Green Economy: Opportunities and Challenges

An International Perspective

Green Economy: Opportunities and Challenges

An International Perspective

Ishwar C. Dhingra
(Ex) Associate Professor in Economics
Shaheed Bhagat Singh College
University of Delhi

Routledge
Taylor & Francis Group
LONDON AND NEW YORK

Manakin
PRESS

First published 2022
by Routledge
2 Park Square, Milton Park, Abingdon, Oxon OX14 4RN

and by Routledge
605 Third Avenue, New York, NY 10158

Routledge is an imprint of the Taylor & Francis Group, an informa business

British Library Cataloguing-in-Publication Data
A catalogue record for this book is available from the British Library

Library of Congress Cataloging-in-Publication Data
A catalog record for this book has been requested

ISBN: 978-1-032-07399-6 (hbk)
ISBN: 978-1-003-20672-9 (ebk)

Typeset in Arial
by Manakin Press Pvt. Ltd.

Manakin
PRESS

Brief Contents

1.	Green Economics: An Introduction	1–18
2.	Global Environmental Issues	19–60
3.	Global Response to Green Economy Initiatives	61–80
4.	Green Economy Strategies	81–100
5.	Guidelines to Green Agriculture	101–146
6.	Guidelines to Green Fisheries	147–156
7.	Guidelines for Green Water Sector	157–174
8.	Guidelines for Green Forest	175–212
9.	Guidelines for Green Renewable Energy	213–228
10.	Guidelines for Green Manufacturing	229–238
11.	Guidelines for Greening Waste Management	239–250
12.	Guidelines for Greening Building	251–258
13.	Guidelines for Green Transport	259–266
14.	Guidelines for Green Tourism	267–278
15.	Guidelines for Green Cities	279–290
16.	Success Stories in Green Economy	291–306
17.	International Initiatives for Green Economy	307–342
18.	Profile of Natural Resources and Issues thereof in India	343–390
19.	Green Economy Issues in India	391–412
	Appendices	413–430
	Index	431–434

Detailed Contents

Preface *xiii*

1. Green Economics: An Introduction **1**

1.1 Introduction 1

1.2 What is a Green Economy? 5

1.3 Pathways to a Green Economy 12

1.4 Approach and Structure—Towards a Green Economy 18

2. Global Environmental Issues **19**

2.1 Introduction 19

2.2 Protection of Environment and Sustainable Development 20

2.3 Seven Biggest Environmental Threats 24

2.4 Environmental Issues 26

2.5 List of Environmental Disasters 31

2.5 List of Environmental Organizations 37

2.6 Outline of Green Topics 39

2.7 Environmental Challenges in a Global Context 45

3. Global Response to Green Economy Initiatives **61**

3.1 Introduction 61

3.2 Principles for a Green Economy 62

3.3 Analysis and Consolidation of Principles 64

3.4 Green Growth: Why it Matters for Developing Countries? 68

3.5 Green Growth for Sustainable and Equitable Development 71

3.6 Patterns of Growth in Developing Countries 72

3.7 A Policy Framework for Greening Growth in Developing Countries 73

3.8 How Can the Global Community Help? 74

3.9 Measuring Progress 77

Appendix: Selected Environmental Chronology 79

4. Green Economy Strategies **81**

4.1 Introduction 81

4.2 A Typology of Green Economy Policies 81

4.3 Recent Green Economy Policy Toolkits 83
4.4 Green Economy and Green Growth Strategies 92
4.5 Use of a Mix of Green Economy Policy Instruments 97

5. Guidelines to Green Agriculture 101
5.1 Introduction 101
5.2 General Background 101
5.3 Conventional/industrial agriculture 104
5.4 The Greening of Agriculture 107
5.5 Challenges and Opportunities 108
5.6 Opportunities 116
5.7 The Case for Greening Agriculture 118
5.8 Investment Priorities for Greening Agriculture 120
5.9 The benefits of Greening Agriculture 131
5.10 Modelling: Future Scenarios for Green Agriculture 137
5.11 Getting There: Enabling Conditions 140

6. Guidelines to Green Fisheries 147
6.1 Introduction 147
6.2 Challenges in Global Fisheries 148
6.3 Opportunities in Global Fisheries 151
6.4 The Cost of Greening Global Fisheries 153
6.5 Enabling Conditions: Institutions, Planning, Policy and Regulatory Reform and Financing 155

7. Guidelines for Green Water Sector 157
7.1 Introduction 157
7.2 The World's Water Resources 157
7.3 Water: A Unique Natural Resources 159
7.4 Challenges and Opportunities 161
7.5 The Economics of Greening Water Use 167
7.6 Enabling Conditions—Overcoming Barriers and Driving Change 169

8. Guidelines for Green Forest 175
8.1 Introduction 175
8.2 World's Last Intact Forests are Becoming Increasingly Fragmented 177
8.3 Vision for the Forest Sector in a Green Economy 179
8.4 Challenges and Opportunities 180
8.5 The Case for Investing in Greening the Forest Sector 188
8.6 Modelling Green Investment in Forests 206
8.7 Impacts of Investment in Reducing Deforestation and in Planted Forest 208

9. Guidelines for Green Renewable Energy **213**
9.1 Introduction 213
9.2 Challenges and Opportunities 214
9.3 Investing in Renewable Energy 219
9.4 Overcoming Barriers: Enabling Conditions 223
10. Guidelines for Green Manufacturing **229**
10.1 Introduction 229
10.2 Green Manufacturing 230
10.3 Challenges—The Risks and Costs of Inaction 230
10.4 Opportunities—Strategic Options for the Manufacturing Sector 235
11. Guidelines for Green Waste Management **239**
11.1 Introduction 239
11.2 Greening the Waste Sector 239
11.3 Challenges in the Waste Sector 241
11.4 Opportunities 247
11.5 Making an Economic Case for Investing in Greening the Waste Sector 249
12. Guidelines for Greening Building **251**
12.1 Introduction 251
12.2 Sizing the Building Sector 251
12.3 Opportunities 254
13. Guidelines for Green Transport **259**
13.1 Introduction 259
13.2 Challenges in the Transport Sector 259
13.3 Opportunities 262
14. Guidelines for Green Tourism **267**
14.1 Introduction 267
14.2 Tourism in a Green Economy 267
14.3 Challenges 268
14.4 Opportunities 270
14.5 Overcoming Barriers: Enabling Conditions 272
14.6 Sum Up 276
Appendix: India's New Tourism Policy 277
15. Guidelines for Green Cities **279**
15.1 Introduction 279
15.2 Green Cities 279
15.3 Challenges 279
15.4 Opportunities 281

15.5 The Case for Green Cities 282
15.6 Greening Urban Sectors 286
15.7 Barriers and Constraints 289

16. Success Stories in Green Economy **291**
16.1 Introduction 291
16.2 Success Stories in Green Initiative 292

17. International Initiatives for Green Economy **307**
17.1 Introduction 307
17.2 UN Conference on Environment and Development 308
17.3 International Environmental Governance 309
17.4 The United Nations Environment Program 309
17.5 Distinctive Doctrines of International Environmental Law 311
17.6 Multilateral Agreements 320
17.7 The Oceans 328
17.8 Biological Diversity 332
17.9 Antarctica 335
17.10 Water Resources 336
17.11 New Multilateral Agreements 336
17.12 From Kyoto to Paris 337
17.13 Marrakech Negotiations 340

18. Profile of Natural Resources and Issues thereof in India **343**
18.1 Introduction 343
18.2 Natural Resources 344
18.3 Land and Soils 345
18.4 Crop Pattern in India 349
18.5 Water Resources 354
18.6 Biodiversity 360
18.7 Forest Resources 360
18.8 Mineral Resources 365
18.9 Conclusion 369
18.10 National Environment Policy, 2006 (NEP) 370
18.11 Five Year Reform Plan (2012–17) 371
18.12 India's Role in International Action (Kyoto Protocol and afterwards) 373
18.13 Carbon Credits and Carbon Trading 379
18.14 Green India Mission 380
18.15 Environmental Accounting System 381
18.16 Land Boom 382
18.17 Issue of Land Acquisition 383

18.18 National Law on Water 387
18.19 Green Technology Fund 389

19. Green Economy Issues in India **391**
19.1 Introduction 391
19.2 Environmental Issues in India 391
19.3 Environmental Damage 399
19.4 Environmental Protection in India 399
19.5 Allocation of Natural Resources 401
19.6 Natural Resource Deficit and Natural Resource Investment Trust (NRIT) 404
19.7 Emission Intensity Cut Leaves Scope for Growth 405

Appendices **413**
A.1 2016 Deemed the Warmest Year on Record. What's in Store for 2017? 413
A.2 Global Green Economy Index (GGEI) 414
A.3 World Green Economy Organisation: A New Approach from the Arab World 416
A.4 Green Climate Fund 417
A.5 Environmental Performance Index 419
A.6 National Air Quality Index (AQI) 420
A.7 Environmental Vulnerability Index 421
A.8 Environmental Sustainability Index 422
A.9 Paris Agreement 422
A.10 Super Typhoon Meranti: 2016's Most Powerful Cyclone 428
A.11 The Dragon's Appetite for Commodities 429

Index **431**

Preface

It is no more quizzical to ask: Why has the Green Economics (GE) emerged as a separate branch of economics. It owes its origin primarily to the growing concern among the policy-makers about the indiscriminate and random destruction of resources, both renewable and non-renewable. Renewable resources get replenished in a cyclical manner. These include such resources as ground water, forests, biodiversity parks, etc. Non-renewable resources are available to mankind only in a limited defined stock. As we make use of these resources to meet our increasing demand for goods and resources, the available stock shrinks and goes on depreciating, with no alternative to replenish them. These include such valuable resources as crude oil, natural gas, minerals, etc. In both cases, unplanned, indiscriminate use of resources may result in irreversible damage. We may reach a situation where all available resources get exhausted, with nothing or little left for use by the succeeding generations. Similar fate awaits us if we resort to indiscriminate exploitation of renewable resources. These undoubtedly get redeemed. But if the pace of redemption falls short of current usage, this may result in a situation called environmental degradation. Economists and policy-makers have been actively engaged in identifying the strategies that may contribute to the setting up of what is called a Green Economy.

The concept of 'Green Economy' extends beyond the frontiers of a 'sustainable economy.' In compilation of the subject-matter we have drawn heavily from the UN reports and other research studies.

Author

THE COLOUR GREEN

The colour green is the colour of balance, harmony and growth

This is the colour of balance and harmony. From a colour psychology perspective, it is the great balancer of the heart and the emotions, creating equilibrium between the head and the heart.

From a meaning of colours perspective, green is also the colour of growth, the colour of spring, of renewal and rebirth. It renews and restores depleted energy. It is the sanctuary away from the stresses of modern living, restoring us back to a sense of well being. This is why there is so much of this relaxing colour on the earth, and why we need to keep it that way.

Green is an emotionally positive colour, giving us the ability to love and nurture ourselves and others unconditionally. A natural peacemaker, it must avoid the tendency to become a martyr.

It loves to observe, and therefore relates to the counsellor, the good listener, the social worker. It loves to contribute to society. It is the charity worker, the good parent and the helpful neighbour.

Being a combination of yellow and blue, green encompasses the mental clarity and optimism of yellow with the emotional calm and insight of blue, inspiring hope and a generosity of spirit not available from other colours.

This is a colour that has a strong sense of right or wrong, inviting good judgment. It sees both sides of the equation, weighs them up, and then usually takes the moral stand in making appropriate decisions. On the negative side, it can be judgmental and over-cautious.

Green promotes a love of nature, and a love of family, friends, pets and the home. It is the colour of the garden lover, the home lover and the good host.

It is generous and loves to share, but it also looks for recognition. It is friendly and can keep confidences.

This colour relates to stability and endurance, giving us persistence and the strength to cope with adversity.

Green is the colour of prosperity and abundance, of finance and material wealth. It relates to the business world, to real estate and property. Prosperity gives a feeling of safety to green.

On the negative, the colour green can be possessive and materialistic, with a need to own people and things.

If your favorite colour is green, it will reflect in your personality! (With apologies to Aarna who associates colour 'green' with 'envy.')

1

Green Economics: An Introduction

1.1 INTRODUCTION

The last two decades have seen the idea of a "green economy" float out of its specialist moorings in environmental economics and into the mainstream of policy discourse. It is found increasingly in the words of heads of state and finance ministers, in the text of G20 communiques, and discussed in the context of sustainable development and poverty eradication.

This recent traction for a green economy concept has no doubt been aided by widespread disillusionment with the prevailing economic paradigm, a sense of fatigue emanating from the many concurrent crises and market failures experienced during the very first decade of the new millennium, including especially the financial and economic crisis of 2008. But at the same time there is an increasing evidence of a way forward, a new economic paradigm—one in which material wealth is not delivered perforce at the expense of growing environmental risks, ecological scarcities and social disparities.

Mounting evidence also suggests that transitioning to a green economy has sound economic and social justification. There is a strong case emerging for a redoubling of efforts by both governments as well as the private sector to engage in such an economic transformation. For governments, this would include levelling the playing field for greener products by phasing out antiquated subsidies, reforming policies and providing new incentives, strengthening market infrastructure and market-based mechanisms, redirecting public investment, and greening public procurement. For the private sector, this would involve understanding and sizing the true opportunity represented by green economy transitions across a number of key sectors, and responding to policy reforms and price signals through higher levels of financing and investment.

1.1.1 From Crisis to Opportunity

Several concurrent crises have unfolded during the last two decades: climate, biodiversity, fuel, food, water, and more recently, in the global financial system. Accelerating carbon emissions indicate a mounting threat of climate change, with potentially disastrous human consequences. The fuel price shock of 2007–2008 and the related skyrocketing food and commodity prices, reflect both structural weaknesses and unresolved risks. Forecasts by the International Energy Agency (IEA) and others of rising fossil fuel demand and energy prices suggest an ongoing dependence as the world economy struggles to recover and grow.

Currently, there is no international consensus on the problem of global food security or on possible solutions for how to nourish a population of 9 billion by 2050. Box 1.1 throws further information on the population challenge. Freshwater scarcity is already a global problem, and forecasts suggest a growing gap by 2030 between annual freshwater demand and renewable supply. The outlook for improved sanitation still looks bleak for over 1.1 billion people and 844 million people still lack access to clean drinking water. Collectively, these crises are severely impacting the possibility of sustaining prosperity worldwide and achieving the Millennium Development Goals (MDGs) for reducing extreme poverty. They are also compounding persistent social problems, such as job losses, socio-economic insecurity, disease and social instability.

The causes of these crises vary, but at a fundamental level they all share a common feature: the gross misallocation of capital. During the last two decades, much capital was poured into property, fossil fuels and structured financial assets with embedded derivatives. However, relatively little in comparison was invested in renewable energy, energy efficiency, public transportation, sustainable agriculture, ecosystem and biodiversity protection, and land and water conservation.

Most economic development and growth strategies encouraged rapid accumulation of physical, financial and human capital, but at the expense of excessive depletion and degradation of natural capital, which includes the endowment of natural resources and ecosystems. By depleting the world's stock of natural wealth—often irreversibly—this pattern of development and growth has had detrimental impact on the well being of current generations and presents tremendous risks and challenges for the future. The recent multiple crises are symptomatic of this pattern.

Box 1.1: Managing the Population Challenge in the Context of Sustainable Development

The link between population dynamics and sustainable development is strong and inseparable, as reflected in Principle 8 of the 1992 Rio Declaration on Environment and Development.

"To achieve sustainable development and a higher quality of life for all people, States should reduce and eliminate unsustainable patterns of production and consumption and promote appropriate demographic policies." Rio Declaration, Principle 8.

This year the world population will reach 7 billion and by mid-century grow to over 9 billion. Contrary to previous projections the most recent population projections expect continued population growth thereafter. Population growth raises the stakes in efforts to reduce poverty. It not only increases the challenge of feeding a growing population, which crucially depends on higher agricultural output, but also requires creation of sufficient employment opportunities, which in turn depend on favorable economic development.

A transition to a green economy can assist in overcoming the contribution that population growth makes to the depletion of scarce natural resources. The world's Least Developed Countries (LDCs) are more strongly affected by environmental degradation than most other developing countries and therefore have much to gain from the transition to a green economy.

In addition, changing spatial distributions of populations, driven both by rural to urban migration and by urban growth, are changing environmental impacts and vulnerabilities. When planned, urbanisation can be a powerful driver of sustainable development. Given that in 2008, the share of the urban population has for the first time exceeded the share of people living in the rural areas at the global level, a transition to a green economy becomes increasingly important. Significantly, in the least developed countries where the majority of people are still living in the rural areas, 2000 to 2010 was the first decade that growth of the urban population outpaced the growth of the rural populations. These types of changes at a societal level can also present opportunities for a green economy to develop.

For example, cities can provide essential services, including health and education, at lower costs per capita due to economies of scale benefits. Efficiencies are also realised in the development of vital infrastructure including housing, water, sanitation and transport. Urbanisation can also reduce energy consumption, particularly in transport and housing, and create interactive spaces that further cultural outreach and exchange. Realisation of these positive benefits requires proactive planning for the future demographic changes.

Forward planning by governments and local authorities can address population dynamics in a proactive way. For example, one tool available to assist countries is to make better use of available population data and conduct a systematic population situation analysis, aiming to highlight how current and projected population trends affect the development of countries. Such analysis provides the necessary foundation to address population dynamics and their links to sustainable development and poverty reduction strategies.

Source: UNFPA

Existing policies and market incentives have contributed to this problem of capital misallocation because they allow businesses to run up significantly, largely unaccounted for, and unchecked social and environmental externalities. To reverse such misallocation requires better public policies, including pricing and regulatory measures, to change the perverse incentives that drive this capital misallocation and ignore social and environmental externalities. At the same time, appropriate regulations, policies and public investments that foster changes in the pattern of private investment are increasingly being adopted around the world, especially in developing countries.

1.1.2 Myths About Green Economy

UNEP's report, *Towards a Green Economy*, aims to debunk several myths and misconceptions about greening the global economy, and provides timely and practical guidance to policy makers on what reforms they need to unlock the productive and employment potential of a green economy.

Perhaps the most prevalent myth is that there is an inescapable trade-off between environmental sustainability and economic progress. There is now substantial evidence that the greening of economies neither inhibits wealth creation nor employment opportunities. To the contrary, many green sectors provide significant opportunities for investment, growth and jobs. For this to occur, however, new enabling conditions are required to promote such investments in the transition to a green economy, which in turn calls for urgent action by policy makers.

A second myth is that a green economy is a luxury only wealthy countries can afford, or worse, a ruse to restrain development and perpetuate poverty in developing countries. Contrary to this perception, numerous examples of greening transitions can be found in the developing world, which should be replicated elsewhere. *Towards a Green Economy* brings some of these examples to light and highlights their scope for wider application.

UNEP's work on green economy raised the visibility of this concept in 2008, particularly through a call for a *Global Green New Deal* (GGND). The GGND recommended a package of public investments and complementary policy and pricing reforms aimed at kick-starting a transition to a green economy, while reinvigorating economies and jobs and addressing persistent poverty. Designed as a timely and appropriate policy response to the economic crisis, the GGND proposal was an early output from the United Nations' *Green Economy Initiative*. This initiative, coordinated by UNEP, was one of the nine Joint Crisis Initiatives undertaken by the Secretary-General of the UN and his Chief Executives Board in response to the 2008 economic and financial crisis.

Towards a Green Economy—the main output of the Green Economy Initiative—demonstrates that the greening of economies need not be a drag on growth. On the contrary, the greening of economies has the potential to be a new engine of growth, a net generator of decent jobs and a vital strategy to eliminate persistent poverty. The report also seeks to motivate policy makers to create the enabling conditions for increased investments in a transition to a green economy in three ways.

First, the report makes an economic case for shifting both public and private investment to transform key sectors that are critical to greening the global economy. It illustrates through examples how added employment through green jobs offsets job losses in a transition to a green economy.

Second, it shows how a green economy can reduce persistent poverty across a range of important sectors—agriculture, forestry, freshwater, fisheries and energy. Sustainable forestry and ecologically friendly farming methods help conserve soil fertility and water resources. This is especially critical for subsistence farming, upon which almost 1.3 billion people depend for their livelihoods.

Third, it provides guidance on policies to achieve this shift by reducing or eliminating environmentally harmful or perverse subsidies, addressing market failures created by externalities or imperfect information, creating market- based incentives, implementing appropriate regulatory frameworks, initiating green public procurement and by stimulating investment.

1.2 WHAT IS A GREEN ECONOMY?

UNEP defines a green economy as one that results in "improved human well-being and social equity, while significantly reducing environmental risks and ecological scarcities." In its simplest expression, a green economy is low-carbon, resource efficient, and socially inclusive. In a green economy, growth in income and employment are driven by public and private investments that reduce carbon emissions and pollution, enhance energy and resource efficiency, and prevent the loss of biodiversity and ecosystem services.

These investments need to be catalysed and supported by targeted public expenditure, policy reforms and regulation changes. The development path should maintain, enhance and, where necessary, rebuild natural capital as a critical economic asset and as a source of public benefits. This is especially important for poor people whose livelihoods and security depend on nature.

The key aim for a transition to a green economy is to enable economic growth and investment while increasing environmental quality and social inclusiveness. Critical to attaining such an objective is to create the conditions for public and private investments to incorporate broader environmental and social criteria. In addition, the main indicators of economic performance, such as growth in Gross Domestic Product (GDP) need to be adjusted to account for pollution, resource depletion, declining ecosystem services, and the distributional consequences of natural capital loss to the poor.

A major challenge is reconciling the competing economic development aspirations of rich and poor countries in a world economy that is facing increasing climate change, energy insecurity and ecological scarcity. A green economy can meet this challenge by offering a development path that reduces carbon dependency, promotes resource and energy efficiency and lessens environmental degradation. As economic growth and investments become less dependent on liquidating environmental assets and sacrificing environmental quality, both rich and poor countries can attain more sustainable economic development.

The concept of a green economy does not replace sustainable development; but there is a growing recognition that achieving sustainability rests almost entirely on getting the economy right. Decades of creating new wealth through a "brown economy" model based on fossil fuels have not substantially addressed social marginalisation, environmental degradation and resource depletion. In addition, the world is still far from delivering on the Millennium Development Goals set for 2015. We need to look at the important linkages between the concept of a green economy and sustainable development.

1.2.1 A Green Economy and Sustainable Development

In 2009, the UN General Assembly decided to hold a summit in Rio de Janeiro in 2012 (Rio+20) to celebrate the 20th anniversary of the first Rio Earth Summit in 1992. Two of the agenda items for Rio+20 were, "*Green Economy in the Context of Sustainable Development and Poverty Eradication*", and "*International Framework for Sustainable Development.*" With the green economy now firmly established on the international policy agenda, it is useful to review and clarify the linkages between a green economy and sustainable development.

Most interpretations of sustainability take as their starting point the consensus reached by the World Commission on Environment and Development (WCED) in 1987, which defined sustainable development as "development that meets the needs of the present without compromising the ability of future generations to meet their own needs."

Economists are generally comfortable with this broad interpretation of sustainability, as it is easily translatable into economic terms: an increase in well-being today should not result in reducing well-being tomorrow. That is, future generations should be entitled to atleast the same level of economic opportunities—and thus atleast the same level of economic welfare—as is available to current generations.

As a result, economic development today must ensure that future generations are left no worse off than current generations. Or, as some economists have succinctly expressed it, per capita welfare should not be declining over time. According to this view, it is the total stock of capital employed by the economic system, including natural capital, which determines the full range of economic opportunities, and thus well-being, available to both current and future generations.

Society must decide how best to use its total capital stock today to increase current economic activities and welfare. Society must also decide how much it needs to save or accumulate for tomorrow, and ultimately, for the well-being of future generations.

However, it is not simply the aggregate stock of capital in the economy that may matter but also its composition, in particular whether current generations are using up one form of capital to meet today's needs. For example, much of the interest in sustainable development is driven by concern that economic development may be leading to rapid accumulation of physical and human capital at the expense of excessive depletion and degradation of natural capital. The major concern is that by irreversibly depleting the world's stock of natural wealth, today's development path will have detrimental implications for the well-being of future generations.

One of the first economic studies to make the connection between this capital approach to sustainable development and a green economy was the 1989 book *Blueprint for a Green Economy*. The authors argued that because today's economies are biased towards depleting natural capital to secure growth, sustainable development is unachievable. A green economy that values environmental assets, employs pricing policies and regulatory changes to translate these values into market incentives, and adjusts the economy's measure of GDP for environmental losses is essential to ensuring the well-being of current and future generations.

As pointed out by the *Blueprint for a Green Economy* authors, a major issue in the capital approach to sustainable development is whether substitution among different forms of capital—human capital, physical capital and natural capital—is possible. A strong conservationist perspective might maintain that the natural component of the total capital stock must be kept intact, as measured in physical terms. However, this may be questioned in practice, especially in the context of developing countries, if natural capital is relatively abundant while physical and human capital needs to be developed to meet other human demands. This type of substitution reflects the unfortunate reality that the creation of physical capital—for example roads, buildings and machinery—often requires the conversion of natural capital. While substitution between natural capital and other forms of capital is often inevitable, there is often room for efficiency gains. There is also a growing recognition of environmental thresholds that would constrain substitution beyond minimum levels needed for human welfare.

Yet, there has always been concern that some forms of natural capital are essential to human welfare, particularly key ecological goods and services, unique environments and natural habitats, and irreplaceable ecosystem attributes. Uncertainty over the true value of these important assets to human welfare, in particular the value that future generations may place on them if they become increasingly scarce, further limits our ability to determine whether we can adequately compensate future generations for today's irreversible losses in such essential natural capital. This concern is reflected in other definitions of sustainable development. For example, in 1991, the World Wide Fund for Nature, the International Union for Conservation of Nature (IUCN), and UNEP interpreted the concept of sustainable development as "improving the quality of human life within the carrying capacity of supporting ecosystems."

As this definition suggests, the type of natural capital that is especially at risk is ecosystems. As explained by Partha Dasgupta: "Ecosystems are capital assets. Like reproducible capital assets ... ecosystems depreciate if they are misused or are overused. But they differ from reproducible capital assets in three ways: (1) depreciation of natural capital is frequently irreversible (or at best the systems take a long time to recover); (2) except in a very limited sense, it isn't possible to replace a depleted or degraded ecosystem by a new one; and (3) ecosystems can collapse abruptly, without much prior warning."

Rising ecological scarcity is an indication that we are irrevocably depleting ecosystems too rapidly, and the consequence is that current and future economic welfare is affected. An important indicator of the growing ecological scarcity worldwide was provided by the Millennium Ecosystem Assessment (MEA) in 2005, which found that over 60 per cent of the world's major ecosystem goods and services covered in the assessment were degraded or used unsustainably.

Some important benefits to humankind fall in this category, including fresh water; capture fisheries; water purification and waste treatment; wild foods; genetic resources; biochemicals; wood fuel; pollination; spiritual, religious and aesthetic values; the regulation of regional and local climate; erosion; pests; and natural hazards. The economic values associated with these ecosystem services, while generally not marketed, are substantial (see Table 1.1).

Table 1.1: Natural capital—Underlying components and illustrative services and values

Biodiversity	Ecosystem goods and services (examples)	Economic values (examples)
Ecosystems (variety and extent/area)	• Recreation • Water regulation • Carbon storage	Avoiding greenhouse gas emissions by conserving forests: US$ 3.7 trillion (NPV)
Species (diversity and abundance)	• Food, fiber, fuel • Design inspiration • Pollination	Contribution of insect pollinators to agricultural output: ~US$ 190 billion/year
Genes (variability and population)	• Medicinal discoveries • Disease resistance • Adaptive capacity	25–50% of the US$ 640 billion pharmaceutical market is derived from genetic resources

One major difficulty is that the increasing costs associated with rising ecological scarcity are not routinely reflected in markets. Almost all the degraded ecosystem goods or services identified by the Millennium Ecosystem Assessment are not marketed. Some goods, such as capture fisheries, fresh water, wild foods, and wood fuel, are often commercially marketed, but due to the poor management of the biological resources and ecosystems that are the source of these goods, and imperfect information, the market prices do not reflect unsustainable use and overexploitation.

Nor have adequate policies and institutions been developed to handle the costs associated with worsening ecological scarcity globally. All too often, policy distortions and failures compound these problems by encouraging wasteful use of natural resources and environmental degradation. The unique challenge posed by rising ecological scarcity and inefficient resource and energy use today is to overcome a vast array of market, policy, and institutional failures that prevents recognition of the economic significance of this environmental degradation.

Reversing this process of unsustainable development requires three important steps. First, as argued by the *Blueprint for a Green Economy* authors, improvements in environmental valuation and policy analysis are required to ensure that markets and policies incorporate the full costs and benefits of environmental impacts. Environmental valuation and accounting for natural capital depreciation must be fully integrated into economic development policy and strategy. As suggested above, the most undervalued components of natural capital are ecosystems and the myriad goods and services they provide. Valuing ecosystem goods and services is not easy, yet it is fundamental to ensuring the sustainability of global economic development efforts.

A major international research effort supported by UNEP, *the Economics of Ecosystems and Biodiversity* (TEEB), is illustrating how ecological and economic research can be used to value ecosystem goods and services, as well as how such valuation is essential for policy making and investments in the environment.

Second, the role of policy in controlling excessive environmental degradation requires implementing effective and appropriate information, incentives, institutions, investments and infrastructure. Better information on the state of the environment, ecosystems and biodiversity is essential for both private and public decision-making that determines the allocation of natural capital for economic development. The use of market-based instruments, the creation of markets, and where appropriate, regulatory measures, have a role to play in internalising this information in everyday allocation decisions in the economy. Such instruments are also important in correcting the market and policy failures that distort the economic incentives for improved environmental and ecosystems management.

However, overcoming institutional failures and encouraging more effective property rights, good governance and support for local communities, is also critical. Reducing government inefficiency, corruption and poor accountability are also important in reversing excessive environmental degradation in many countries. But there is also a positive role for government in providing an appropriate and effective infrastructure through public investment, protecting critical ecosystems and biodiversity conservation, creating new incentive mechanisms such as payment for ecosystem services, fostering the technologies and knowledge necessary for improving ecosystem restoration, and facilitating the transition to a low-carbon economy.

Third, continuing environmental degradation, land conversion and global climate change affect the functioning, diversity, and resilience of ecological systems and the goods and services they supply. The potential long-term impacts of these effects on the health and stability of ecosystems are difficult to quantify and value. Increasing collaboration between environmental scientists, ecologists and economists will be required to assess and monitor these impacts. Such interdisciplinary ecological and economic analysis is also necessary to identify and assess the welfare consequences for current and future generations from increasing ecological scarcity. Further progress in reversing unsustainable development calls for more widespread interdisciplinary collaboration to analyse complex problems of environmental degradation, biodiversity loss and ecosystem decline.

Interdisciplinary research also needs to determine the thresholds that should govern the transformation of specific types of natural capital into other forms of capital. For example, how much forest land is allowed for conversion into farmland, industrial use or urban development in a given area? How much underground water is allowed for extraction each year? How much and what fish species can be caught in a given season? Which chemicals should be banned from production and trading? And more important, what are the criteria for setting these thresholds? Once these standards are established, incentive measures at national or international levels can be devised to ensure compliance.

The other key to balancing different forms of capital recognises that substitutability is a characteristic of current technologies. Investing in changing and substituting these technologies can lead to new complementarities. Most renewable energy sources, such as wind turbines or solar panels, considerably reduce the amount of natural capital that is sacrificed in their construction and the lifetime of their operation, compared to fossil fuel burning technologies. Both of these types of solutions—setting thresholds and altering technologies—are important for achieving a green economy.

In sum, moving towards a green economy must become a strategic economic policy agenda for achieving sustainable development. A green economy recognises that the goal of sustainable development is improving the quality of human life within the constraints of the environment, which include combating global climate change, energy insecurity, and ecological scarcity. However, a green economy cannot be focused exclusively on eliminating environmental problems and scarcity. It must also address the concerns of sustainable development with intergenerational equity and eradicating poverty.

1.2.2 A Green Economy and Eradicating Poverty

Most developing countries, and certainly the majority of their populations, depend directly on natural resources. The livelihoods of many of the world's rural poor are also intricately linked with exploiting fragile environments and ecosystems. Well over 600 million of the rural poor currently live on lands prone to degradation and water stress, and in upland areas, forest systems, and drylands that are vulnerable to climatic and ecological disruptions. The tendency of rural populations to be clustered on marginal lands and in fragile environments is likely to be a continuing problem for the foreseeable future, given current global rural population and poverty trends. Despite rapid global urbanisation, the rural population of developing regions continues to grow, albeit at a slower rate in recent decades. Furthermore, around three-quarters of the developing world's poor still live in rural areas, which means about twice as many poor people live in rural rather than in urban areas.

The world's poor are especially vulnerable to the climate-driven risks posed by rising sea levels, coastal erosion and more frequent storms. Around 14 per cent of the population and 21 per cent of urban dwellers in developing countries live in low elevation coastal zones that are exposed to these risks. The livelihoods of billions—from poor farmers to urban slum dwellers—are threatened by a wide range of climate-induced risks that affect food security, water availability, natural disasters, ecosystem stability and human health. For example, many of the 150 million urban inhabitants, who are likely to be at risk from extreme coastal flooding events and sea level rise, are likely to be the poor living in cities in developing countries.

As in the case of climate change, the link between ecological scarcity and poverty is well established for some of the most critical environmental and energy problems. For example, for the world's poor, global water scarcity manifests itself as a water poverty problem. One-in-five people in the developing world lacks access to sufficient clean water, and about half the developing world's population, 2.6 billion people, do not have access to basic sanitation. More than 660 million of the people without sanitation live on less than US$ 2 a day, and more than 385 million on less than US$ 1 a day. Billions of people in developing countries have no access to modern energy services, and those consumers who do have access often pay high prices for erratic and unreliable services. Among the energy poor are 2.4 billion people who rely on traditional biomass fuels for cooking and heating, including 89 per cent of the population of Sub-Saharan Africa; and 1.6 billion people who do not have access to electricity.

Thus, finding ways to protect global ecosystems, reduce the risks of global climate change, improve energy security, and simultaneously improve the livelihoods of the poor are important challenges in the transition to a green economy, especially for developing countries.

A transition to a green economy can contribute to eradicating poverty. A number of sectors with green economic potential are particularly important for the poor, such as agriculture, forestry, fishery and water management, which have public goods qualities. Investing in greening these sectors, including through scaling up microfinance, is likely to benefit the poor in terms of not only jobs, but also secure livelihoods that are predominantly based on ecosystem services. Enabling the poor to access microinsurance coverage against natural disasters and catastrophes is equally important for protecting livelihood assets from external shocks due to changing and unpredictable weather patterns.

However, it must be emphasised that moving towards a green economy will not automatically address all poverty issues. A pro-poor orientation must be superimposed on any green economy initiative. Investments in renewable energy, for example, will have to pay special attention to the issue of access to clean and affordable energy. Payments for ecosystem services, such as carbon sequestration in forests, will need to focus more on poor forest communities as the primary beneficiaries. The promotion of organic agriculture can open up opportunities, particularly for poor small- scale farmers who typically make up the majority of the agricultural labour force in most low-income countries, but will need to be complemented by policies to ensure that extension and other support services are in place.

In sum, the top priority of the UN MDGs has been eradicating extreme poverty and hunger, including halving the proportion of people living on less than US$ 1 a day by 2015. A green economy must not only be consistent with that objective, but must also ensure that policies and investments geared towards reducing environmental risks and scarcities are compatible with ameliorating global poverty and social inequity.

1.3 PATHWAYS TO A GREEN ECONOMY

If the desirability of moving to a green economy is clear to most people, the means of doing so is still a work in progress for many. We look at the theory of greening, the practice and the enabling conditions required for making such a transition. However, before embarking on this analysis, let us frame the dimensions of the challenge.

1.3.1 How Far is the World from a Green Economy?

Over the last quarter of a century, the world economy has quadrupled, benefiting hundreds of millions of people. However, 60 per cent of the world's major ecosystem goods and services that underpin livelihoods have been degraded or used unsustainably *(Millennium Ecosystem Assessment 2005)*. This is because the economic growth of recent decades has been accomplished mainly through drawing down natural resources, without allowing stocks to regenerate, and through allowing widespread ecosystem degradation and loss.

For instance, today only 20 per cent of commercial fish stocks, primarily low priced species, are underexploited; 52 per cent are fully exploited with no further room for expansion; about 20 per cent are overexploited; and 8 per cent are depleted. Water is becoming scarce and water stress is projected to increase with water supply satisfying only 60 per cent of world demand in 20 years. Agriculture saw increasing yields primarily due to the use of chemical fertilisers, yet has resulted in declining soil quality, land degradation, and deforestation—which resulted in 13 million hectares of forest lost annually over 1990–2005. Ecological scarcities are seriously affecting the entire gamut of economic sectors that are the bedrock of human food supply (fisheries, agriculture, freshwater, and forestry) and a critical source of livelihoods for the poor. At the same time, ecological scarcity and social inequity are clear indicators of an economy that is not sustainable.

For the first time in history, more than half of the world population lives in urban areas. Cities now account for 75 per cent of energy consumption and of carbon emissions. Rising and related problems of congestion, pollution and poorly provisioned services affect the productivity and health of all, but fall particularly hard on the urban poor. With approximately 50 per cent of the global population now living in emerging economies that are rapidly urbanising and developing, the need for green city planning, infrastructure and transportation is paramount.

The transition to a green economy will vary considerably among nations, as it depends on the specifics of each country's natural and human capital and on its relative level of development. As demonstrated graphically, there are many opportunities for all countries in such a transition. Some countries have attained high levels of human development, but often at the expense of their natural resource base, the quality of their environment, and high greenhouse gas (GHG) emissions. The challenge for these countries is to reduce their per capita ecological footprint without impairing their quality of life.

Other countries still maintain relatively low per capita ecological footprints, but need to deliver improved levels of services and material well-being to their citizens. Their challenge is to do this without drastically increasing their ecological footprint. One of these two challenges affects almost every nation, and globally, the economy is still very far from being green.

1.3.2 Enabling Conditions for a Green Economy

To make the transition to a green economy, specific enabling conditions will be required. These enabling conditions consist of national regulations, policies, subsidies and incentives, as well as international market and legal infrastructure, trade and technical assistance. Currently, enabling conditions are heavily weighted towards, and encourage, the prevailing brown economy, which depends excessively on fossil fuels, resource depletion and environmental degradation.

For example, price and production subsidies for fossil fuels collectively exceeded US$ 650 billion in 2016. This high level of subsidisation can adversely affect the adoption of clean energy while contributing to more greenhouse gas emissions. In contrast, enabling conditions for a green economy can pave the way for the success of public and private investment in greening the world's economies. At a national level, examples of such enabling conditions are: changes to fiscal policy, reform and reduction of environmentally harmful subsidies, employing new market-based instruments, targeting public investments to green key sectors, greening public procurement, and improving environmental rules and regulations, as well as their enforcement. At an international level, there are also opportunities to add to market infrastructure, improve trade and aid flows and foster greater international cooperation.

At the national level, any strategy to green economies should consider the impact of environmental policies within the broader context of policies to address innovation and economic performance. In this view, government policy plays a critical role within economies to encourage innovation and growth. Such intervention is important as a means for fostering innovation and for choosing the direction of change.

For some time, economists such as Kenneth Arrow have shown that competitive firms and competitive markets do not necessarily produce the optimal amount of innovation and growth within an economy. Public intervention within an economy is therefore critically important for these purposes. This is because industries in competitive markets have few incentives to invest in technological change or even in product innovation, as any returns would be immediately competed away. This is one of the best-known examples of market failure in the context of competitive markets, and provides the rationale for various forms of interventions.

Examples of spurring growth and innovation can be seen from histories of many recently emerged economies. In the 1950s and 1960s, the Japanese and South Korean governments chose the direction of technological change through importing the technology of other countries. This changed in the 1970s when these economies shifted to aggressive policies for encouraging energy-efficient innovation. Shortly afterwards, Japan was one of the leading economies in the world in terms of Research and Development (R&D)

investment in these industries. This pattern of directed spending and environmental policies is being repeated today across much of Asia. The cases of South Korea and China in particular are illustrative, where a large proportion of their stimulus packages was directed at a "green recovery" and has now been instituted into longer-term plans for retooling their economies around green growth.

Thus, moving towards a green development path is almost certainly a means for attaining welfare improvements across a society, but it is also often a means for attaining future growth improvement. This is because a shift away from basic production modes of development based on extraction and consumption and towards more complex modes of development can be a good long-term strategy for growth. There are several reasons why this shift might be good for long-term competitiveness as well as for social welfare.

First, employing strong environmental policies can drive inefficiencies out of the economy by removing those firms and industries that only exist because of implicit subsidies in under-priced resources. The free use of air, water and ecosystems is not a value-less good for any actor in an economy and amounts to subsidising negative net worth activities. Introducing effective regulation and market-based mechanisms to contain pollution and limit the accumulation of environmental liabilities drives the economy in a more efficient direction.

Second, resource pricing is important not just for the pricing of natural capital and services, but also for pricing of all the other inputs within an economy. An economy allocates its efforts and expenditures according to relative prices, and under-priced resources result in unbalanced economies. Policy makers should be targeting the future they wish their economies to achieve, and this will usually require higher relative prices on resources. An economy that wishes to develop around knowledge, R&D, human capital and innovation should not be providing free natural resources.

Third, employing resource pricing drives investments into R&D and innovation. It does so because avoiding costly resources can be accomplished by researching and finding new production methods. This will include investment in all of the factors (human capital and knowledge) and all of the activities (R&D and innovation) listed above. Moving towards more efficient resource pricing is about turning the economy's emphasis towards different foundations of development.

Fourth, these investments may then generate innovation rents. Policies that reflect scarcities that are prevalent in the local economy can also reflect scarcities prevalent more widely. For this reason, a solution to a problem of resource scarcity identified locally (via R&D investments) may have applicability and hence more global marketability. The first solution to a widely experienced problem can be patented, licensed and marketed widely.

Fifth, aggressive environmental regulation may anticipate future widely-experienced scarcities and provide a template for other jurisdictions to follow. Such policy leadership can be the first step in the process of innovation, investment, regulation and resource pricing described above.

In sum, the benefits from a strong policy framework to address market failures and ecological scarcities will flow down the environment pathway that comes from altering the direction of an economy. Policies and market-based mechanisms that enhance perceived resource prices creates incentives to shift the economy onto a completely different foundation—one based more on investments in innovation and its inputs of human capital, knowledge, and research and development.

1.3.3 How to Measure Progress Towards a Green Economy?

It is difficult, if not impossible, to manage what is not measured. Notwithstanding the complexity of an overall transition to a green economy, appropriate indicators at both a macroeconomic level and a sectoral level will be essential to informing and guiding the transition.

To complicate matters, conventional economic indicators, such as GDP, provide a distorted lens for economic performance, particularly because such measures fail to reflect the extent to which production and consumption activities may be drawing down natural capital. By either depleting natural resources or degrading the ability of ecosystems to deliver economic benefits, in terms of provisioning, regulating or cultural services, economic activity is often based on the depreciation of natural capital.

Ideally, changes in stocks of natural capital would be evaluated in monetary terms and incorporated into national accounts. This is being pursued in the ongoing development of the *System of Environmental and Economic Accounting* (SEEA) by the UN Statistical Division, and the World Bank's adjusted net national savings methods. The wider use of such measures would provide a better indication of the real level and viability of growth in income and employment. *Green Accounting* or *Inclusive Wealth Accounting* are available frameworks that are expected to be adopted by a few nations[1] initially and pave the way for measuring the transition to a green economy at the macroeconomic level.

1.3.4 How Might a Green Economy Perform Over Time?

The macroeconomic *Threshold 21* (T21) model is used by UNEP to explore the impacts of investments in greening the economy against investments in

1. World Bank, together with UNEP and other partners, have announced a global project on Ecosystem Valuation and Wealth Accounting which will enable a group of developing and developed nations to test this framework and evolve a set of pilot national accounts that are better able to reflect and measure sustainability concerns.

business as usual. The T21 model measures results in terms of traditional GDP as well as its affects on employment, resource intensity, emissions, and ecological impacts.

The T21 model was developed to analyse strategies for medium to long-term development and poverty reduction, most often at the national level, complementing other tools for analysing short-term impacts of policies and programmes. The model is particularly suited to analysing the impacts of investment plans, covering both public and private commitments. The global version of T21 models the world economy as a whole to capture the key relationships between production and key natural resource stocks at an aggregate level.

The T21 model reflects the dependence of economic production on the traditional inputs of labour and physical capital, as well as stocks of natural capital in the form of resources, such as energy, forest land, soil, fish and water. Growth is thus driven by the accumulation of capital—whether physical, human or natural - through investment, also taking into account depreciation or depletion of capital stocks. The model is calibrated to reproduce the past 40-year period of 1970-2010; simulations are conducted over the next 40-year period, 2010-2050. Business-as-usual projections are verified against standard projections from other organisations, such as the United Nations Population Division, World Bank, OECD, the International Energy Agency, and the Food and Agriculture Organisation.

The inclusion of natural resources as a factor ofproduction distinguishes T21 from all other global macroeconomic models. Examples of the direct dependence of output (GDP) on natural resources are the availability of fish and forest stocks for the fisheries and forestry sectors, as well as the availability of fossil fuels to power the capital needed to catch fish and harvest timber, among others. Other natural resources and resource efficiency factors affecting GDP include water stress, waste recycle and reuse and energy prices.

Based on existing studies, the annual financing demand to green the global economy was estimated to be in the range US$ 1.05 to US$ 2.59 trillion. To place this demand in perspective, it is about one-tenth of total global investment per year, as measured by global Gross Capital Formation. Taking an annual level of US$ 1.3 trillion (2 per cent of global GDP) as a reference scenario, varying amounts of investment in the 10 sectors covered in this report were modelled to determine impact on growth, employment, resource use and ecological footprint. The results of the model, suggest that over time investing in a green economy enhances long-term economic performance. Significantly, it does so while enhancing stocks of renewable resources, reducing environmental risks, and rebuilding capacity to generate future prosperity.

1.4 APPROACH AND STRUCTURE—TOWARDS A GREEN ECONOMY

UNEP focuses on 10 key sectors considered to be driving the defining trends of the transition to a green economy. These trends include increasing human well-being and social equity, and reducing environmental risks and ecological scarcities. Across many of these sectors, greening the economy can generate consistent and positive outcomes for increased wealth, growth in economic output, decent employment and reduced poverty.

In Part I of its report the UNEP focuses on the sectors derived from natural capital—agriculture, fishing, forests and water. These sectors have a material impact on the economy as they form the basis for primary production, and because the livelihoods of the rural poor depend directly upon them. The analysis looks at the principal challenges and opportunities for bringing more sustainable and equitable management to these sectors, and reviews investment opportunities to restore and maintain the ecosystem services that underpin these sectors. In so doing, the report highlights several sector-specific investment opportunities and policy reforms that are of global importance as they appear replicable and scalable in the goal to transition to a green economy.

In Part II, the UNEP report focuses on those sectors that may be characterised as "built capital" traditionally considered the brown sectors of the economy. In these sectors—such as transportation, energy and manufacturing—the report finds large opportunities for energy and resources savings. These savings, it is argued, can be scaled up and become drivers of economic growth and employment, as well as having important equity effects in some cases. Resource efficiency is a theme that has many dimensions as it cuts across energy efficiency in manufacture and habitation, materials efficiency in manufacture, and better waste management.

Finally, Part III focuses on enabling conditions for ensuring a successful transition to a green economy. These include appropriate domestic fiscal measures and policy reforms, international collaboration through trade, finance, market infrastructure, and capacity building support. Much has been said about the potential for a green economy to be used as a pretext for imposing aid conditionalities and trade protectionism. It is argued that to be green, an economy must not only be efficient, but also fair. Fairness implies recognising global and country level equity dimensions, particularly in assuring a just transition to an economy that is low-carbon, resource efficient, and socially inclusive. These enabling conditions for a fair and just transition are described and addressed at length in the report before conclusions, along with the steps necessary to mobilise finance at scale for a global transition to a green economy.

■ ■ ■

2

Global Environmental Issues

2.1 INTRODUCTION

Environment has been defined as the aggregate of all the external conditions and influences affecting the life and development of an organism. It comprises the whole range of external influences acting on an organism, of both the physical and the biological forces of nature surrounding the human individual. All living beings are part of balanced and interacting eco-system: They draw sustenance from the solid, liquid and aerial resources on the earth. They undergo passage through the phases of reproduction, sustenance and extinction. The ultimate source of energy for the entire ecosystem is the solar energy. The ecosystem consists of subsystems such as sea, forests, water reservoirs, plants, trees, insects, animals in a forest—all of them are inter-connected in a network of devourer-devoured relationship. None of the elements in the network is useless or purposeless, each of them helps to regulate the balance of the subsystem. Bio-gas-chemical cycles are essential for the process of reproduction in the physical environment. The process of reproduction is held in balance as long as the natural cycles, their interrelations and the hierarchy of the food-chain are not disturbed.

There are many different models of the relationship between the environment and the economy. The simple model depicted in Fig. 2.1 illustrates the four functions of the environment in supporting economic activity and the effects of this activity on the environment. These four functions are life-support, supply of natural resources, absorption of waste products and supply of amenity services. The economy is represented in Fig. 2.1 by households consuming goods and services, and firms producing with natural resources provided by the environment, with labour and man-made capital provided by households.

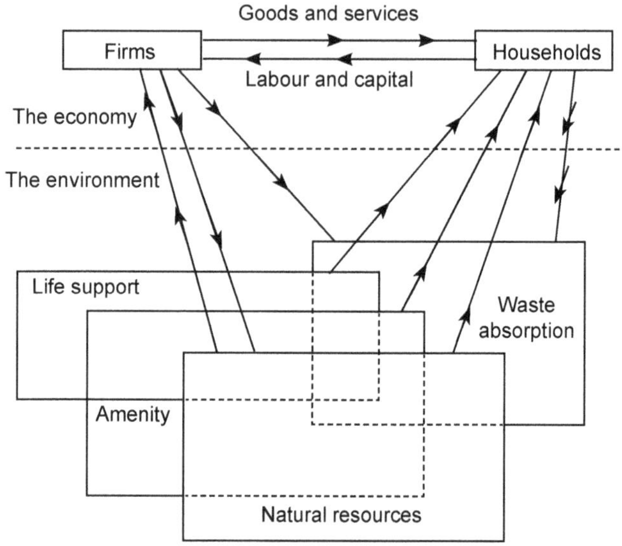

Fig. 2.1

Man's desire for more joy and comfort has led him to exploit nature's free goods to the extent of reducing its natural capacities for self-stabilisation. As countries industrialise, they increase their global footprint, which is the pressure they put on the earth's resources. The footprints of Europe and Japan are about 4.7 global hectares per person, and the USA's is 9.7. India's footprint at present is 0.8 and China's has reached 1.6. Mankind's overall global footprint, which was about 60% of the global biocapacity in the 1960s, has already reached 130%. This gets reflected in growing environmental problems. Environmental problems centre on human activities resulting in pollution of the atmosphere, oceans and land. These range from the global (greenhouse warming and ozone depletion) to the regional (acid rain and desertification), national (deforestation) and local (soil erosion, contamination of fresh water resources and urban pollution). The relevance of such concerns and the priority attached to each varies between the developing and developed countries.

2.2 PROTECTION OF ENVIRONMENT AND SUSTAINABLE DEVELOPMENT

The protection of the environment is an essential part of development. Without adequate environmental protection, development is undermined. There are two ways in which this can happen.

1. Environment quality—water that is safe and plentiful and air that is healthy—is itself part of the improvement in welfare that development

attempts to bring. If the benefits from rising incomes are offset by the costs imposed on health and the quality of life by pollution, this cannot be called development.

2. Environmental damage can undermine future productivity. Soils that are degraded, aquifers that are depleted, and ecosystems that are destroyed in the name of raising incomes today can jeopardise the prospects for earning income tomorrow.

In view of these considerations, economists have evolved the concept of sustainable development. This term was popularised by the 1987 Brundtland Report, *Our Common Future*, although it had been around earlier in the decade. This Report by the UN-created World Commission on the Environment and Development was significant in so far as it was the first such initiative to involve social scientists. The Brundtland Report and other independent writers highlighted the accelerating rate of decline in the land and water resources of the UDCs. These also highlighted the need for re-estimating growth rates on the basis of net national product, where net takes into account depreciation of natural resources.

The concept of sustainable development was also paraphrased for the first time in these writings. Today, there are a large number of definitions of sustainable development (some think the number is over 100).

Sustainable development requires that in the process of economic growth we maintain our natural resources and environment intact and use/harvest only that much quantity which is regenerated naturally, that is, we, live on the 'flows' and keep the 'stock' of natural resources and environment intact. In the context of sustainability, two questions are important: (*i*) How long will our natural resources last given current consumption patterns? and (*ii*) How should we manage our natural resources and environment do that the future generations have access to the same quality of life as the present generation?

The sustainability condition can be summarized as follows:

$$K \geq K^{-*};\ H \geq H^{-*};\ SC \geq SC^{-*};\ \text{and}\ N \geq N^{*}$$

where K is man-made capital, H is human capital, SC is social capital, N is natural capital and * is some predetermined threshold level of all the forms of capital expressed in monetary terms.

For attaining and maintaining strong sustainable development the following three principles can be specified:

1. Renewable resources must be harvested at or below the growth rate for some predetermined stock of the resource.
2. As non-renewable resources are depleted, renewable substitutes must be developed so as to maintain the flow of goods and services over time.

3. Pollution emissions should be limited to the assimilative capacity of the environment.

A recent commercial advertisement by Mitsubhish Electricals sums it well as follows:

Technology + Ecology = Sustainability

2.2.1 Effects of Economic Growth on Environment

Economic growth can cause environmental problems but can also, with the right policies and institutions, help address them. Three patterns, as follows, emerge:

1. Some problems decline as income increases. This is because increasing income provides the resources for public services such as sanitation and rural electricity. When individuals no longer have to worry about day-to-day survival, they can devote resources to profitable investments in conservation. These positive relations between economic growth and environment quality must not be underestimated.
2. Some problems initially worsen but then improve as incomes rise. Most forms of air and water pollution fit into this category, as do some types of deforestation and encroachment on natural habitats. There is nothing automatic about this improvement; it occurs only when governments deliberately introduce policies to ensure that additional resources are devoted to dealing with environment problems.
3. Some indicators of environmental stress worsen as incomes increase. Emissions of carbon and of nitrogen oxides and municipal wastes are current examples. In these cases abatement is relatively expensive and the costs associated with the emissions and wastes are not yet perceived as high—often because they are borne by someone else. The key is, once again, policy. More generally, individuals and firms have few incentives to cut back on wastes and emissions, and until such incentives are put into place—through regulation, charges, or other means—damage will continue to increase. The experience with the turnarounds achieved in other forms of pollution, however, shows what may be possible once a policy commitment is made.

Studies suggest that developing economies now follow a sequence in which pollution rises in the early stages of growth when incomes are low. As per capita gross domestic product continues to rise, pollution reaches a peak and eventually decreases, even as GDP continues to rise. This phenomenon can be represented as an inverted U-shaped curve, known as an *environmental Kuznets curve,* illustrated in Fig. 2.2.

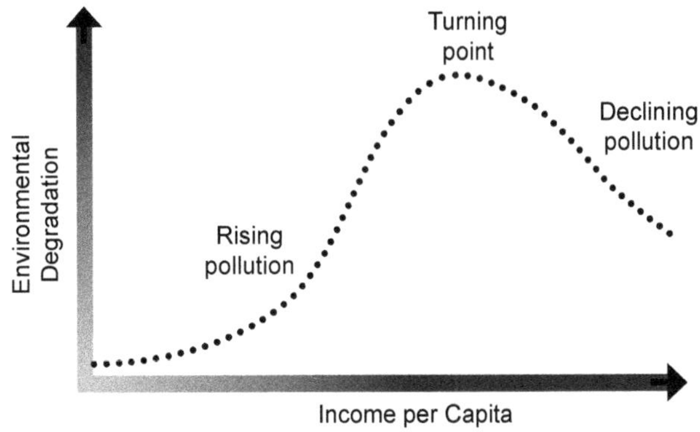

Fig. 2.2

2.2.2 Policies for Environment

Two broad sets of policies are needed to attack the underlying causes of environmental damage:

1. ***Policies Built on the Positive Links*** These seek to harness the positive links between development and the environment by correcting or preventing policy failures, improving access to resources and technology, and promoting equitable income growth. Such policies include the following:

(*i*) Removing subsidies that encourage excessive use of fossil fuels, irrigation water, and pesticides and excessive logging.

(*ii*) Clarifying rights to manage and own land, forests and fisheries.

(*iii*) According higher priority to provision of sanitation and clean water, education (especially for girls), family planning service, and agricultural extension, credit, and research.

(*iv*) Taking measures to empower, educate, and involve farmers, local communities, indigenous people, and women so that they can make decisions and investments in their long-term interests.

2. ***Policies Targeted at Specific Environmental Problems*** Specific policies are required to induce or require resource users to take account of the spillover effects that their actions have on the rest of society. Policies designed to change behaviour are of two broad types: those based on incentives, which tax or charge polluters according to the amount of damage they do, and those based on quantitative restrictions, which provide no such flexibility.

The appropriate choice among varied instruments of policy will depend on circumstances. However, several lessons, as follows, can be drawn from recent experience.

(*i*) Trade-offs between income and environmental quality need to be carefully assessed, taking long-term, uncertain, and irreversible impacts into account. Carefully balancing costs and benefits is especially important for developing countries, where resources are scarce and where basic needs still must be met.

(*ii*) Standards and policies need to be realistic and consistent with the monitoring and enforcement capacity and the administrative traditions of the country.

(*iii*) Blunt and more self-enforcing policies are likely to be attractive in developing countries. Policies need to work with the grain of the market rather than against it using incentives rather than regulations where possible.

(*iv*) Governments need to build constituencies for change—to curb the power of vested interests, to hold institutions accountable, and to increase willingness to pay the costs of protection. Local participation in setting and implementing environmental policies and investments will yield high returns.

(*v*) The costs of protecting and improving the environment are high in absolute terms, but they are modest in comparison with their benefits and with the potential gains from economic growth.

2.3 SEVEN BIGGEST ENVIRONMENTAL THREATS

Examining the various threats to the Earth's environment must include the human impact on the planet. Catch phrases such as carbon footprint, global warming, deforestation, and other commonly used terms have become the everyday jargon for those concerned about the environment.

1. Human Population and Pollution

A growing world population might seem like an obvious threat to the environment that goes far beyond the debatable theory of global warming. The bigger threat is far more complex and directly linked not to the controversial idea of a carbon footprint, but to the unique system of supply and demand.

Consumers place more and more demands on the earth's natural resources as the population increases year after year. These demands leave pollution and waste in the wake of human daily activity. Compound this with each world government doing its own brand of commerce, many without environmental consciences, and you get the formula for environmental chaos and disaster.

A prime example of higher consumption demands can be found in the fishery industry, where the world's marine life is being harvested with few to no renewable methods in place. Consumers are also responsible via industry for hundreds of hazardous chemicals used in the production of various products. Heavy metals continue to contaminate land, water and air.

The power of a consumer can be mighty when each person in the world realizes that action can be taken and changes made by carefully choosing how each consumer dollar is spent.

2. Earth Changes

The last major climate change was an ice age and the world is in the final stages of that event. The result is a rise in temperatures and the melting of glaciers and even the polar ice cap. Many highly respected scientists disagree that global warming is the result of human caused pollution any more than it can cause powerful hurricanes, tsunamis, earthquakes, floods, and even solar flares. This school of thought views earth changes as being the result of the natural processes found in an evolving living planet and its sun. While the cause of global warming remains controversial, both sides agree that it's a very real environmental threat to the world as you know it.

3. Deforestation

When a region loses its biodiversity, it becomes more vulnerable to other environmental elements. Deforestation disrupts the natural balance of ecological systems in the area where the trees have been harvested far beyond. Food production can be impacted due to drought and erosion directly linked to the loss of forests.

4. Ozone Deterioration

Chemicals and chlorofluorocarbons pollutants are created by industry and agriculture. They have a negative impact on ozone layer. The lack of strict enforcement of laws to prevent the use of such pollutants compounds the situation. World governments that continue to allow various pollutants into the environment impede the recovery of the ozone layer.

5. Acid Rain

Acid rain is created by excessive sulfuric and nitric acid being pumped into the atmosphere, rivers, oceans, and land. While some acid rain is the byproduct of the natural processes of decaying vegetation and volcanic activity, the current crisis comes directly from the burning of fossil fuels. Water becomes toxic when acid rain imbues the oceans or lakes with an absorption quality that cause the water to absorb soil-based aluminum and poisons the aquatic plant and marine life.

6. Dead Zones in the Ocean

Another harmful source of excessive nitrogen being dumped into the oceans can be traced back to agricultural practices of over-fertilization of crops, lawns and gardens. The end result has been the creation of over 160 dead zones throughout the world's oceans.

The oceans' eco-systems are dependent upon the natural process of organic ocean matter known as phytoplankton, which is found on ocean surfaces. This eventually breaks down and filters to the bottom of the ocean floor where it's broken down further by ocean floor bacteria. This process is called bacterial respiration.

When too much nitrogen feeds the phytoplankton, like any fertilized crop, it begins to overproduce. The bacteria are unable to break down the plankton fast enough and the chemical processes that convert carbondioxide into oxygen can't keep up. The oxygen is used up quicker than it can be produced. The plankton chokes out the flow of water and oxygen so that marine and plant life die from the lack of oxygen.

7. Species Extinction

An alarming rate of species extinction is happening worldwide. As of 2017, the rate of loss is estimated to be more than 1,000 times the normal rate. Greater preservation tactics and strategies are needed with laws put into place to protect species. Once more, manmade pollution is the culprit along with land encroachment by developers. Both causes are created by consumer demands as people branch out into areas that were once remote habitats for various species.

An example of successful endangered species preservation is the American national symbol, the bald eagle. In the 1960s, there were fewer than 470 eagle nestlings. As of 2016, there were over 7,000 nestlings in the United States. This increase in the bald eagle population demonstrates how threatened species can be brought back from the brink of extinction. More and more animals and other forms of wildlife are being added to the endangered species list each year. It makes sense to become better land stewards, instead of encroaching on forests and wetlands.

While there are many other threats to the environment that have a significant impact, these are certainly the seven biggest environmental threats facing the world today.

2.4 ENVIRONMENTAL ISSUES

Environmental issues are harmful effects of human activity on the biophysical environment. Environmental protection is a practice of protecting the natural environment on individual, organisational or governmental levels, for the benefit of both the environment and humans. Environmentalism, a social and environmental movement, addresses environmental issues through advocacy, education and activism.

The carbondioxide equivalent of greenhouse gases (GHG) in the atmosphere has already exceeded 400 parts per million (NOAA) (with total "long-term" GHG exceeding 455 parts per million). (Intergovernmental Panel on Climate Change Report) This level is considered a tipping point. "The amount of greenhouse gas in the atmosphere is already above the threshold that can potentially cause dangerous climate change. We are already at risk of many areas of pollution—It's not next year or next decade, it's now." Report from the UN Office for the Coordination of Humanitarian Affairs (OCHA) says that climate disasters are on the rise. Around 70 percent of disasters are now climate related—up from around 50 percent from two decades ago.

These disasters take a heavier human toll and come with a higher price tag. In the last decade, 2.4 billion people were affected by climate related disasters, compared to 1.7 billion in the previous decade. The cost of responding to disasters has risen manifold between 1992 and 2017. Destructive sudden heavy rains, intense tropical storms, repeated flooding and droughts are likely to increase, as will the vulnerability of local communities in the absence of strong concerted action. (OCHA) "Climate change is not just a distant future threat. It is the main driver behind rising humanitarian needs and we are seeing its impact. The number of people affected and the damages inflicted by extreme weather has been unprecedented."

Environment destruction caused by humans is a global problem, and this is a problem that is on-going every day. By the year 2050, the global human population is expected to grow by 2 billion people, thereby reaching a level of 9.6 billion people (Living Blue Planet). The human effects on Earth can be seen in many different ways. The main one is the temperature rise, and according to the report "Our Changing Climate", the global warming that has been going on for the past 50 years is primarily due to human activities. Since 1895, the U.S. average temperature has increased from 1.3 °F to 1.9 °F, with most of the increase taken place since around year 1970.

Major current environmental issues may include climate change, pollution, environmental degradation, and resource depletion etc. The conservation movement lobbies for protection of endangered species and protection of any ecologically valuable natural areas, genetically modified foods and global warming.

2.4.1 Scientific Grounding

The level of understanding of Earth has increased markedly in recent times through science especially with the application of the scientific method. Environmental science is now a multi-disciplinary academic study taught and researched at many universities. This is used as a basis for addressing environmental issues.

Large amounts of data have been gathered and these are collated into reports, of which a common type is the State of the Environment publications. A recent major report was the Millennium Ecosystem Assessment, with input from 1200 scientists and released in 2005, which showed the high level of impact that humans are having on ecosystem services.

2.4.2 Organisations

Environmental issues are addressed at a regional, national or international level by government organisations.

The largest international agency, set up in 1972, is the United Nations Environment Programme. The International Union for Conservation of Nature brings together 83 states, 108 government agencies, 766 Non-governmental organisations and 81 international organisations and about 10,000 experts and scientists from countries around the world. International non-governmental organisations include Greenpeace, Friends of the Earth and World Wide Fund for Nature. Governments enact environmental policy and enforce environmental law and this is done to differing degrees around the world.

2.4.3 List of Environmental Issues

This is an alphabetical list of environmental issues, harmful aspects of human activity on the biophysical environment. As such, they relate to the anthropogenic effects on the natural environment, which are loosely divided into causes, effects and mitigation, noting that effects are interconnected and can cause new effects.

Issues

- **Human overpopulation**—Biocapacity; Carrying capacity; Exploitation; Industrialisation; I = PAT; Land degradation; Land reclamation; Optimum population; Overshoot (population); Population density; Population dynamics; Population growth; Projections of population growth; Total fertility rate; Urbanisation; Waste; Water conflict; Water scarcity; Overdrafting.
- **Hydrology**—Environmental impacts of reservoirs; Tile drainage; Hydrology (agriculture); Flooding; Landslide.
- **Intensive farming**—Agricultural subsidy; Environmental effects of meat production; Intensive animal farming; Intensive crop farming; Irrigation; Monoculture; Nutrient pollution; Overgrazing; Pesticide drift; Plasticulture; Slash and burn; Tile drainage.
- **Land use**—Built environment; Desertification; Habitat fragmentation; Habitat destruction; Land degradation—Land pollution; Lawn-environmental concerns; Urban heat island; Urban sprawl.
- **Nanotechnology**—Impact of nanotechnology.
- **Natural disasters.**

- **Nuclear issues**—Nuclear fallout; Nuclear meltdown; Nuclear power; Nuclear weapons; Nuclear and radiation accidents; Nuclear safety; High-level radioactive waste management.
- **Ocean trash.**

Effects

- **Climate change**—Global warming; Global dimming; Fossil fuels; Sea level rise; Greenhouse gas; Ocean acidification; Shutdown of thermohaline circulation; Environmental impact of the coal industry; Urban Heat Islands; Flooding.
- **Environmental degradation**—Habitat destruction; Invasive species.
- **Environmental health**—Air quality; Asthma; Birth defect; Developmental disability; endocrine disruptors; Environmental impact of the coal industry; Environmental impact of nanotechnology; Electromagnetic field; Electromagnetic radiation and health; Indoor air quality; Lead poisoning; Leukemia; Nanotoxicology; Nature deficit disorder; One Health; Sick Building Syndrome; Environmental impact of hydraulic fracturing.
- **Environmental issues with energy**—Environmental impact of the coal industry; Environmental impact of the energy industry; Environmental impact of hydraulic fracturing.
- **Environmental issues with war**—Agent Orange; Depleted Uranium; Military Superfund site (Category only).
- **Scorched earth**—War and environmental law; Unexploded ordnance
- **Overpopulation**—Burial; Overpopulation in companion animals; Tragedy of the commons; Gender Imbalance in Developing Countries; Sub-replacement fertility levels in developed countries.
- **Genetic engineering**—Genetic pollution; Genetically modified food controversies.
- Pollution—Nonpoint source pollution; Point source pollution.
- **Air pollution**—Environmental impact of the coal industry; Environmental impact of hydraulic fracturing; Indoor air quality; Smog; Tropospheric ozone; Volatile organic compound Atmospheric particulate matter CFC; Biological effects of UV exposure; Light pollution; Visual pollution.
- Noise pollution.
- **Soil pollution**—Alkali soil; Brownfield; Residual Sodium Carbonate Index; Soil conservation; Soil erosion; Soil contamination; Soil salination; Superfund; Superfund sites; Space debris; Interplanetary contamination; Ozone depletion.
- **Water pollution**—Acid rain; Agricultural runoff; Algal bloom; Environmental impact of the coal industry; Environmental impact of

hydraulic fracturing; Eutrophication; Fish kill; Groundwater pollution; Groundwater recharge; Marine debris; Marine pollution; Mercury in fish; Microplastics; Nutrient pollution; Ocean acidification; Ocean dumping; Oil spills.

- **Soda lake**—Ship pollution; Thermal pollution; Urban runoff; Waste-water.
- **Resource depletion**—Exploitation of natural resources; Overdrafting (groundwater); Overexploitation.
- **Consumerism**—Consumer capitalism; Planned obsolescence; Over-consumption.
- **Fishing**—Blast fishing; Bottom trawling; Cyanide fishing; Ghost nets; Illegal, unreported and unregulated fishing; Overfishing; Shark finning; Whaling.
- **Logging**—Clearcutting; Deforestation; Illegal logging.
- **Mining**—Acid mine drainage; Environmental impact of hydraulic fracturing; Mountaintop removal mining; Slurry impoundments.
- **Water (depletion)**—Anoxic waters; Aral Sea; California Water Wars; Dead Sea; Lake Chad; Water scarcity.
- **Toxicants**—Agent Orange; Asbestos; Beryllium; Bioaccumulation; Biomagnification; Chlorofluorocarbons (CFCs); Cyanide; DDT; Endocrine disruptors; Explosives; Environmental impact of the coal industry; Herbicides; Hydrocarbons; Perchlorate; Pesticides; PBDE; Persistent organic pollutant; PBBs; PBDEs; Toxic heavy metals; PCB; Dioxin; Polycyclic aromatic hydrocarbons; Radioactive contamination; Volatile organic compounds.
- **Waste**—Electronic waste; Great Pacific Garbage Patch; Illegal dumping; Incineration; Litter; Waste disposal incidents; Marine debris; Medical waste; Landfill; Leachate; Toxic waste; Environmental impact of the coal industry; Exporting of hazardous waste.

Mitigation

- Conservation
  - o Ecosystems—Anoxic waters; Biodiversity; Biosecurity; Coral bleaching; Edge effect; Habitat destruction; Habitat fragmentation; In-situ leach
  - o Fishing—Blast fishing; Bottom trawling; By-catch; Cetacean bycatch; Gillnetting; Illegal, unreported and unregulated fishing; Environmental effects of fishing; Marine pollution; Overfishing; Whaling
  - o Forests—Clearcutting; Deforestation; Illegal logging
  - o Natural resources—Resource depletion; Exploitation of natural resources

  - o Species—Endangered species; Genetic diversity; Habitat destruction; Holocene extinction; Invasive species; Poaching; Pollinator decline; Species extinction; Threshold host density; Wildlife trade; Wildlife disease
  - o Energy conservation—Efficient energy use
  - o Renewable energy—Renewable energy commercialization
  - o Recreation—Protected areas
  - o Water conservation
- Disaster mitigation
- **Environmental law**—Environmental crime; Environmental justice; Polluter pays principle; Precautionary Principle; regulatory capture.

2.5 LIST OF ENVIRONMENTAL DISASTERS

Below is reproduced list of environmental disasters. In this context it is an annotated list of specific events caused by human activity that results in a negative effect on the environment.

2.5.1 Environmental Disasters by Category

This list is incomplete; it can always be expanded.

Agricultural

- Mismanagement and shrinking of the Aral Sea
- Salinity in Australia
- Salinization of the Fertile Crescent
- The Dust Bowl in Canada and the United States (1934–1939)
- The Great sparrow campaign; sparrows were eliminated from Chinese farms, which caused locusts to swarm the farms and contributed to a famine which killed 38 million people
- Africanized bees, known colloquially as "killer bees"
- "Dirty dairying" in New Zealand.

Biodiversity

- Chestnut blight
- Extinction of American megafauna
- Extinction of Australian megafauna
- Deforestation of Easter Island
- Destruction of the old growth forests
- Rabbits in Australia
- Red imported fire ants
- Dutch Elm Disease

- Devil facial tumour disease
- Reduction in the number of the American Bison
- Introduction of the Nile perch into Lake Victoria in Africa, decimating indigenous fish species
- The Saemangeum Seawall
- Emerald Ash Borer
- Environmental threats to the Great Barrier Reef.

Box 2.1: Water

The importance of water in the 21st century—comparable to that of oil in the 20th—can hardly be overstated. Yet some strategic experts continue to underestimate it. The reality is that oil has alternatives like natural gas, wind, solar, and nuclear energy. By contrast, for industry and agriculture as much as for drinking and sanitation, the only alternative to water, as former Slovenian president Danilo Türk once put it, is water.

The same is true for trade. Consider the Rio Chagres. While it may not be widely known, it is vitally important, as it feeds the Panama Canal, through which 50% of the trade between Asia and the Americas flows. There is no risk of the natural depletion of the river flow for the next 100 years, but in the event of a security crisis in Central America, it could be taken over by rogue forces. The impact on the global economy would be enormous.

In the last three years, the Islamic State (IS) captured the Tabqa, Tishrin, Mosul, and Fallujah dams on the Tigris and Euphrates Rivers. IS subsequently lost control of all of them, but not before using them to flood or starve downstream populations, to pressure them to surrender.

This tactic is not exclusive to IS. Extremist groups in South Asia have also threatened to attack water infrastructure. And of course state actors, too, can use water resources to gain a strategic advantage.

The consensus on the need to protect water resources and installations in conflict zones is clear. What is less clear is how to do it. Unlike medicines and food packets, water cannot be airdropped into conflict zones. And UN peacekeeping forces are badly overstretched.

The International Committee of the Red Cross does negotiate safe passage for technicians to inspect and repair damage to water pipes and storage systems in Iraq, Syria and Ukraine; but each passage needs to be negotiated with governments in conflict and rebel commanders—a long and cumbersome process. A better approach would be for great powers, with their considerable influence, to negotiate short-term ceasefires in areas experiencing protracted conflict, specifically to repair and restore water systems.

To pave the way for such an approach, however, the UN Security Council will have to declare water a "strategic resource of humanity" and adopt a resolution to protect water resources and installations, similar to Resolution 2286, adopted May, 2016 to protect medical facilities in armed conflicts.

In the longer term, countries that share riparian systems will need to establish regional security arrangements to preserve and protect their resources. With collaborative management underpinning collective protection, water, often a source of competition and conflict, could become a facilitator of peace and cooperation.

- 2006 Zakouma elephant slaughter
- Invasive species in New Zealand
- The loss of Biodiversity of New Zealand
- Ghost nets
- Grounding of SS Makambo on Lord Howe Island
- Shark finning
- Decline of vultures in India due to Diclofenac leading to increased incidence of rabies
- Extinction of the Tasmanian Tiger.

Human Health

- Introduction of the Bubonic Plague (the Plague of Justinian) in Europe from Africa in the 7th century resulting in the death of upto 60% (100 million) of the population.
- Introduction of the Bubonic Plague (the Black Death) in Europe from Central Asia in the 14th century resulting in the death of up to 60% (200 million) of the population and recurring until the 18th century.
- Introduction of infectious diseases by Europeans causing the death of indigenous people during European colonization of the Americas
- Health effects arising from the September 11 attacks
- Goiânia accident, human deaths resulting from dismantling a scrapped medical machine containing a source of radioactivity.

Industrial

Coordinates of the Industrial Environmental Disasters found on this page, shown in Google. Complete with the Wikipedia descriptions listed below built into each location.

- Spring Valley, a neighborhood in Washington, D.C. which was used as a chemical weapons testing ground during World War I.
- Minamata disease—mercury poisoning in Japan (1950s and 1960s)
- Ontario Minamata disease in Canada.
- Itai-itai disease, due to cadmium poisoning in Japan.
- Love Canal toxic waste site.
- Seveso disaster (1976), chemical plant explosion, caused highest known exposure to 2,3,7,8-tetrachlorodibenzo-p-dioxin (TCDD) in residential populations.
- Times Beach, Missouri (1983) the town was completely evacuated due to a dioxin contamination.
- Bhopal disaster (December 3, 1984, India), leak of methyl isocyanate that took place in 1984 resulted in more than 22,000 deaths.

- Sandoz chemical spill into the Rhine river (1986).
- United States Environmental Protection Agency Superfund sites in the United States.
- AZF Explosion at a Toulouse chemical factory (2001).
- 2005 Jilin chemical plant explosions.
- The Sydney Tar Ponds and Coke Ovens sites in the city of Sydney, Nova Scotia, Canada, known as the largest toxic waste site in North America.
- Release of lead dust into Esperance Harbour.
- Release of cyanide, heavy metals and acid into the Alamosa River, Colorado from the Summitville mine, causing the death of all aquatic life 17 miles downstream.
- Release of 20,000 gallons of lethal chemicals (metam sodium, tradename Vapam) into the Upper Sacramento River near Dunsmuir, causing the death of all aquatic life within a 38-mile radius.
- Release of CFCs resulting in ozone depletion.
- Release of sulfur dioxide after a fire at the Al-Mishraq plant in Iraq.
- The Phillips Disasters.
- Vila Parisi (Brazil).
- Health issues on the Aamjiwnaang First Nation due to chemical factories.
- Environmental issues with the Three Gorges Dam.
- Kingston Fossil Plant coal fly ash slurry spill.
- The Great Smog in London in 1952.

Mining

- 1947 Centralia mine disaster, Illinois
- Centralia mine fire, Pennsylvania, 1962
- Phosphate mining in Nauru
- Phosphate mining in St. Pierre Island
- Talvivaara gypsum pond leak, Finland, 2012
- Mount Polley mine disaster, British Columbia, 2014
- Rio Doce, Mariana. Iron mining residue spill which spread for over 2 states, Minas Gerais and Espirito Santo, all the way to the Atlantic sea, Brazil, 2015.

Coal Mining

- Martin County sludge spill
- Mountaintop removal mining

Oil Industry

- Environmental issues in the Niger Delta relating to the oil industry
- Lago Agrio oil field issues
- Exxon Valdez oil spill
- Arctic Refuge drilling controversy
- Deepwater Horizon oil spill
- Sidoarjo mud flow triggered by Lapindo Brantas gas exploration in 2006; East Java, Indonesia
- Leaded gasoline introduced 1920s; phased out globally by 2012.

Nuclear

November 1951 nuclear test at the Nevada Test Site, from Operation Buster, with a yield of 21 kilotons. It was the first U.S. nuclear field exercise conducted on land; troops shown are 6 mi (9.7 km) from the blast.

- Chernobyl disaster in 1986 in Chernobyl, Ukraine, "killed at least 4056 people and damaged almost $7 billion of property". Radioactive fallout from the accident concentrated near Belarus, Ukraine and Russia and at least 350,000 people were forcibly resettled away from these areas. After the accident, "traces of radioactive deposits unique to Chernobyl were found in nearly every country in the northern hemisphere".
- Fukushima Daiichi nuclear disaster: Following an earthquake, tsunami, and failure of cooling systems at Fukushima I Nuclear Power Plant and issues concerning other nuclear facilities in Japan on March 11, 2011, a nuclear emergency was declared. This was the first time a nuclear emergency had been declared in Japan, and 140,000 residents within 20 km of the plant were evacuated. Explosions and a fire have resulted in dangerous levels of radiation, sparking a stock market collapse and panic-buying in supermarkets.
- Mayak nuclear waste storage tank explosion, (Chelyabinsk, Soviet Union, 29 September 1957), 200+ people died and 270,000 people were exposed to dangerous radiation levels. Over thirty small communities had been removed from Soviet maps between 1958 and 1991.
- Windscale fire, United Kingdom, October 8, 1957. Fire ignites plutonium piles and contaminates surrounding dairy farms.
- Soviet submarine K-431 accident, August 10, 1985 (10 people died and 49 suffered radiation injuries).
- Soviet submarine K-19 accident, July 4, 1961. (8 deaths and more than 30 people were over-exposed to radiation).
- Nuclear testing at Moruroa and Fangataufa in the Pacific Ocean.
- Fallout from the Castle Bravo nuclear test at Bikini Atoll in the Marshall Islands.

- The health of Downwinders
- Atomic bombings of Hiroshima and Nagasaki Within the first two to four months of the bombings, the acute effects killed 90,000–166,000 people in Hiroshima and 60,000–80,000 in Nagasaki, with roughly half of the deaths in each city occurring on the first day.
- Hanford Nuclear, 1986—The U.S. government declassifies 19,000 pages of documents indicating that between 1946 and 1986, the Hanford Site near Richland, Washington, released thousands of US gallons of radioactive liquids. Radioactive waste was both released into the air and flowed into the Columbia River (which flows to the Pacific Ocean). In 2014, the Hanford legacy continues with billions of dollars spent annually in a seemingly endless cleanup of leaking underground.

Air/land/water

- Proliferation of plastic shopping bags.

Air

- The Donora Smog of 1948 in Donora, Pennsylvania in the United States
- The Great Smog of 1952, which killed 4,000 Londoners
- The 1983 Melbourne dust storm
- The 1997 Southeast Asian haze
- The 2005 Malaysian haze
- The 2006 Southeast Asian haze
- Yokkaichi asthma in Japan
- Health problems due to the Jinkanpo Atsugi Incinerator in Japan
- Kuwaiti oil fires.

Land

- The Dust Bowl of Canada and the United States
- Contaminated soils in Mapua, New Zealand due to the operation of an agricultural chemicals factory
- Basin F, a disposal site in the United States for contaminated liquid wastes from the chemical manufacturing operations of the Army and its lessee Shell Chemical Company
- Exide lead contamination in southeast Los Angeles County, California, United States, from a battery recycling plant that emitted lead, arsenic and other dangerous pollutants
- 2006 Côte d'Ivoire toxic waste dump
- Nigeria gully erosion crisis.

Water

- Sandoz chemical spill, severely polluting the Rhine in 1986
- Selenium poisoning of wildlife due to farm runoff used to create Kesterson National Wildlife Refuge, and the artificial wetland
- The Jiyeh Power Station oil spill in the Mediterranean region
- Effects of polluted water in the Berkeley Pit in the United States
- Ignition and conflagration (13 times from 1868 to 1969) of the Cuyahoga River in Ohio, United States
- Cheakamus River derailment which polluted a river with caustic soda
- Draining and development of the Everglades
- Loss of Louisiana Wetlands due to Mississippi River levees, saltwater intrusion through manmade channels, timber harvesting, subsidence, and hurricane damage
- Lake Okeechobee is heavily polluted and during extreme events releases large volumes of polluted water into the St. Lucie River estuary and the Caloosahatchee River estuary.

Marine

- Coral bleaching
- Gulf of Mexico Dead Zone due to high-nutrient fertilizer runoff from the Midwest that is drained through the Mississippi River.
- The artificial Osborne Reef off the coast of Fort Lauderdale, Florida in the United States
- Dumping of conventional and chemical munitions in Beaufort's Dyke, a sea trench between Northern Ireland and Scotland
- Marine debris
- Environmental threats to the Great Barrier Reef
- Nurdles, plastic pellet typically under 5mm in diameter
- Friendly Floatees
- The Great Pacific Garbage Patch
- Minamata disease, mercury poisoning in Japan
- Mercury in fish
- Ocean acidification due to anthropogenic greenhouse gas emissions.

2.6 LIST OF ENVIRONMENTAL ORGANISATIONS

An environmental organisation is an organisation coming out of the conservation or environmental movements that seeks to protect, analyse or monitor the environment against misuse or degradation from human forces.

In this sense the environment may refer to the biophysical environment, the natural environment or the built environment. The organisation may be a charity, a trust, a non-governmental organisation or a government organisation. Environmental organisations can be global, national, regional or local.

History

Founded on 28 May, 1892 in San Francisco, California, Sierra Club was one of the first large-scale environmental preservation organisations in the world.

Politics

Most organisations are in some way try to be involved in policy-making to exert more influence. Green politics is the political ideology which places a high importance on environmental goals and Green parties have formed to implement environmental policy at a government level.

Issues in Focus

Audubon Center at Bent of the River. The Audubon Society is an environmental organisation dedicated to the protection of birds.

Some of the environmental issues that are of interest to environmental organisations are pollution, waste, resource depletion, human overpopulation and climate change.

Notable global environmental organisations are the United Nations Environment Programme (UNEP), World Wide Fund for Nature, the Intergovernmental Panel on Climate Change (IPCC), World Nature Organisation (WNO), Greenpeace and Friends of the Earth.

Counter-movements

In order to counterbalance the sometimes progress—captalism and economy-skeptic attitude and influence of environmental organisations, many interest groups and associations have been formed, most often industry-supported, seldom of purely ideological nature.

Solutions

Sustainability is the key to prevent or reduce the effect of environmental issues. There is now clear scientific evidence that humanity is living unsustainably, and that an unprecedented collective effort is needed to return human use of natural resources to within sustainable limits. For humans to live sustainably, the Earth's resources must be used at a rate at which they can be replenished.

Concerns for the environment have prompted the formation of Green parties, political parties that seek to address environmental issues. Initially, these were formed in Australia, New Zealand and Germany but are now present in many other countries.

Film and Television

There are an increasing number of films being produced on environmental issues, especially on climate change and global warming. Al Gore's 2006 film An Inconvenient Truth gained commercial success and a high media profile.

2.7 OUTLINE OF GREEN TOPICS

Environmental issues have increasingly become a topic in film and television.

Within the last twenty years, commercially successful films with an environmentalism theme have been released theatrically and made by the major Hollywood studios. The Annual Environmental Media Awards have been presented by the Environmental Media Association (EMA) since 1991 to the best television episode or film with an environmental message.

Some notable films with an environmental message include:

A documentary film is a film whose author(s), at the time of its production, believed its information to be true.

Table 2.1: List of documentary films about the environment

Title	Theme(s) and subtheme(s)	Author(s)	Year(s)
The 11th Hour	Various themes	Nadia Conners, Leonardo DiCaprio, and Leila Conners Petersen	2007
Angry Inuk	Hunting: seal hunting	Alethea Arnaquq-Baril	2016
Ashes to Honey	Nuclear power: radiation		2011
Baraka	Various themes	Constantine Nicholas and Genevieve Nicholas	1992
Before Vanishing	Water pollution: Barada (river)		2005
The Bell of Chernobyl	Chernobyl nuclear disaster	Rollan Sergienko	1987
Bhopal Express	Air pollution: from a pesticide factory	Prasoon Pandey and Piyush Pandey	1999
Biùtiful cauntri	Pollution: toxic waste dumping in Southern Italy		2007
Black Hole	Coal mining: Australia		2015
Blue Vinyl	Pollution: production and disposal of polyvinyl chloride	Daniel B. Gold and Judith Helfand	2002
Burning the Future: Coal in America	Mining: mountaintop removal mining	David Novack and Richard Hankin	2008
The Burning Season (1994)	Forests: Amazon rainforest	William Mastrosimone	1994
The Burning Season (2008)	Forests: deforestation in Indonesia	Cathy Henkel	2008
Butterfly	Environmentalism: resisting clear-cutting		2000
Cane Toads: The Conquest	Introduced species: cane toads in Australia	Mark Lewis	2010
Cane Toads: An Unnatural History	Introduced species: cane toads in Australia	Mark Lewis	1988

Title	Theme(s) and subtheme(s)	Author(s)	Year(s)
The Chances of the World Changing	Endangered species: turtles		2006
Chasing Ice	Global warming	James Balog	2012
Chernobyl Heart	Chernobyl disaster: acute radiation syndrome		2003
Children of Tsunami: No More Tears	Natural disasters: 2004 Indian Ocean earthquake and tsunami		2005
The City Dark	Light pollution		2011
Climate Refugees	Climate change: effects on natural resources	Michael P. Nash	2010
Clouds of Smoke	Global warming		
The Cove	Hunting: dolphin drive hunting	Mark Monroe	2009
Cowspiracy: The Sustainability Secret	Animal husbandry: policies of environmental organisations		2014
Crude (2007)	Petroleum industry: peak oil and climate change	Richard Smith	2007
Crude (2009)	Petroleum industry: pollution and class action	Joe Berlinger	2009
A Crude Awakening: The Oil Crash	Peak oil		2006
Dead Ahead: The Exxon Valdez Disaster	Petroleum industry: Exxon Valdez oil spill		1992
Deadly Deception: General Electric, Nuclear Weapons and Our Environment	Energy: nuclear weapons		1991
A Delicate Balance—The Truth	Diet: effects on health and the natural environment	Adrianna Scheibner and Aaron Scheibner	2008
Demain (English: Tomorrow)	Sustainable communities	Cyril Dion and Mélanie Laurent	2015
Dreamland	Dams: Kárahnjúkar Hydropower Plant in Iceland		2009
Dust to Dust: The Health Effects of 9/11	Air pollution: from September 11 attacks		2006
Earth	Environmental change: effects on three particular species	Alastair Fothergill, Mark Linfield, and Leslie Megahey	2007
Earth Days	Environmentalism: environmental movement in the United States	Robert Stone	2009
Eco-Pirate: The Story of Paul Watson	Hunting: anti-whaling		2011
Elemental	Water pollution: Ganges River		2012
The End of Eden	Environmental degradation: cattle ranching	Rick Lomba	1986
Fight for the Planet	Global warming		2009
Flow: For Love of Water	Water privatization: big business and the natural environment		2008
Food, Inc.	Agriculture: corporate farming		2009
Forbidden Forest	Forests: sustainable forest management		2004
Forests for the 21st Century	Forests: reforestation		2009

Title	Theme(s) and subtheme(s)	Author(s)	Year(s)
The Forgotten District	Forests: ecotourism in Belize		2008
The Fourth Revolution: Energy	Energy: renewable energy	Carl-A. Fechner	2010
Fractured Land	Environmentalism: awareness of hydraulic fracturing		2015
The Future of Food	Agriculture: genetically modified food	Deborah Koons Garcia	2004
Global Warming: The Signs and The Science	Global warming		2005
GMO OMG	Genetically modified organisms		2013
The Great Global Warming Swindle	Global warming: climate change denial		2007
The Great Warming	Global warming: politics of global warming		2006
Haida Gwaii: On the Edge of the World	General environment of Haida Gwaii		2015
Hinterland Who's Who	Wildlife	Environment Canada Wildlife Service and the National Film Board of Canada	1960s, 1970s, 2000s
Home	Various themes	Yann Arthus-Bertrand	2009
How to Change the World	Environmentalism: Greenpeace	Jerry Rothwell	2015
Ice and the Sky	Global warming: Antarctic	Luc Jacquet	2015
The Idiot Cycle	Chemical industry: genetically modified organisms	Emmanuelle Schick Garcia	2009
If a Tree Falls: A Story of the Earth Liberation Front	Forests: eco-terrorism	Marshall Curry and Matthew Hamacheck	2011
If You Love This Planet	Energy: nuclear weapons		1982
An Inconvenient Truth	Global warming	Al Gore	2006
An Inconvenient Truth...Or Convenient Fiction?	Global warming	Steven F. Hayward	2007
Journey to the Safest Place on Earth	Radioactive waste	Edgar Hagen	2013
Koyaanisqatsi	Various themes		1982
The Last Paradises: On the Track of Rare Animals	Endangered species	Eugen Schuhmacher	1967
Material Of The Future	Plastic		2014
The Messenger	Songbirds: bird conservation	Su Rynard	2015
Millennium: Tribal Wisdom and the Modern World (series of videos)	Various themes		1992
Minamata: The Victims and Their World	Water pollution: mercury poisoning	Noriaki Tsuchimoto	1971
Mountaintop Removal	Mining: mountaintop removal mining		2007
No Impact Man	Sustainability: sustainable living		2009
No Pressure	Climate change mitigation	Richard Curtis and Franny Armstrong	2010

Title	Theme(s) and subtheme(s)	Author(s)	Year(s)
Oil on Ice	Petroleum industry: Arctic Refuge drilling controversy	Stephen Most	2004
Oil Sands Karaoke	Petroleum industry: oil sands of Alberta		2013
On the Line	Petroleum industry: Enbridge Northern Gateway Pipelines	Frank Wolf	2011
Pandora's Promise	Nuclear power and climate change	Robert Stone	2013
Polluting Paradise	Waste management: in Turkey		2012
The Power of Community: How Cuba Survived Peak Oil	Peak oil: anti-consumerism		2006
Powerful: Energy for Everyone	Energy: renewable energy	David Chernushenko	2010
The Redwoods	Forests: protection of redwoods		1967
Renewal	Environmentalism: environmental movement in the United States		2008
The Return of Navajo Boy	Uranium mining in the United States		2000
Revenge of the Electric Car	Sustainable design: electric cars	Chris Paine and P.G. Morgan	2011
Revolution	Various themes	Rob Stewart	2012
Rokkasho Rhapsody	Nuclear power: radiation		2006
Semper Fi: Always Faithful	Water pollution: Camp Lejeune water contamination		2011
Sharkwater	Endangered species: sharks	Rob Stewart	2007
The Shiranui Sea	Water pollution: mercury poisoning	Noriaki Tsuchimoto	1975
Shored Up	Weather: Hurricane Sandy: rising sea level		2013
Silent Storm	Nuclear weapons testing: radioactive contamination	Peter Butt	2003
Spoiled the Movie	Energy: environmental impact of the energy industry, problems with alternative energy, etc.	Kevin Miller	2011
The Story of Stuff	Product lifecycle management: anti-consumerism	Annie Leonard	2007
The Superior Human?	Speciesism	Samuel McAnallen	2012
Surviving the Tsunami - My Atomic Aunt	Nuclear power: Fukushima Daiichi nuclear disaster		2013
That Should Not Be: Our Children Will Accuse Us	Agriculture: agrochemicals		2008
The Toxic Clouds of 9/11: A Looming Disaster	Air pollution: from September 11 attacks		2006
Toxic Legacy	Air pollution: from September 11 attacks		2006
Trash Inc: The Secret Life of Garbage	Waste management: landfills		2010
The True Cost	Various themes		2015
Ultimate Tornado	Tornadoes in the United States		2006
Unacceptable Levels	Environmental toxicology	Edward Brown	2013
Under the Dome	Air pollution in China	Chai Jing	2015

Title	Theme(s) and subtheme(s)	Author(s)	Year(s)
Vanishing of the Bees	Honey bees: colony collapse disorder	Maryam Henein, George Langworthy, and James Erskine	2009
Vanishing Point	Climate change: Arctic sea ice decline	Marina Endicott and others	2012
Virunga	Wildlife conservation: Democratic Republic of the Congo	Orlando von Einsiedel	2014
Voices of Transition	Agriculture: energy and resource scarcity	Nils Aguilar	2012
Walking Through A Minefield	Uranium mining in Australia	Cathy Henkel	1999
Watch	Forests: clearcutting	Briana Waters	2001
Watermark	Water	Jennifer Baichwal and Edward Burtynsky	2013
What a Way to Go: Life at the End of Empire	Various themes: peak oil, climate change, overpopulation, and species extinction	Timothy S. Bennett	2007
What Is the Electric Car?	Sustainable design: electric cars		2010
White Light/Black Rain: The Destruction of Hiroshima and Nagasaki	Atomic bombings of Hiroshima and Nagasaki: acute radiation syndrome	Steven Okazaki	2007
Who Killed the Electric Car?	Sustainable design: electric cars and roles of industries and governments	Chris Paine	2006
Wind Over Water: The Cape Wind Story	Energy: Cape Wind	Ole Tangen Jr.	2003
Windfall	Wind power: environmental impact of wind power		2010
The World According to Monsanto	Agriculture: Monsanto	Marie-Monique Robin	2008

Table 2.2: List of fictional films about the environment. Some fictional films are based on true events.

Title	Theme(s) and subtheme(s)	Author(s)	Year(s)
2012	Natural disasters: 2012 phenomenon	Harald Kloser and Roland Emmerich	2009
The Age of Stupid	Climate change	Franny Armstrong	2009
Aftermath	Nuclear warfare	Christian McDonald	2012
Avatar	Natural resources	James Cameron	2009
Birdemic: Shock and Terror	Global warming	James Nguyen	2008
Captain Planet, Ted Turner's animated television series	Various themes		
Carface	Petroleum industry: car culture		2015
The China Syndrome	Energy: nuclear meltdown	Mike Gray, T.S. Cook, and James Bridges	1979
A Civil Action	Water pollution: by trichloroethylene	Steven Zaillian	1998

Title	Theme(s) and subtheme(s)	Author(s)	Year(s)
Climate Change Denial Disorder	Climate change denial	Nicol Paone	2015
The Day After Tomorrow	Climate change	Roland Emmerich and Jeffrey Nachmanoff	2004
Denmark	Pollution: water pollution		2010
Earthquake	Earthquake in Los Angeles	George Fox and Mario Puzo	1974
Earthquake in Chile	Earthquake in Chile	Helma Sanders-Brahms	1975
Erin Brockovich	Energy development: water pollution	Susannah Grant	2000
FernGully: The Last Rainforest	Forests: rainforests		1992
Fire Down Below	Pollution: from mining		1997
Fly Away Home	Wildlife: Canada goose	Bill Lishman, Robert Rodat, and Vince McKewin	1996
H_2O	Water supply: Kaveri River water dispute in India	Upendra	2000
The Happening	Toxicology	M. Night Shyamalan	2008
Happy Feet	Environment of Antarctica	Warren Coleman, John Collee, George Miller, and Judy Morris	2006
Hell and High Water	Atomic bomb explosion in the Cold War	Jesse L. Lasky, Jr., Samuel Fuller, and David Hempstead	1954
An Inconvenient Penguin	Invasive species		2008
The Lorax	Forests	Dr. Seuss	2012
Muro Ami	Fishing: illegal, unreported and unregulated fishing	Marilou Diaz-Abaya	1999
On Deadly Ground	Pollution: from petroleum industry	Ed Horowitz and Robin U. Russin	1994
Once Upon a Forest	Forests		1993
Plastic Bag	Plastic bags	Ramin Bahrani and Jenni Jenkins	2009
Safe	Multiple chemical sensitivity	Todd Haynes	1995
San Andreas	Earthquake in California		2015
Sharknado	Severe weather and sharks	Thunder Levin	2013
Sizzle	Global warming	Randy Olson	2009
Soylent Green	Natural resources	Richard Fleischer	1973
Space Tourists	Pollution: space debris		2002
Star Trek IV: The Voyage Home	Wildlife: Humpback whale		1986
Super Comet: After The Impact	Impact event		2007
Tentacles	Wildlife: gigantic octopus	Steven W. Carabatsos, Tito Carpi, Jerome Max, and Sonia Molteni	1977
WALL-E	Waste management		2008

Many anime movies by Hayao Miyazaki carry a hidden environmentalist message. The best-known is Nausicaä of the Valley of the Wind, but also well known is Pom Poko as well as Princess Mononoke, which is based on a conflict between technology and nature.

Television

Title	Theme(s) and subtheme(s)	Producer(s)	Year(s)
Saving Britain's Past	Cultural heritage	Mary Sackville-West	2009
The Truth About Wildlife	Wildlife: United Kingdom	Simon Willis	2011
Years of Living Dangerously	Climate change	James Cameron, Jerry Weintraub and Arnold Schwarzenegger	2014

2.8 ENVIRONMENTAL CHALLENGES IN A GLOBAL CONTEXT

1. Environmental challenges in Europe and in the rest of the world are intertwined.

There is a two-way relationship between Europe and the rest of the world. Europe is contributing to environmental pressures and accelerating feedbacks in other parts of the world through its dependence on fossil fuels, mining products and other imports. Conversely, in a highly interdependent world, changes in other parts of the world are increasingly felt closer to home, both directly through the impacts of global environmental changes, or indirectly through intensified socio-economic pressures.

Climate change is an obvious example. Most of the growth in global greenhouse gas emissions is projected to occur outside Europe, as a result of increasing wealth in populous emerging economies. In spite of successful efforts to reduce emissions and a decreasing share in the global total, European societies continue to be major emitters of greenhouse gases.

Many of the countries that are most vulnerable to climate change are outside the European continent, others are our direct neighbours. Often these countries are highly dependent on climate-sensitive sectors such as farming and fishing. Their adaptive capacity varies, but is often rather low, in particular due to persistent poverty. The links between climate change, poverty and political and security risks and their relevance for Europe have been extensively analysed.

Biodiversity has continued to decline globally despite a few encouraging achievements and increased policy action. The global rate of species extinction is escalating and is now estimated to be up to 1000 times the natural rate. Evidence is growing that critical ecosystem services are under great pressure globally. According to one estimate, approximately one quarter of the potential net primary production has been converted by humans, either through direct cropping (53%), land-use-induced productivity changes (40%) or human-induced fires (7%). While such figures should be treated with caution, they do give an indication of the substantial impact of humans on natural ecosystems.

Loss of biodiversity in other regions of the world affects European interests in several ways. It is the world's poor that bear the brunt of biodiversity loss, as they are usually most directly reliant on functioning ecosystem services. Increases in poverty and inequality are likely to further fuel conflict and instability in regions that are already characterised by often fragile governance structures. Moreover, reduced genetic variety in crops and cultivars implies future losses of economic and social benefits for Europe in such critical areas as food production and modern healthcare.

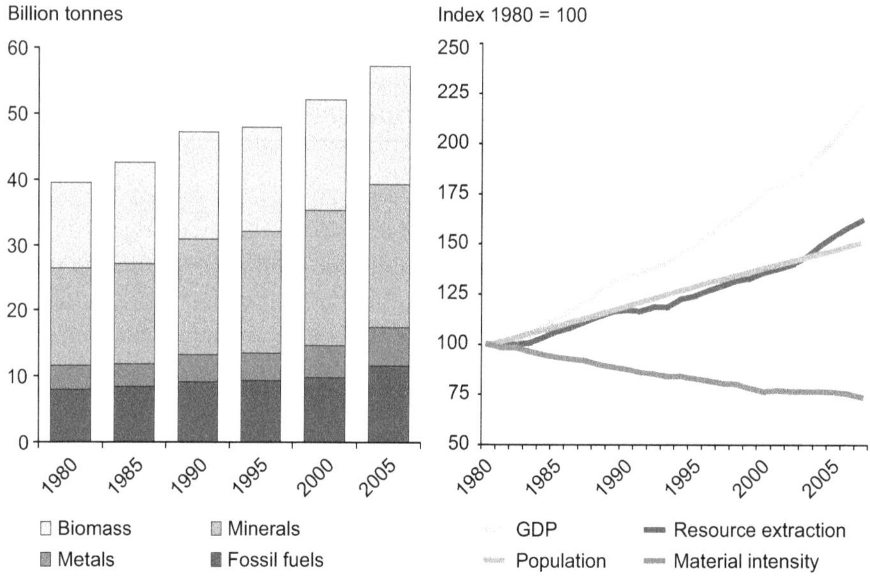

Fig. 2.3: Global extraction of natural resources from ecosystems and mines, 1980 to 2005/2007. *Source:* SERI Global Material Flow Database.

Global extraction of natural resources from ecosystems and mines grew more or less steadily over the past 25 years, from 40 billion tonnes in 1980 to 58 billion tonnes in 2005. Resource extraction is unevenly distributed across the world, with Asia accounting for the largest share in 2005 (48% of total tonnage, compared with Europe's 13%). Over this period, a relative decoupling of global resource extraction and economic growth took place: resource extraction increased by roughly 50% and world economic output (GDP) by about 110%.

Nonetheless, resource use and extraction is still increasing in absolute terms, outweighing gains in resource efficiency. Such a composite indicator does not, however, reveal information on specific resource developments. Global food, energy and water systems appear to be more vulnerable and fragile than thought a few years ago, the factors responsible being increased demand, decreased supply, and supply instabilities. Over-exploitation, degradation and loss of soils are relevant concerns in this regard. With global competition and increased geographic and corporate concentration of supplies for some resources, Europe faces increasing supply risks.

In spite of general progress in the area of environment and health in Europe, the global human toll of environmental health impacts remains deeply worrying. Unsafe water, poor sanitation and hygiene conditions, urban outdoor air pollution, indoor smoke from solid fuels and lead exposure and global climate change account for nearly a tenth of deaths and disease burden globally, and around one quarter of deaths and disease burden in children under 5 years of age. It is again poor populations in low latitudes that are affected most heavily.

Table 2.3: Death and DALYs (disability-adjusted life years) attributable to five environmental risks, by region, 2004

Risk	World	Low and middle income	High income
Percentage of deaths			
Indoor smoke from solid fuels	3.3	3.9	0.0
Unsafe water, sanitation, hygiene	3.2	3.8	0.1
Urban outdoor air pollution	2.0	1.9	2.5
Global climate change	0.2	0.3	0.0
Lead exposure	0.2	0.3	0.0
All five risks	8.7	9.6	2.6
Percentage of DALYs			
Indoor smoke from solid fuels	2.7	2.9	0.0
Unsafe water, sanitation, hygiene	4.2	4.6	0.3
Urban outdoor air pollution	0.6	0.6	0.8
Global climate change	0.4	0.4	0.0
Lead exposure	0.6	0.6	0.1
All five risks	8.0	8.6	1.2

Source: World Health Organisation.

Many low and middle-income countries now face a growing burden from new risks to health, while still fighting an unfinished battle with the traditional risks to health. The World Health Organisation (WHO) forecasts that between 2006 and 2015, deaths from non-communicable diseases could increase worldwide by 17%. The greatest increase is projected for the African region (24%) followed by the eastern Mediterranean region (23%). Europe is likely to be faced with the increased problem of emerging or re-emerging infectious diseases that are critically influenced by changes in temperature or precipitation, habitat loss and ecological destruction. In an increasingly urbanised world, which is tightly linked by long-distance transport, the incidence and distribution of infectious diseases affecting humans is likely to increase.

2. Links between environmental challenges are particularly apparent in Europe's direct neighbourhood

Europe's direct neighbourhood—the Arctic, the Mediterranean and the eastern neighbours—is worth particular attention here due to the strong socio-economic and environmental links and the importance of these regions in EU external policy. Furthermore, some of the world's largest reservoirs of natural resources are in these regions, which is of immediate relevance to a resource-scarce Europe.

Box 2.2: Global Sea-level Rise and Ocean Acidification

During the 20th century, global sea level rose by an average of 1.7 mm/year. This was due to an increase in the volume of ocean water as a consequence of temperature rise, although inflow of water from melting glaciers and ice sheets is playing an increasing role. In the past 15 years, sea-level rise has been accelerating and averaged about 3.1 mm/year, based on data from satellites and tide gauges, with a significantly increasing contribution from the ice sheets of Greenland and Antarctica. Sea level is projected to rise considerably during this century and beyond.

In 2007, the IPCC presented a projected rise of 0.18 to 0.59 m above the 1990 level by the end of the century. However, since 2007, reports comparing the IPCC projections with observations show that sea level is currently increasing at an even greater rate than indicated by these projections. Recent estimates suggest, in case of unabated greenhouse gas emissions, a projected global average sea-level rise of about 1.0 m or possibly (although unlikely) even up to 2.0 m, by 2100.

Ocean acidification is a direct consequence of CO_2 emissions to the atmosphere. The oceans have already taken up around a third of the CO_2 produced by humankind since the industrial revolution. While this has limited the amount of CO_2 in the atmosphere somewhat, it has come at the price of a significant change to ocean chemistry. Evidence indicates that ocean acidification is likely to become a serious threat to many organisms and will have implications for food webs and ecosystems, for example, tropical coral reefs.

It is expected that, at atmospheric carbon dioxide concentrations above 450 ppm, large areas of the polar oceans will probably become corrosive to shells of key marine calcifiers, an effect that will be strongest in the Arctic. Already, loss of shell weight in planktonic Antarctic calcifiers has been observed. The rate of change in ocean chemistry is high, and faster than previous ocean acidification-driven extinctions in the Earth's history .

Source: EEA.

These regions are also home to some of the world's richest and yet most fragile natural environments which are facing multiple threats. At the same time, concerns remain related to many transboundary issues like water management and air pollution deposition shared between Europe and its neighbours. Some of the main environmental challenges in these regions include:

- *The Arctic*—European activities, such as those resulting in long-range emission of air pollution, black carbon and greenhouse gas emissions, leave a considerable footprint in the Arctic. At the same time what happens in the Arctic also influences Europe's environment because the Arctic plays a key role, for example, in the context of climate change and related sea-level rise projections. Furthermore, multiple pressures on Arctic ecosystems have resulted in biodiversity loss across the region. Such changes have global repercussions because of the loss of key ecosystem functions and are creating additional challenges for the people living in the Arctic as changing seasonal patterns affect hunting and food provision.

- *Eastern neighbours*—EU neighbours to the east face many environmental challenges affecting human health and ecosystems. The EEA's fourth assessment report of Europe's environment summarises key environmental issues across the pan-European region, including countries in Eastern Europe, the Caucasus and Central Asia. It focuses on the challenges posed by air and water pollution, climate change, biodiversity loss, pressures on the marine and coastal environment, consumption and production patterns, and assesses sectoral developments that drive environmental change across the region.

Box 2.3: The European Neighbourhood Policy

The European Neighbourhood Policy (ENP) aims to strengthen cooperation between the EU and its neighbours. It is a dynamic and evolving platform for dialogue and action based on joint responsibility and ownership. In recent years, the ENP has been further strengthened through initiatives such as the Eastern Partnership, Black Sea Synergy and Union for the Mediterranean.

Within the ENP, relevant EU instruments—the EU maritime policy, the Water Framework Directive and the development of a Shared Environmental Information System (SEIS)—are gradually being implemented beyond EU borders to help streamline environmental efforts. International legal instruments have also been developed and gradually implemented to address common transboundary issues—such as the UN LRTAP Convention or transboundary water convention, covering also the eastern neighbours.

For the Mediterranean, the Horizon 2020 initiative (k) supports the riparian countries in addressing the priority issues of dealing with industrial emissions, municipal waste and wastewater treatment to reduce pollution of the Mediterranean.

Within the Arctic, a number of environmental treaties and conventions, as well as shipping and industrial regulations, provide a backdrop for policy deliberations in the context of the EU's Arctic policy: while the EU has taken the first steps towards an Arctic Policy, no comprehensive policy approach exists at present, several EU policies—such as the EU's agricultural policy, fisheries policy, maritime policy, environmental and climate policy or energy policy—affect the Arctic environment both directly and indirectly.

However, it is worth noting here that environmental trend analyses covering Europe's neighbouring regions often lack reliable data and indicators that are comparable over time and space. Better and more targeted information to underpin environmental analysis and assessment is needed.

The EEA—within the framework of the European Neighbourhood Policy, and in cooperation with the countries and main partners in the regions—is implementing a series of activities that aim to strengthen existing environmental monitoring, data and information management.

Source: EEA.

- *The Mediterranean*—Located at the crossroad of three continents this is one of the richest 'eco-regions' and yet one of the most vulnerable natural environments in the world. The recent report on the State of the Environment and Development in the Mediterranean presents the major impacts of climate change, the characteristics of the natural resources and environment in the region, and the challenges linked to their conservation. In particular, some of the main pressures from human activities are identified (such as tourism, transport, and industry) and their impacts on coastal and marine ecosystems are assessed, together with considerations about their environmental sustainability.

While Europe is contributing directly and indirectly to some of the environmental pressures in these regions, it is also in an unique position to cooperate to improve their environmental conditions, particularly through fostering technology transfer and helping to build institutional capacity. These dimensions are increasingly reflected in European neighbourhood policy priorities.

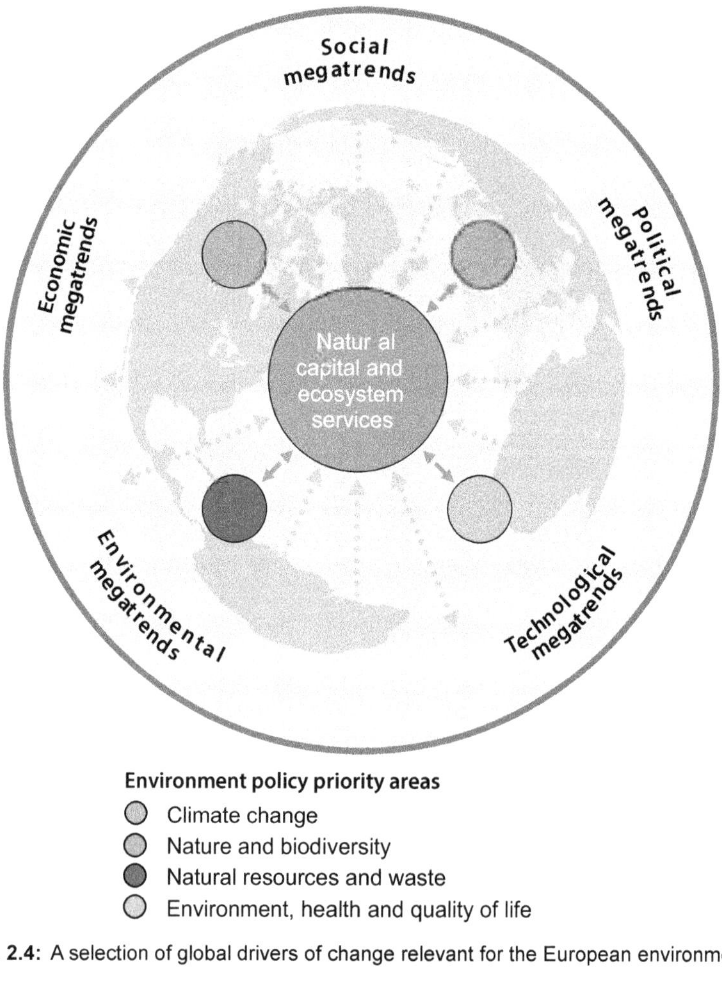

Fig. 2.4: A selection of global drivers of change relevant for the European environment.

3. Environmental challenges are closely connected with global drivers of change

A range of unfolding trends are shaping the future European and global context, and many of these are outside the realm of Europe's direct influence. Related global megatrends cut across social, technological, economic, political and even environment dimensions. Key developments include changing demographic patterns or accelerating rates of urbanisation, ever faster technological changes, deepening market integration, evolving economic power shifts or the changing climate.

4. A selection of global megatrends

- Increasing global divergence in population trends: populations ageing, growing and migrating
- Living in an urban world: spreading cities and spiralling consumption
- Changing patterns of global disease burdens and the risk of new pandemics
- Accelerating technologies: racing into the unknown
- Continued economic growth
- Global power shifts: from a uni-polar to a multi-polar world
- Intensified global competition for resources
- Decreasing stocks of natural resources
- Increasing severity of the consequences of climate change
- Increasingly unsustainable environmental pollution load
- Global regulation and governance: increasing fragmentation, but converging outcomes.

Table 2.4: Population of the world and different regions, 1950, 1975, 2005 and 2050 according to different growth variants

Region	Population in millions			Population in 2050			
	1950	1975	2005	Low	Medium	High	Constant
World	2 529	4 061	6 512	7 959	9 150	10 461	11 030
More developed regions	812	1 047	1 217	1 126	1 275	1 439	1 256
Less developed regions	1 717	3 014	5 296	6 833	7 875	9 022	9 774
Africa	227	419	921	1 748	1 998	2 267	2 999
Asia	1 403	2 379	3 937	4 533	5 231	6 003	6 010
Europe*	547	676	729	609	691	782	657
Latin America and Caribbean	167	323	557	626	729	845	839
Northern America	172	242	335	397	448	505	468
Oceania	13	21	33	45	51	58	58
Europe (EEA-38)	419	521	597	554	628	709	616

Note: *Europe (in UN terminology) includes all EEA member countries (except Turkey) and EEA cooperating countries, as well as Belarus, Republic of Moldova, Russian Federation, Ukraine.

Source: United Nations Population Division.

In 1960, the world's population was 3 billion. Today, it is about 6.8 billion. The United Nations Population Division expects this growth to continue and that the global population will exceed 9billion by 2050, according to the 'medium growth variant' of their population estimate. However, uncertainties are apparent, and forecasts depend on several assumptions, including for fertility rates. As such, by 2050, the world population could thus exceed 11 billion or be limited to 8 billion. The implications of this uncertainty for global resource demands are huge.

In contrast to the global trend, European populations are expected to decline and age significantly. In its neighbourhood, population decline is particularly dramatic in Russia and large parts of Europe. At the same time, northern African countries along the southern Mediterranean are witnessing strong population growth. In general, the wider region of Northern Africa and the Middle East has experienced the highest rate of population growth of any region in the world over the past century.

The regional distribution of population growth, the age structure, and migration between regions are also important. Ninety percent of the population growth since 1960 has been in countries classified as 'less developed' by the United Nations. Meanwhile, the world is urbanising at an unprecedented rate. By 2050, about 70% of the global population is likely to live in cities, compared with less than 30% in 1950. Population growth is now largely an urban phenomenon concentrated in the developing world, particular Asia, which is estimated to be home to more than 50% of the global urban population by 2050.

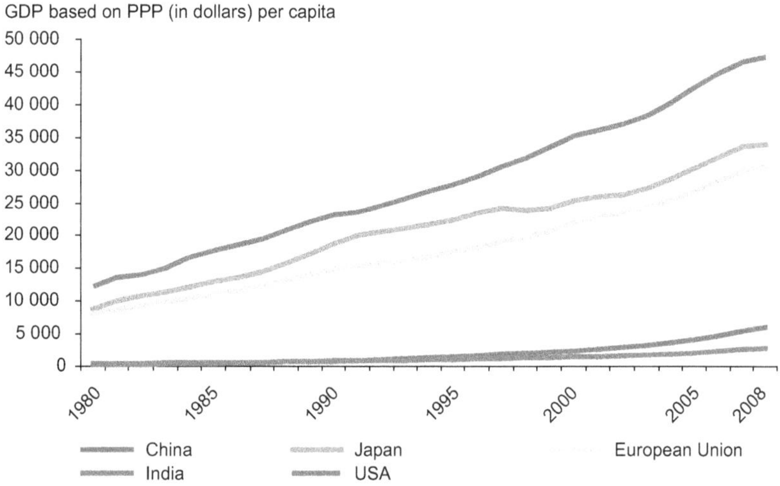

Fig. 2.3: Growth of GDP per capita in the EU-27, USA, China, Japan and India, 1980 to 2008. *Source:* International Monetary Fund.

Global integration of markets, shifts in global competitiveness and changing global spending patterns comprise another complex set of drivers. As a result of liberalisation and due to the lowering of transport and communication costs, international trade over the past half-century has grown rapidly: global exports grew in value from US$296 billion in 1950 to more than US$8 trillion (measured in relation to 'purchasing power parity') in 2005, and their share of global GDP rose from around 5% to close to 20%. Similarly, remittances sent home from emigrant workers often represent a large source of income for developing countries. For some countries remittances exceeded a quarter of the respective GDP in 2008 (forexample, 50% in Tajikistan, 31% in Moldova, 28% in the Krygyz Republic, and 25% in Lebanon).

Aided by globalisation, many countries have been able to lift larger proportions of their populations out of poverty. Global economic growth and trade integration have fuelled long-term shifts in international competitiveness, characterised by a high growth of productivity in emerging economies. The number of middle-income consumers world-wide is growing rapidly, particularly in Asia. The World Bank has estimated that, by 2030, there could be 1.2 billion middle-income consumers (C) in the emerging and developing economies of today. Already in 2010, the economies of the BRIC countries—Brazil, Russia, India and China—are expected to contribute almost half of global consumption growth.

Large differences in individual wealth accumulation are expected to persist between developed economies and key emerging economies. Yet the world's economic balance of power is changing. Large shifts in purchasing power towards middle-income economies and middle-income consumers are underway, creating significant consumer markets in emerging markets that are likely to fuel future global resource demands, again particularly in Asia. According to one estimate, the BRIC countries together could match the G7 share of global GDP by the 2040s.

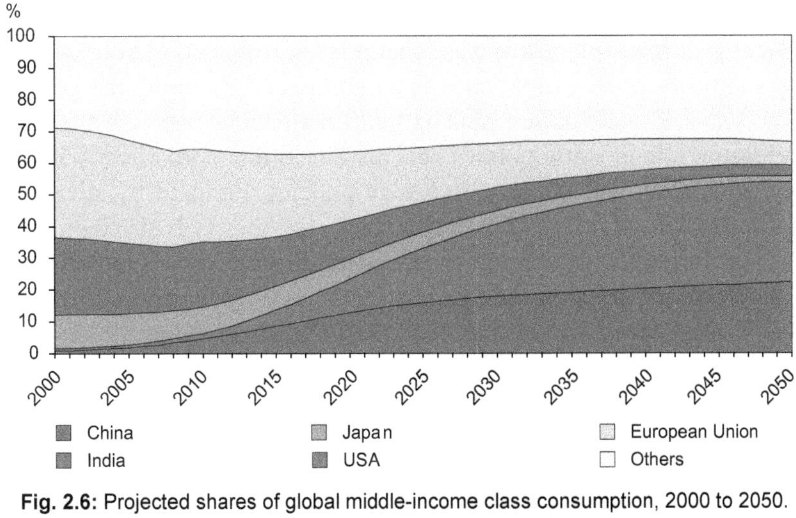

Fig. 2.6: Projected shares of global middle-income class consumption, 2000 to 2050.

Source: Kharas.

A number of critical uncertainties are, however, embedded in those projections. Examples include uncertainties about the degree to which Asia might integrate economically, the impact of population ageing and the capacity to strengthen private investment and education. In the context of greater interconnectedness of markets and a higher susceptibility to risks of market failures, global regulatory regimes are likely to expand in the future, yet their contours and thus their role are unpredictable.

Furthermore, the speed and scope of scientific and technological progress influences key socio-economic trends and drivers. Eco-innovation and eco-friendly technologies are of key relevance in this regard; European companies are already relatively well-positioned in global markets. Supporting policies are relevant both in terms of facilitating market entry of new eco-innovations and technologies as well as increasing global demand.

In the longer-term perspective, developments and technology convergence in nanoscience and nanotechnologies, biotechnologies and life-sciences, information and communication technologies, cognitive sciences and neuro-technologies are expected to have profound effects on economies, societies and the environment. They are likely to open up completely new options for mitigating and remedying environmental problems including, for example, new pollution sensors, new types of batteries and other technologies for energy storage, and lighter and more durable materials for cars, buildings or aircrafts.

However, these technologies also give rise to concerns about detrimental effects on the environment, given the scale and level of complexity of their interactions. The existence of unknown, even unknowable, impacts poses a great challenge to risk governance. Rebound effects might also jeopardize environmental and resource-efficiency achievements.

As a result of demographic and economic power shifts, the contours of the global governance landscape are changing. A diffusion of political power towards multiple poles of influence is on-going, and changing the geo-political landscape. Private actors such as multi-national enterprises are playing an increasing role in world politics, and are becoming more directly involved in the formulation and implementation of policies. Fostered by advances in communications and information technology, civil society is also increasingly taking part in global negotiation processes of all kind. The interdependence and complexity of decision-making is growing as a result, giving rise to new modes of governance and posing new questions about responsibility, legitimacy and accountability.

2.8.1 Environmental Challenges may Increase Risks to Food, Energy and Water Security on a Global Scale

Global environmental challenges, such as impacts of climate change, loss of biodiversity, over-use of natural resources and environmental and health issues, are critically linked to issues of poverty and the sustainability of ecosystems, and consequently, issues of resource security and political stability. This adds pressure and uncertainty to the overall competition for natural resources, which might intensify as a consequence of increased demands, decreased supplies and decreased stability of supplies. Ultimately, this further increases pressure on ecosystems globally, and especially their capacity to ensure continued food, energy and water security.

According to the Food and Agriculture Organisation of the United Nations (FAO), demand for food, feed and fibres could grow by 70% by 2050. The fragility of global food, water and energy systems has become apparent over recent years. For example, arable land per person declined globally from 0.43 ha in 1962 to 0.26 ha in 1998. The FAO expects this value to fall further by 1.5% per year between now and 2030, if no major policy changes are initiated.

Similarly, the International Energy Agency (IEA) expects global demand for energy to rise by 40% over the next 20 years if no major policy changes are implemented. The IEA has repeatedly warned about an impeding global energy crisis due to rising long-term demand. Massive and continuous investments are needed in energy efficiency, renewable energies and new infrastructures to achieve the transition to a low-carbon, resource efficient energy system that is compliant with long-term environmental objectives.

Seeking to carry forward the Gujarat-model of irrigation reform in the country and to quickly complete the task of aquifer mapping to quantify available groundwater resources, the Centre has constituted a seven-member committee to restructure the central water commission (CWC) and the central ground water board (CGWB)—the two key central bodies that manage water resources in India.

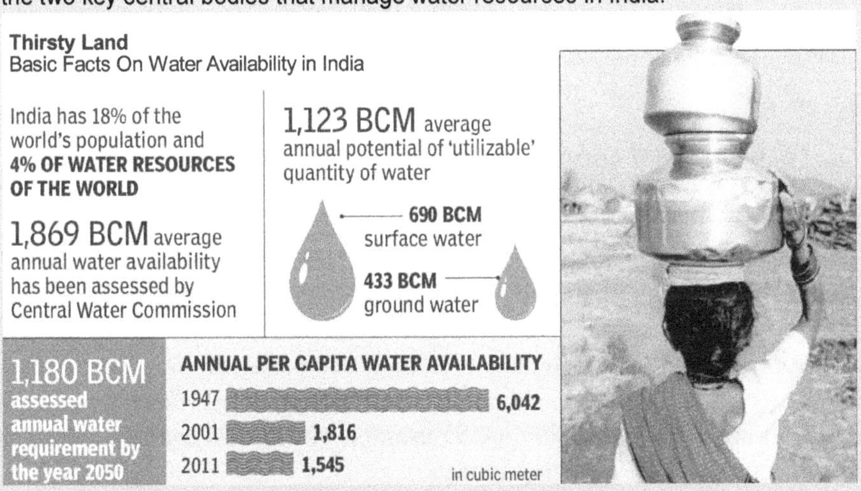

Andhra Pradesh has created history by completing the river linking project to divert about 80 TMC of water from Godavari to Krishna. The linkage of Godavari water with Krishna basin is being made possible by the 'Pattiseema' lift irrigation scheme on Godavari river in West Godavari district.

But it could be water shortages that will hit hardest over the coming decades. One estimate suggests that in just 20 years, global demand for water could be 40% higher than today, and more than 50% higher in the most rapidly developing countries. Furthermore, according to a recent estimate prepared by the Secretariat of the Convention of Biological Diversity, the flow in more than 60 per cent of the large river systems in the world has been heavily altered. Limits of ecological sustainability of water availability for abstraction have thus been reached, and upto 50% of the world could be living in areas with high water stress by 2030, while more than 60% could still lack improved access to sanitation.

Water infrastructure systems are often old and there is a lack of information about actual performance and losses. One estimate foresees an average annual investment need of US$772 billion for maintaining water and wastewater services around the world by 2015. Here, potential for ripple effects for food and energy supply exist, for example, cutting agricultural output which could result in decreasing overall social resilience.

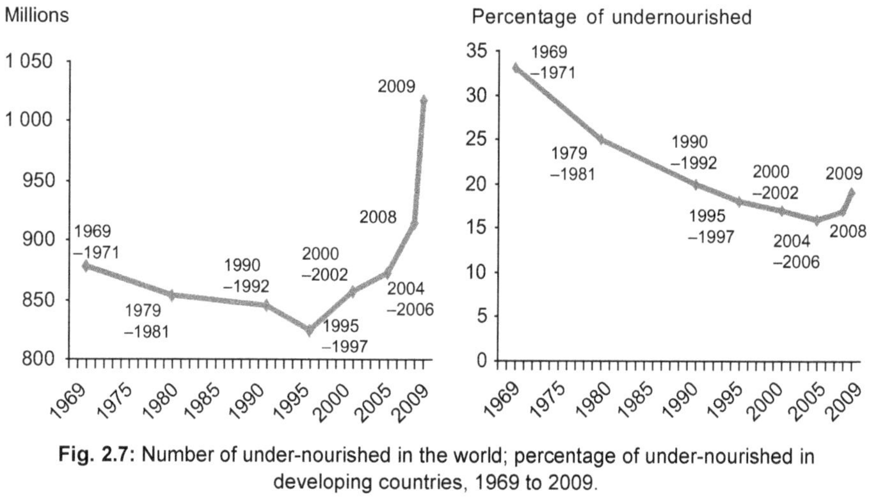

Fig. 2.7: Number of under-nourished in the world; percentage of under-nourished in developing countries, 1969 to 2009.

Source: Food and Agriculture Organisation of the United Nations.

Already today, in many parts of the world, non-renewable resource use is close to its limit and potentially renewable resources are being used beyond their reproductive capacity. This kind of dynamics can also be recognised in Europe's neighbouring regions with their comparatively rich natural capital. Water resource over-exploitation, combined with insufficient access to safe drinking water and sanitation, for example, are critical challenges both in Eastern Europe and the Mediterranean.

At the global level, poverty and social exclusion are further exacerbated by ecosystem degradation and changes in the climate. Globally, efforts to alleviate extreme poverty were reasonably effective until the 1990s. However, the recurring food and economic crises throughout 2006 to 2009 have magnified the trend of increasing under-nourishment rates around the world. The number of undernourished rose, for the first time, to more than 1 billion in 2009 and the proportion of undernourished in developing countries, which was declining quite rapidly, has risen in the past few years.

Resource over-exploitation and changes in the climate aggravate threats to natural capital. They also affect quality of life, potentially undermining social and political stability. Furthermore, the livelihoods of billions of people are inevitably linked with the sustainability of local ecosystem services. Combined with demographic pressures, decreasing socio-ecological resilience can add a new dimension to the environment and security debate, as conflict around scarcer resources is likely to intensify and add to migration pressures.

Box 2.4: Towards Identifying Environmental Thresholds and Planetary Boundaries

Earth system scientists are trying to understand the complexity of the interactions in bio-geophysical processes that determine the Earth's capacity for self-regulation. In this regard ecologists have observed thresholds in a range of essential ecosystem processes, that when crossed cause the functioning of an ecosystem to fundamentally change.

More recently, a group of scientists have proposed a number of planetary boundaries within which humanity must stay to avoid catastrophic environmental change. They suggest that three critical boundaries have already been exceeded; the rate of biodiversity loss, climate change and human interference with the nitrogen cycle, but acknowledge that there are serious knowledge gaps and uncertainties.

The attempt to identify and quantify such planetary boundaries has started a broader debate about the feasibility of such an undertaking, and whether it is meaningful to calculate a global rate for processes some of which are inherently localised, for example nitrate levels and the loss of biodiversity. While the general value of such a scientific exercise can be acknowledged, concerns have been raised about the scientific justification, the possibility of choosing exact values that are non-arbitrary and the problems of reducing the complexity of interactions into single boundary values.

Problems might arise with regard to balancing limits with ethical and economic issues and confusing values with targets. Some argue that the setting of quantitative boundaries might delay effective action and contribute to the degradation of the environment up to the point of no return.

Source: EEA.

2.8.2 Global Developments may Increase Europe's Vulnerabilities to Systemic Risks

Since many of the global drivers of change operate beyond Europe's direct influence, Europe's vulnerability to external change could increase markedly, particularly accentuated by developments in its direct neighbourhood. Being a resource-scarce continent and neighbour to some of the world's regions most prone to global environmental change, active engagement and cooperation with these regions can help address the range of problems that Europe is facing.

Many key drivers operate on a global scale and are likely to unfold over decades rather than years. In a recent assessment, the World Economic Forum warned about a higher level of systemic risk due to the increase in interconnections among various risks. Furthermore, the assessment emphasised that unexpected, sudden changes in external conditions are inevitable in a highly inter-linked world. While sudden changes can have huge impacts, the biggest risks may be from slow failures which unfold their full damage potential over decades and may be seriously underestimated in their potential economic impact and societal cost. The continued over-exploitation of natural capital is an example for a slow failure.

Box 2.5: Tipping points: Risks of Large Scale (Non-linear) Climate Change

What are tipping points? If a system has more than one equilibrium state, transitions to structurally different states are possible. If and when a tipping point is passed, the development of the system is no longer determined by the time-scale of the pressure, but rather its internal dynamics, which can be much more rapid than the original pressure.

A variety of tipping points have been identified, some of which have potentially significant consequences for Europe—however, it is worth noting that these may unfold on very different, and sometimes very long, time-scales.

One of the potential large-scale changes likely to affect Europe is the deglaciation of the West Antarctic Ice Sheet (WAIS) and Greenland ice sheet (GIS)—there is already evidence of accelerated melting of the GIS. Sustained 1–2°C, respectively 3–5°C, global warming above 1990 temperatures could be tipping points beyond which at least partial deglaciation of respectively GIS and WAIS and a significant rise in sea level will follow.

There is less confidence about other non-linear effects, for example, what may happen with ocean circulation. Parts of the Atlantic meridianal overturning circulation exhibits considerable seasonal and decadal variability, but data do not support a coherent trend in the overturning circulation. A slow-down of the meridianal overturning circulation may temporarily counteract global warming trends in Europe, but may have unexpected and serious consequences elsewhere.

Other examples of possible tipping points are the accelerated emission of methane from permafrost melting, destabilisation of hydrates on the ocean floor, and rapid climate-driven transitions from one ecosystem type to another. The understanding of these processes is as yet limited and the chance of major implications in the current century is generally considered to be low.

Source: EEA.

Such systemic risks—whether they manifest themselves as sudden changes or slow failures—include the potential damage to, or even full failure of, an entire system, for example a market or an ecosystem, as opposed to effects on individual elements only. The interconnectedness between drivers and risks highlighted here are relevant in this regard: while these links can lead to higher robustness when risk sharing is distributed across a greater number of elements in the system, they can also lead to greater fragility. Failure in one critical link can have cascading effects, often as a consequence of decreased system diversity and governance gaps.

A key related risk is that of accelerating global environmental feedback mechanisms and their direct and indirect impacts on Europe. Since the Millennium Ecosystem Assessment and the IPCCFourth Assessment Report, scientific assessments have warned that environmental feedback mechanisms are increasing the likelihood of large scale non-linear changes in key Earth system components. With increasing global temperatures, for example, there is an increasing risk of passing tipping points that may trigger large-scale, non-linear changes.

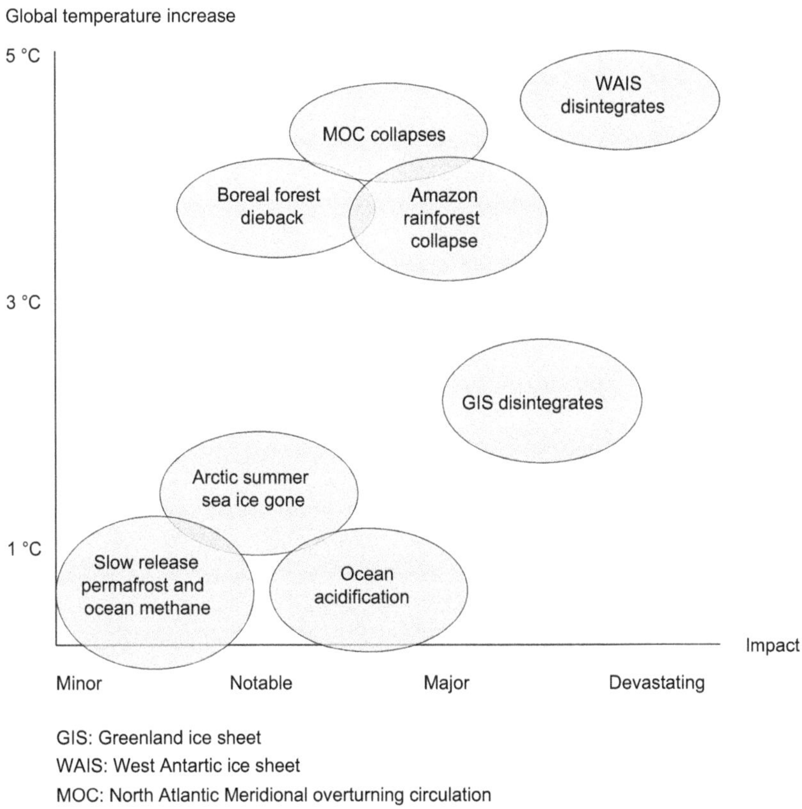

Fig, 2.8: Estimated global warming at which the onset of the events could occur versus their impact.

Note: The shapes and sizes of the ovals do not represent uncertainties in the impact and temperature onset of eventualities. These uncertainties may be significant.

Systemic risks have the potential, if they are not properly addressed, to inflict devastating damage on the vital systems, natural capital and infrastructures on which our well-being depends both at a local and at a global scale. Thus, joint efforts are required to tackle some of the causes of systemic risks, develop adaptive management practices and strengthen resilience in view of increasingly pressing environmental challenges.

■ ■ ■

3

Global Response to Green Economy Initiatives

3.1 INTRODUCTION

We stand at a critical moment in Earth's history, a time when humanity must choose its future. The identification of the 'green economy' as one of the key themes for the UN Conference on Sustainable Development 2012 (UNCSD) or 'Rio+20', represented a significant opportunity to define a new global economic paradigm, but it also entailed a risk that previously 'hard-won' global agreements on sustainable development might be lost or obscured in the pursuit of new agenda. There are also as yet few definitions as to what constitutes a 'green economy', which make some actors weary about its potential application.

Recognizing the challenges of this 'new' agenda, it is therefore critical that any global agreements that advance progress towards a green economy are governed by an over-arching set of principles that have common currency among governments and stakeholders alike. Arriving at an all-encompassing definition of a green economy may be both laborious and constraining, so the application of broader principles may ultimately prove more helpful.

There is range of literature that can be drawn upon in the identification of some common guiding principles for the green economy. The following document aims to combine some of the most prominent existing principles relating to sustainable development and the green economy into a cohesive guiding tool. Fifteen principles have been identified that represent a consolidation of existing international agreements and more radical and forward-thinking proposals, cutting across The Stockholm Declaration, the Rio Declaration, The Johannesburg Declaration, The Earth Charter, The One

Planet Living Principles, The Green Economy Coalition, the TUC 'Just Transition' principles, and The New Economics Foundation.

One of the goals of the Rio+20 Conference is to secure renewed political commitment for sustainable development. To enhance political will to achieve an ambitious outcome from Rio+20, the international community will have to agree on a common ethical framework of shared values and principles. This document offers State and non-State actors an overview the kinds of principles that might constitute that framework.

3.2 PRINCIPLES FOR A GREEN ECONOMY

1 Equitable Distribution of Wealth

Promote the equitable distribution of wealth within nations and among nations, to reduce disparities between rich and poor, and achieve social and economic justice, within a sustainable and fair share of the world's resources and leaving sufficient space for wildlife and wilderness.

2. Economic Equity and Fairness

Guided by the principle of common but differentiated responsibilities, create economic partnerships that would transfer substantial financial and technological assistance to less developed countries, to help minimize the gap between the developed and developing world and support the environmental sustainability of both.

3. Intergenerational Equity

Environmental resources and ecosystems must be carefully managed and safeguarded so as to enhance the value of environmental assets for future generations, thereby equitably meeting their needs and allowing them to flourish.

4. Precautionary Approach

Science should be utilized to enhance social and environmental outcomes, through the identification of environmental risk. Scientific uncertainty of environmental impacts shall not lead to avoidance of measures to prevent environmental degradation. The 'burden of proof' should lie with those claiming that there will not be significant environmental impacts.

5. The Right to Development

Human development in harmony with the environment is fundamental to the achievement of sustainable development, so that individuals and societies are empowered to achieve positive social and environmental outcomes.

6. Internalization of Externalities

Building true social and environmental value should be the central goal of policy. To this end, market prices must reflect real social and environmental costs and benefits, so that that the polluter bears the cost of pollution. Tax regimes and regulatory frameworks should be used to 'tilt the playing field', making 'good' things cheap and 'bad' things very expensive.

7. International Cooperation

The application of environmental standards within nation States must be undertaken in a cooperative manner with the international community, based on an understanding of the possible impact on the development potential of other States. Environmental measures relating to trade should avoid unfair protectionism, but overall should ensure that trade supports sustainable resource use, environmental protection and progressive labor standards, promoting a 'race to the top' rather than the bottom.

8. International liability

Acknowledging that actions within national boundaries can cause environmental impacts beyond national jurisdictions, requiring cooperation in the development of international law that allows for independent judicial remedies in such cases.

9. Information, Participation and Accountability

All citizens should have access to information concerning the environment, as well as the opportunity to participate in decision-making processes. To ensure that environmental issues are handled with the participation of all concerned citizens, institutions at all levels (national and international) must be democratic and accountable, and make use of tools that enable civil society to hold them to account. In this regard, the access to justice by citizens for redress and remedy in environmental matters is a cornerstone of enhancing accountability.

10. Sustainable Consumption and Production

Introduce sustainable production and consumption with sustainable and equitable resource use. Reduce and eliminate unsustainable patterns of production and consumption, *i.e.* reduce, reuse, and recycle the materials used, acknowledge the scarcity of the Earth resources and implement activities accordingly.

11. Strategic, Co-Ordinated and Integrated Planning to Deliver Sustainable Development, the Green Economy and Poverty Alleviation

An integrated approach must be adopted at all levels to expedite the achievement of socio-economic and environmental sustainability through strategic planning with civil society and stakeholders, and across all relevant government departments.

12 Just Transition

There will be costs in making the transition to a low carbon, green economy in the pursuit of sustainable development. Some States and actors are better able to bear those costs than others and are more resilient to transitional changes. In the process of change, the most vulnerable must be supported and protected—developing countries must have access to appropriate financial and technical assistance, citizens and communities must also have access to new skills and jobs.

13 Redefine Well-being

GDP is an inadequate tool for measuring social wellbeing and environmental integrity. Many socially and environmentally damaging activities enhance GDP—such as fossil fuel exploitation and financial speculation. Human well-being and quality of life, and environmental health should be the guiding objectives of economic development.

14 Gender Equality

Gender equality and equity are prerequisites to the transition to a green economy and the achievement of sustainable development. Women have a vital role to play as agents of change for environmental management and development—their actions must be rewarded accordingly and their skills enhanced.

15 Safeguard Biodiversity and Prevent Pollution of Any Part of the Environment

Protect and restore biodiversity and natural habitats as integral to development and human well-being, and develop a system of governance that protects the resilience of ecosystems to prevent irreversible damage.

3.3 ANALYSIS AND CONSOLIDATION OF PRINCIPLES

3.3.1 Equitable Distribution of Wealth

Promote the equitable distribution of wealth within nations and among nations, to reduce disparities between rich and poor, and achieve social and economic justice, within a sustainable and fair share of the world's resources and leaving sufficient space for wildlife and wilderness.

3.3.2 Economic Equity and Fairness

Guided by the principle of common but differentiated responsibilities, create economic partnerships that would transfer substantial financial and technological assistance to less developed countries, to help minimize the gap between the developed and developing world and support the environmental sustainability of both.

Stockholm Declaration	Rio Declaration	Johannesburg Declaration
Man has the fundamental right to freedom, equality and adequate conditions of life, in an environment of a quality that permits a life of dignity and well-being, and he bears a solemn responsibility to protect and improve the environment for present and future generations. In this respect, policies promoting or perpetuating apartheid, racial segregation, discrimination, colonial and other forms of oppression and foreign domination stand condemned and must be eliminated.	All States and all people shall cooperate in the essential task of eradicating poverty as an indispensable requirement for sustainable development, in order to decrease the disparities in standards of living and better meet the needs of the majority of the people of the world.	14. The rapid integration of markets, mobility of capital and significant increases in investment flows around the world have opened new challenges and opportunities for the pursuit of sustainable development. But the benefits and costs of globalization are unevenly distributed, with developing countries facing special difficulties in meeting this challenge. 12. The deep fault line that divides human society between the rich and the poor and the ever-increasing gap between the developed and developing worlds pose a major threat to global prosperity, security and stability.
Earth Charter	**One Planet Living**	**The New Economics Foundation**
3b. Promote social and economic justice, enabling all to achieve a secure and meaningful livelihood that is ecologically responsible. **10a.** Promote the equitable distribution of wealth within nations and among nations.	One planet living is a positive vision of the world where we live happy, healthy lives within a sustainable and fair share of the world's resources and have left sufficient space for wildlife and wilderness. **Principle 9:** Equity and the local economy Thriving, diverse and resilient local economies support fair employment, inclusive communities and international fair trade.	Redistribution of both income and wealth would create value as resources are moved from those who do not need them to those who do. More equal societies are happier societies. By focusing on fairness we reap both social and economic benefits, as we no longer have to pay such a high price for the social 'ills' associated with high levels of inequality.

3.3.3 Intergenerational Equity

Environmental resources and ecosystems must be carefully managed and safeguarded so as to enhance the value of environmental assets for future generations, thereby equitably meeting their needs and allowing them to flourish.

In the face of pressing economic and environmental challenges, national and international efforts to promote green growth as a new source of growth have been intensifying in recent years. Building on this momentum can help to accelerate progress towards sustainable development and poverty reduction through, for example, more sustainable use of natural resources, efficiencies in the use of energy, and valuation of ecosystem services.

Stockholm Declaration	Rio Declaration	Johannesburg Declaration
Principle 9 Environmental deficiencies generated by the conditions of under-development and natural disasters pose grave problems and can best be remedied by accelerated development through the transfer of substantial quantities of financial and technological assistance as a supplement to the domestic effort of the developing countries and such timely assistance as may be required. **Principle 20** Environmental technologies should be made available to developing countries on terms which would encourage their wide dissemination without constituting an economic burden on the developing countries.	**Principle 7** States shall cooperate in a spirit of global partnership to conserve, protect and restore the health and integrity of the Earth's ecosystem. In view of the different contributions to global environmental degradation, States have common but differentiated responsibilities. The developed countries acknowledge the responsibility that they bear in the international pursuit to sustainable development in view of the pressures their societies place on the global environment and of the technologies and financial resources they command. **Principle 9** States should cooperate to strengthen endogenous capacity-building for sustainable development by improving scientific understanding through exchanges of scientific and technological knowledge, and by enhancing the development, adaptation, diffusion and transfer of technologies, including new and innovative technologies.	**18.** At the same time, we will work together to help one another gain access to financial resources, benefit from the opening of markets, ensure capacity-building, use modern technology to bring about development and make sure that there is technology transfer, human resource development, education and training to banish underdevelopment forever.

Earth Charter	One Planet Living	The New Economics Foundation
2(b) Affirm that with increased freedom, knowledge, and power comes increased responsibility to promote the common good. **7(c)** Promote the development, adoption, and equitable transfer of environmentally sound technologies. **8(a)** Support international scientific and technical cooperation on sustainability, with special attention to the needs of developing nations. **10(b)** Enhance the intellectual, financial, technical, and social resources of developing nations, and relieve them of onerous international debt.	**Principle 9:** Equity and local economy As part of national, regional and project level sustainable development planning, governments and civil society consider what citizens need and what local renewable and waste resources are available which could meet those needs. Where it is necessary to develop new techniques, services and technologies to utilise these resources and meet these needs, governments must incentivise and encourage research and development, and support technology transfer with commensurate financial support for developing countries	Ecological debt trading, obliging rich countries to pay poorer countries for overusing resources and over-pollution. This would have significant effects on development through redistributing economic wealth to poorer countries.

Stockholm Declaration	Rio Declaration	Johannesburg Declaration
Principle 2 The natural resources of the earth, including the air, water, land, flora and fauna and especially representative samples of natural ecosystems, must be safeguarded for the benefit of present and future generations through careful planning or management, as appropriate.	**Principle 3** The right to development must be fulfilled so as to equitably meet developmental and environmental needs of present and future generations.	**37.** From the African continent, the cradle of humankind, we solemnly pledge to the peoples of the world and the generations that will surely inherit this Earth that we are determined to ensure that our collective hope for sustainable development is realized.
Earth Charter	**One Planet Living**	**The New Economics Foundation**
4. Secure Earth's bounty and beauty for present and future generations. (a) Recognize that the freedom of action of each generation is qualified by the needs of future generations. (b) Transmit to future generations values, traditions, and institutions that support the long-term flourishing of Earth's human and ecological communities.	Intergenerational equity is implicit in one planet living, where we live well within a sustainable and fair share of the earth's resources today so that resources can be safeguarded for future generations and the natural world.	By focusing on building real value within environmental limits our natural environment can be protected for both current and future generations. Realising the Great Transition would see value rise steadily across three dimensions: for individuals, through a more equal distribution of incomes and assets; for society, by reducing the costs associated with inequality; for future generations, by protecting the natural environment and so reducing climate change- related costs.

Green growth is a matter of both economic policy and sustainable development policy. It tackles two key imperatives together: the continued inclusive economic growth needed by developing countries to reduce poverty and improve well-being; and improved environmental management needed to tackle resource scarcities and climate change. When green growth began to be promoted through the 2008–9 economic stimulus packages, some governments approached it from a short-term growth perspective—the potential to boost jobs and incomes through increased investment in some green (notably low-carbon) technologies. Others approach green growth from an environmental perspective—the potential to internalise environmental externalities by mainstreaming sustainable development requirements into economic decision-making, notably through resource pricing and land use/infrastructure choices. A third imperative, of equity and inclusion, has more recently been expressed, especially by developing countries—the notion that green growth should serve those excluded by the current economic system. The informal economy is very large in many developing countries and its potentials and hazards need to inform any transition to green growth in order to deliver more and better jobs and resilient livelihoods for poor people. Thus there is growing convergence around the notion that the current economic system is not only unsustainable and inefficient in its resource use, but moreover is inequitable in its distribution of costs and benefits.

This is as much an issue of bringing bad practice to a halt as one of opening up new green opportunities. Moreover, it will require systemic adjustments to better link economic, environmental and social policies and institutions—as far as possible identifying synergies, but also being clear about trade-offs and uncertainties, and the political economy of the changes required in varying contexts. As such, the formation of national green growth policy frameworks will be a critically important exercise. The OECD has prepared the *Green Growth and Developing Countries Report* to examine how developing countries might go about this and to discuss the elements of a practical policy framework that developing countries can use to make the transition to green growth. It explores policy instruments as part of a green growth framework while taking into account differences in natural resource endowments, levels of socio-economic development, sources of economic growth, and institutional capacity. Such a framework involves not just environmental policies, but also a broad range of economic and social policies. It will take significant long-term investment and innovation, necessary to avoid locking in inefficient and costly technology and infrastructure. For such investments and policies to deliver sustainable and equitable outcomes, appropriate governance arrangements must be in place and to facilitate this, capacity development is needed.

3.4 GREEN GROWTH: WHY IT MATTERS FOR DEVELOPING COUNTRIES?

Developing countries are the key to achieving global green growth in two major ways. Firstly, the potential economic and social impacts of environmental degradation are particularly important for developing countries. They are the most vulnerable to climate change and tend to be more dependent than advanced economies on the exploitation of natural resources for economic growth. In addition many developing countries face severe economic, social and ecological threats from energy, food and water insecurity to climate change and extreme weather risks. They also face risks from premature deaths due to pollution, poor water quality and diseases associated with a changing climate. All of these factors undermine their development. Secondly, although today most developing countries contribute only minor shares to global greenhouse gas (GHG) emissions compared to the OECD and major emerging economies (Fig.3.1), they will increase their emissions if they follow conventional economic growth patterns (Fig. 3.2). Increasingly developing countries are becoming sources of global economic growth, emissions and, with these, more intensive use of natural resources.

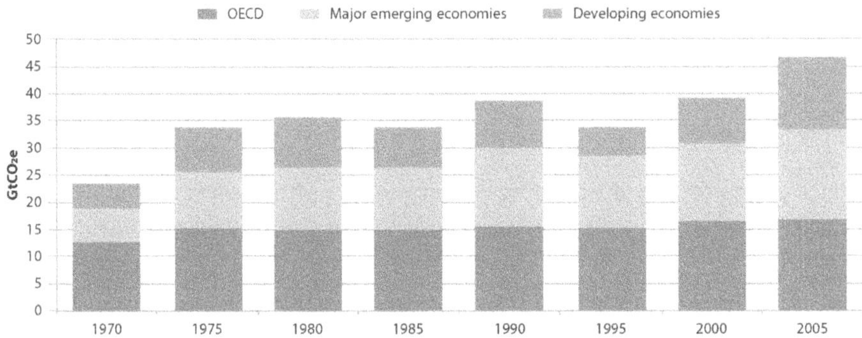

$GtCO_2e$ = Giga tonnes of CO_2 equivalent.

Source: OECD Environmental Outlook Baseline; output from ENV-Linkages.

Fig. 3.1: GHG emissions: 1970–2005

In order to tackle many of the growth and development challenges mentioned above without compromising future growth and poverty reduction goals, the concept of green growth has emerged as a new approach to reframe the conventional growth model and to re-assess many of the investment decisions in meeting energy, agriculture, water needs and the resource demands of economic growth. OECD defines green growth as a means to foster economic growth and development while ensuring that natural assets continue to provide the resources and environmental services on which our well-being relies. However, developing countries will likely interpret green growth in different ways and the concept has generated some concerns (Box 3.1).

Box 3.1: What are the Concerns from Developing Countries About the Concept of Green Growth?

The concept of green growth is generating a diversity of political positions, from enthusiastic to cautious, reflecting variously a lack of clarity and experience, differential opportunities apparent to specific countries, and the risks of international green growth policy regimes disfavouring a given country. For example, emerging economies describe the opportunities offered by green growth in the most enthusiastic terms, and many of them have access to relevant funds and technologies that can realise these opportunities. By 2008, China had already become the largest producer of clean technology in financial terms, accounting for 1.4% of its GDP. By comparison, there is caution on the part of many low-income countries (LICs). LICs are only just beginning to assess the opportunities, threats and indeed meaning of a green economic pathway. However, the policy ideas and technologies are neither easily accessible nor entirely relevant to their national developmental needs. There is a strongly adverse political reaction against the green growth concept in only a few countries.

Developing countries also have some concerns about the specificities of green growth. Some issues relate to the international dimensions of green growth such as the risks of green protectionism and green conditionality for Official Development Assistance. Other critical issues include:

Will green growth help address poverty and other development priorities? The green growth policy recommendations being discussed—with their emphasis on low-carbon and high-technology–do not obviously tackle equity problems at either the national or global level, notably the problem of the lack of inclusion of many poor countries and people within the informal economy in economic decision-making and in major economic opportunities. Not enough attention has been paid to the potential of more efficient use of natural capital. Furthermore, a number of governments are concerned that the focus on green growth could undermine the Rio Principles, particularly the principle of common but differentiated responsibilities.

Will green growth efforts be impeded by high-cost barriers? The high initial costs for the transition to green growth appear to be beyond the reach of many developing countries, *e.g.* solar power for rural communities. Even basic technologies are still lacking in most developing countries, particularly in the fields of wastewater treatment, household and hazardous waste management, energy efficiency and integrated water resource management. In addition there is a concern that developing countries' own technologies, including indigenous approaches, will not be able to compete, and they will need to import technologies from other countries: exchange of scientific and technical knowledge and removing the barriers constituted by intellectual property rights are of great importance if a genuine transfer of green technologies is to take place between developed and developing countries.

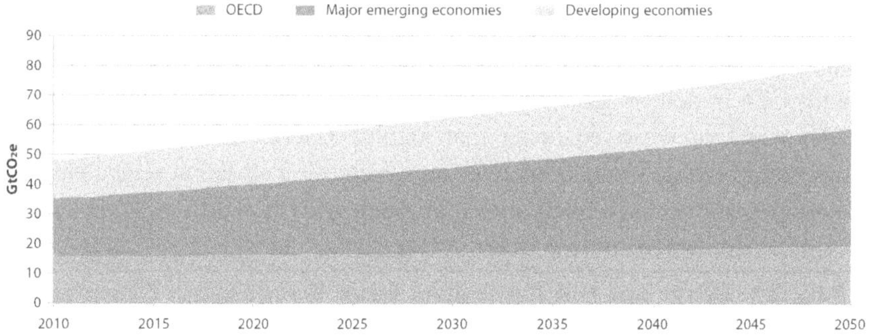

$GtCO_2e$ = Giga tonnes of CO_2 equivalent.

Source: OECD Environmental Outlook Baseline; output from ENV-Linkages.

Fig. 3.2: GHG emissions by region: Baseline, 2010–2050.

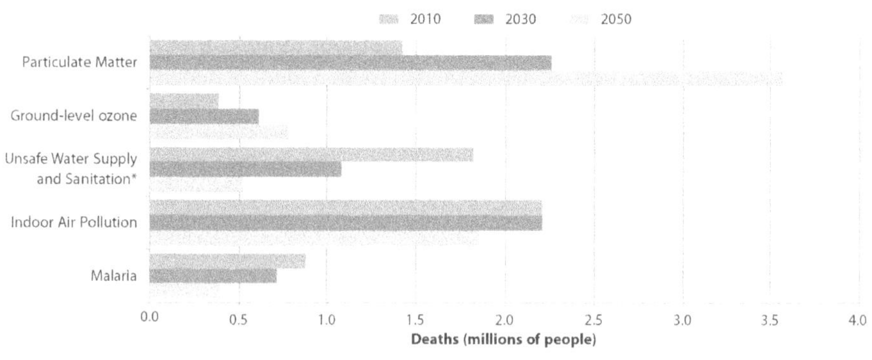

* Child mortality only.

Source: OECD Environmental Outlook Baseline; output from IMAGE (PBL).

Fig. 3.3: Global premature deaths from selected environmental risks: Baseline, 2010 to 2050.

Irrespective of the levels of economic development and to some extent of their political position, the majority of developing country governments have recently embarked on domestic processes to identify and develop particular areas of opportunity and comparative advantage. These domestic actions include carbon taxes, green energy funds, payment for ecosystem services schemes, renewable energy initiatives, sustainable public procurement initiatives and natural resource management initiatives. However there are few holistic or system-wide "green growth" policies, strategies and institutional systems in place. Some developing country strategies stand out, including Cambodia's Green Growth Road Map and Ethiopia's National Development Plans. For most countries, however, national sustainable development strategies go some way to forming an integrated green growth policy framework, though green growth has rarely been addressed in mainstream economic, budget and fiscal policies.

3.5 GREEN GROWTH FOR SUSTAINABLE AND EQUITABLE DEVELOPMENT

Green growth is about reconciling and reinforcing various aspects of economic, environmental and social policies. This is achieved by taking into account the full value of natural capital and recognising its essential role in economic growth. A green growth model promotes a cost-effective and resource efficient way of guiding sustainable production and consumption choices and could lead to the following outcomes if designed and implemented effectively. Therefore we begin by recalling a generic set of green growth outcomes that developing countries are increasingly intending to pursue.

Box 3.2. Green Growth Outcomes

Economic

1. Increased and more equitably distributed GDP–production of conventional goods and services
2. Increased production of unpriced ecosystem services (or their reduction prevented)
3. Economic diversification, *i.e.* improved management of economic risks
4. Innovation, access and uptake of green technologies, *i.e.* improved market confidence

Environmental

5. Increased productivity and efficiency of natural resource use
6. Natural capital used within ecological limits
7. Other types of capital increased through use of non-renewable natural capital
8. Reduced adverse environmental impact and improved natural hazard/risk management

Social

9. Increased livelihood opportunities, income and/or quality of life, notably of the poor
10. Decent jobs that benefit poor people created and sustained
11. Enhanced social, human and knowledge capital
12. Reduced inequality.

3.6 PATTERNS OF GROWTH IN DEVELOPING COUNTRIES

The 2000s saw rapid increases in per capita income in many developing countries. This, however, has been accompanied in many cases by increased inequality within countries. Several global factors underlie these trends: the development of globally integrated supply chains and associated massive increase in global labour supply, as well as soaring demand for commodities, particularly fossil fuels and industrial metals. Accordingly developing countries can be grouped into three "clusters": fuel exporters, non-fuel commodity exporters and manufacturing exporters. Growth in developing countries oriented towards manufacturing, fuel and other extractive industries is often carbon and natural-resource intensive. Moreover, developing countries are particularly vulnerable to the impacts of climate change and resource depletion. These tendencies will have significant economic, social and health consequences, including reductions in agricultural yield, strains on freshwater availability, extreme weather events and premature deaths from uncontrolled pollution.

Developing economies, therefore, need to embrace policies that help achieve concrete and measurable progress towards the twin objectives of stimulating more inclusive economic growth and promoting environmental sustainability. The OECD predicts that emerging economies and developing countries will grow substantially in the next 40 years (see Fig. 3.4). Their taking a greener growth path is therefore essential for global green growth and environmental sustainability.

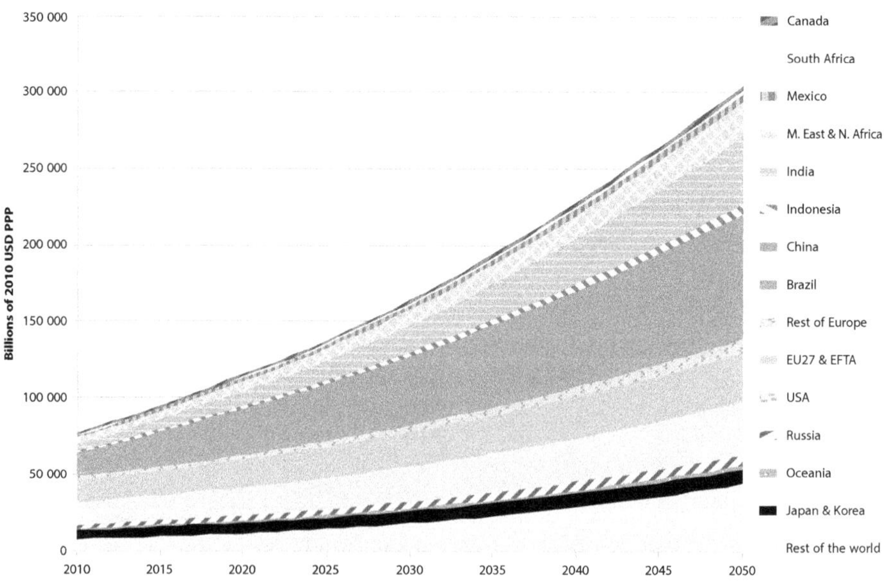

Fig. 3.4: Trend in real gross domestic product (GDP) valued at constant PPP prices by region.

There are opportunities for synergies between environmental and economic sustainability, especially for developing countries which can factor environmental issues into their investment decisions on infrastructure and can further develop agriculture and other natural resources to improve livelihoods, create jobs, and reduce poverty. Green growth provides an opportunity for emerging-market economies and developing countries to leapfrog unsustainable and wasteful production and consumption patterns. While advanced economies are somewhat constrained by the path dependency of sunk capital, adequate financing and capacity would offer developing economies the opportunity to build the infrastructure and networks needed to support sustainable development.

The growth trajectory measured in terms of CO_2 intensity differs markedly across the three clusters of low-and middle-income economies (fuel-based economies, non-fuel commodity exporters and manufacturing exporters). When we exam the relationship between real per capita GDP and per capita CO_2 emissions over the period of 1990–2009 for each cluster, we see significant differences. These two indicators tend to move in a linear fashion in the case of fuel-based economies and non-fuel commodity exporters, though much steeper in the former group. On the other hand, in the case of manufacturing exporters there appear to be two divergent patterns. One is a higher-intensity trend line represented by China, Malaysia, Thailand, Viet Nam and probably South Africa. Another is a lower-intensity trend line followed by the other economies, including among others Brazil, Colombia, Costa Rica and Dominican Republic. The degree of CO_2 intensity looks similar between this group of manufacturing exporters and non-fuel commodity exporters.

3.7 A POLICY FRAMEWORK FOR GREENING GROWTH IN DEVELOPING COUNTRIES

Green growth is understood by many countries to be about mainstreaming inclusive economic development in ways that incorporate, create and sustain environmental and social values. It entails changing and improving the performance of government, formal and informal economic actors and consumers and will require systematic adjustments to mainstream policy and governance systems—in other words, getting economic governance right for sustainable development.

As presented in Fig. 3.5, there are three dimensions which a national government should examine when a green growth strategy is to be developed and deployed:

- a national green growth plan to create enabling conditions;
- green growth mainstreaming mechanisms to ensure opportunities are explored through existing economic activities;

- green growth policy instruments to tap specific opportunities within spatial and resource systems.

Fig. 3.5: Green Growth Framework for Developing Countries.

3.7.1 Dimension 1—National Green Growth Plans that Generate Enabling Conditions

Green growth will only be possible if the broad governance and policy environment is conducive and stakeholders are confident. The most obvious requirement is some kind of co-ordinating national green growth plan that better integrates diverse institutional roles and creates enabling conditions for green growth. Six economy-wide policy challenges need to be addressed in the context of developing a national green growth plan, in each of which governments can exercise leadership. Broadly speaking, they encompass the enabling conditions for economic growth, reinforced by enabling conditions for inclusiveness and tempered by the enabling conditions for environmental protection.

Achieving this depends largely on building on, and "wiring together", existing policies, initiatives and institutions so that they work better together, with a focus on those which have proven effective for mainstreaming sustainable development to date. This will also ensure that the most promising policy options can be quickly mobilised and scaled up, avoiding legislative and other delays.

3.8 HOW CAN THE GLOBAL COMMUNITY HELP?

Creating a global architecture which is conducive to green growth will require further strengthening of arrangements for managing access to the global commons, maintaining the quality of global public goods, increased co-operation in the field of science and technology, provision of finance to support action by developing countries, and facilitating the diffusion of clean technologies. Increased efforts to boost global trade and investment flows would also help to underpin sustained growth. At the same time, there is a need for increasing vigilance around the potential spill-over effects of OECD countries' policy measures on developing countries and the potential for incoherent policy which undermines development prospects in low-income countries.

An international enabling environment will help send the right policy and market signals. It will create momentum for green growth by setting environmental regulations and standards, stimulate the demand and supply of environmental goods and services, catalyse green growth efforts in energy, transportation, agriculture and other sectors. It will make it easier to access financing, technology and innovations. Furthermore, an international enabling environment for green growth will facilitate the international exchange of knowledge about green growth issues or the cooperation in science, technology, and innovation. Effective and comprehensive knowledge sharing platforms are important for the international transfer of knowledge, good practices and technologies to developing countries.

There are several areas where the OECD countries and the international community could help to create such a conducive environment for green growth—these include:

- **Enhancing capacities. Capacity is central to green growth**—capacity to adapt and deploy green technologies, to undertake environmental risk assessment, to make the economic case for greening growth, to incorporate environmental issues into decision making, and to coordinate across sectors and institute environmental fiscal reform. The international community can support developing countries to build and enhance these capacities. Efforts could largely focus on three issues: capacities to integrate green growth considerations into national development planning, national budgetary processes and key economic sectors.
- **Strengthening Official Development Assistance (ODA).** ODA is essential in creating enabling conditions for green growth "targeting areas where incentives for private investment are limited and flows are scarce, including essential infrastructure and human and institutional capacity building." Donor agencies have many decades of experience stimulating and promoting sustainable development in developing countries and have been scaling up assistance in support of the environment (Fig. 3.6). ODA finances major projects in renewable energy, sustainable agriculture and low-carbon transportation networks. It also supports smaller but potentially catalytic efforts such as feasibility studies, pilot projects and technical training. Green growth thinking

should be fully incorporated into development co-operation. Development assistance to green growth should be further strengthened by ensuring that climate proofing and disaster risk reduction approaches are mainstreamed into aid-funded public investment. Furthermore, investments funded by development co-operation in public infrastructure and services should crowd in and leverage private investment in sustainable production systems and value chains. The Aid-for-Trade Initiative, which supports developing countries in the areas of trade infrastructure development, productive capacity building, and implementation of trade agreements, should focus on greening trade-related investments and building capacities for new trade regulations such as green or low carbon certification.

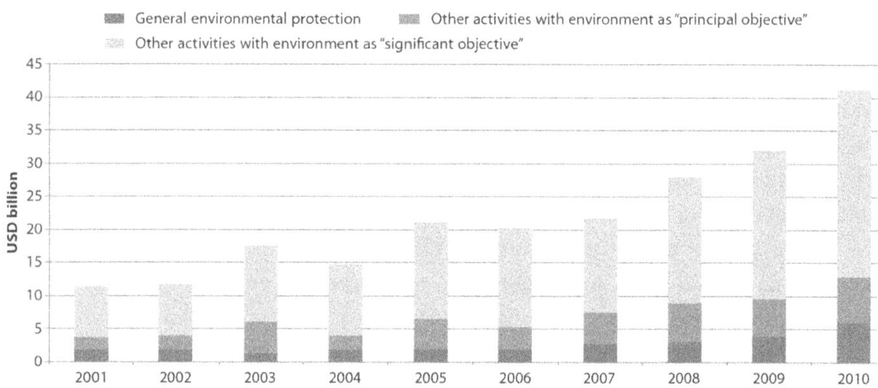

Fig. 3.6: Environment aid from OECD Development Assistance Committee (DAC) members (2001–2010), commitment, constant 2009 prices.

Source: OECD DAC Creditor Reporting System.

- **Promoting innovation and accelerating the dissemination of green technologies.** OECD countries can assist developing countries in promoting green growth by removing existing barriers to the dissemination and transfer of technologies and to trade. In addition, support to developing countries for science and development in green technologies helps to bring forward innovation and technology advancements. It reduces the cost and risks of private sector investment in new technologies. To accelerate the diffusion of innovation, new mechanisms that enhance technology transfer are currently developed, such as through voluntary patent pools and other collaborative mechanisms for easing the application of Intellectual Property Rights (IPRs). To scale up the deployment of green technologies in developing countries, in particular to the least-developed countries, multilateral action is needed to reduce the cost of green technologies for these countries. International co-operation to cover licensing fees, buy out patents of key technologies, and facilitate developing countries and their firms to build more systematically on the knowledge resulting from basic research undertaken by public institutes are some critical steps. Academic partnership and cross border higher education exchange programs can also facilitate technology transfer and lead to spillovers in local innovation systems.

- **Facilitating trade of environmental goods and services.** New opportunities can be created for green growth in developing countries by building international markets for ecosystem services. OECD countries can adopt policy approaches to the pricing of environmental externalities that allocate adequate portions of the consequent revenues to paying for, and thereby creating demand for the production of, the ecosystem services expected from developing countries. The international community can also help developing countries to participate in international trade by facilitating the harmonisation of international standards. This includes environmental, sanitary and phyto-sanitary measures, and certification schemes.
- **Developing coherent policies.** Policy coherence for development (PCD) can play an important part in creating an international enabling environment for green growth. In the long run, the implementation of effective green growth policy mixes will depend on political leadership, widespread public awareness and acceptance that changes are both necessary and affordable. In order to maximise opportunities from PCD, OECD countries have to make sure that their policies in areas other than development co-operation support do not undermine the efforts to improve natural resource management, agriculture, economic development, and sustainability in developing countries. In particular developed countries need to ensure that their green growth policies do not promote the off-shoring of high carbon or pollution intensive components to developing countries. Integrated into a multi-sectoral, whole-of-government approach, green growth policies can be cross-cutting to maximise synergies, generate cobenefits with other sectors, and positively influence poverty alleviation.

3.9 MEASURING PROGRESS

Relevant information and statistics provide the foundation for policies that promote green growth, and are critical to monitoring progress and gauging results. Measuring progress towards green growth in developing countries requires some special considerations. The OECD has developed a measurement framework for green growth that provides developed and developing countries with a way to organise thinking about indicators and to identify relevant, succinct and measurable statistics. The framework reflects the integrated nature of green growth and describes four of the main aspects that need to be monitored: (*i*) the environmental and resource productivity of the economy; (*ii*) the natural asset base; (*iii*) the environmental quality of life, and (*iv*) economic opportunities and policy responses that arise.

Monitoring progress towards green growth in developing countries requires some special considerations. Developing countries face many different challenges—poverty, weak institutional capacity, food insecurity, gender inequality, poor infrastructure—and consequently have policy priorities that differ from developed countries. But the OECD measurement framework is a robust tool that can be adapted to different national circumstances and priorities. It gives countries the flexibility to focus on the indicators that reflect their own green growth objectives, such as building economic and environmental resilience and ensuring that growth is inclusive.

One of the biggest obstacles to establishing a monitoring framework for green growth in developing countries is overall statistical capacity. Faced with other pressing priorities, developing countries have experienced difficulties in mobilising the capacity and resources necessary to collect, produce, analyse and disseminate relevant information to support policy development.

Putting in place a monitoring framework is an integral component of any green growth strategy. For developing countries compiling a set of indicators to monitor progress on green growth need not increase the statistical burden if synergies are maximised. Initiatives to modernise and improve national statistical systems, such as the Modernisation of the Barbados Statistical Service Project (MBSS) and the National Strategy for the Development of Statistics (NSDS) process under the Partnership in Statistics for Development in the 21st Century (PARIS21), provide an important opportunity to mainstream green growth, particularly environmental considerations, into economic and social information systems.

APPENDIX

Selected Environmental Chronology

1909 – International Joint Commission established
1919 – Convention for the Protection of Migratory Birds
1971 – Canada the second country to create a federal Department of the Environment
1972 – United Nations Conference on the Human Environment, Stockholm, Sweden
1972 – Establishment of United Nations Environment Program (UNEP)
1972 – Great Lakes Water Quality Agreement (renegotiated 1978, 1987)
1979 – Geneva Convention on Long Range Transboundary Air Pollution
1980 – Brandt Report
1983 – United Nations General Assembly set up world commission on Environment and Development (the Brundtland Commission)
1984 – Senate Standing Committee Report: "Soil at Risk"
1985 – The Macdonald Commission: "Report of the Royal Commission on the Economic Union and Development Prospect for Canada"
1985 – The Vienna Convention for the Protection of the Ozone Layer
1985 – The Helsinki Agreement
1985 – Inquiry on Federal Water Policy: "Currents of Change"
1987 – National Task Force on Environment and Economy Report
1987 – The Brundtland Report: "Our Common Future"
1988 – The Montreal Protocol on Substances that Deplete the Ozone Layer
1988 – Toronto Conference on the Changing Atmosphere
1989 – Toronto Economic Summit, Hague Declaration, Noordwijk Declaration, Paris Summit, Langkawi Declaration
1988 – The Canadian Environmental Protection Act
1989 – Basel Convention on the control of Transboundary Movement of Hazardous Waste
1989 – Forestry Act established by Forestry Canada
1990 – Houston Summit, Dakar Resolution on the Environment
1990 – Establishment of the International Institute for Sustainable Development in Winnipeg, Manitoba
1990 – House of Commons Standing Committee (HCSC) on the Environment Report: "Our Changing Atmosphere"
1990 – G-7 Summit Houstan, Texas. Mr. Mulroney calls for a instrument to promote sustainable development and conservation of forests
1990 – Canada's Green Plan (11 December, 1990)

1990 – CCME (Canadian Council Of Ministers for the Environment): National Packaging Protocol and Phase 1 of the Management Plan for Nitrogen Oxide and Volatile Organic Compounds

1991 – London Summit

1991 – A Summary of Sustainable Fisheries Activities in Canada

1991 – HCSC on the Environment Report: "Our Changing Atmosphere

1991 – Canada-United States Air Quality Accord

1992 – Senate Approval of Bill C-78 (C-13) Canadian Environmental Assessment Act

1992 – United Nations Conference on Environment and Development (UNCED): Convention on Climate Change, Convention on Biological Diversity, Agenda 21, Rio Declaration, Forestry Agreement, Ocean Conference.

■ ■ ■

4

Green Economy Strategies

4.1 INTRODUCTION

Following Rio+20, the UN Division for Sustainable Development began publishing a new series of guidebooks on the green economy. These guidebooks provide practitioners and other stakeholders with concise resource guides to the green economy and related concepts such as green growth and low-carbon development. They aim to enhance our understanding of the green economy and further clarify the concept. To do so, they endeavour to summarise the rapidly expanding literature from experts and practitioners and emerging international experience to shed new light on what we mean when we talk about the green economy.

A number of developing and developed countries have also released national green economy strategies for the first time, some of which aim to identify initial green economy policy options (or 'low-hanging fruit') that exploit policy synergies that deliver outcomes across all three—economic, social and environmental—dimensions in the short to medium term. For developing countries, these often focus on development pathways that would reduce carbon emissions, increase climate resilience, and tap into emerging climate financing.

4.2 A TYPOLOGY OF GREEN ECONOMY POLICIES

A typology of green economy policies has been developed by the UN—which draws upon the categories used in recent publications by a number of leading international organisations and experts. This typology of green economy policies is set out in Table 4.1 below and proposes six categories that cover the breadth of green economy and complementary policy measures addressing all three dimensions of sustainable development and using an intuitive format based around "6 Is": Internalising; Incentivising; Institutions; Investment; Information; and Inclusion.

Table 4.1: Typology of green economy policy measures

Policy Category: 6 'Is'	Policy Sub-Categories
Internalising (externalities)	1. Taxes, charges, fees, levies on 'bads' (i.e. pollution, resource use or proxy)
	2. Cap-and-trade permit or certificate systems
Incentivising	3. Investment incentives - low-interest loans; micro-financing; tax exemptions etc.
	4. Subsidies, feed-in tariffs and other direct support for 'goods'
	5. Removing policy-induced distortions and perverse incentives (e.g. harmful subsidies)
	6. Leveraging finance - PPPs, long-term guarantees, phased out support, removal of barriers to FDI, lower administrative burden, credit guarantees
Institutions	7. Regulations—norms, standards, info disclosure, labelling, prohibitions, fines and enforcement, mandatory targets
	8. Property right and access right laws, including IPR
	9. Governance and institutional capacities—accountability, transparency, enforcement, anticorruption
	10. Integrated planning, decision-making and resource management—EIA/SEA, IWRM, ICZM, LCA, MCA/CBA, disaster preparedness, other diagnostic tools
Investment (in natural capital, agriculture, human capital, infrastructure, and innovation).	11. Sustainable public procurement
	12. Investment in natural capital - PES, protected areas, direct management and rehabilitation
	13. Investment in sustainable agriculture
	14. Investment in human capital - capacity building, training, skills
	15. Investment in infrastructure - energy, water, transport, waste, ICT
	16. Investment in innovation - R&D, deployment, information sharing
Information	17. Voluntary approaches - information provision, labelling, CSR, targets, agreements, educational initiatives
	18. Measuring progress - green accounting, green targets and indicators, carbon inventories
Inclusion	19. Labour market policies - skills (re-)training, job search assistance, income support and benefits
	20. Social protection floors - unemployment insurance and pensions, cash transfers, compensation for price increases, health care

Within these six categories, Table 4.1 also outlines a consolidated list of 20 green economy and complementary policy sub-categories that were identified through the desktop review of green economy publications. In many cases, this involved intuitively grouping specific and related policy measures together to streamline and avoid duplication (*for example* "taxes, charges and fees" are grouped together). This typology enables us to explore the range of policy measures and most common instruments that are being proposed by practitioners and experts to transition towards greener economies.

Table 4.2 below uses this green economy policy typology based on six categories and 20 policy subcategories to provide an overview of the array of green economy policy instruments that were included in recent green economy studies. Based on this analysis, it can be seen that a wide variety of measures are being proposed in most studies, from economic instruments and regulatory measures, to information-based measures and social policies. The most common green economy policy measures that were proposed across all of the studies were measures for internalising externalities (such as taxes and cap-and-trade systems) and regulatory measures (such as standards, labelling, prohibitions and compliance). Other policies that were proposed in most of the publications included investment in infrastructure (such as sustainable energy, water, transport and waste) as well as investment in innovation (through measures such as funding for R&D and deployment). Green economy policy measures that were the least common across the various studies included measures such as property rights and access rights, integrated planning and resource management, sustainable public procurement and social protection policies.

4.3 RECENT GREEN ECONOMY POLICY TOOLKITS

For governments to implement green economy measures will require much more than simply selecting a policy instrument or instruments from a list of green economy policy measures. It will require them to establish clear goals and targets, to analyse synergies and tradeoffs between different policy options for achieving their goals, to effectively design and implement priority actions, and to monitor progress and review their policies over time through an ongoing adaptive management process. This will be particularly challenging for developing countries that may lack the institutional capacity, adequate data and information, and access to financing and other resources that are required for successful policy development and implementation.

Recognising that a broader framework is required, a number of international organisations have recently begun to elaborate green economy policy toolkits, expanding upon published lists and typologies of green economy policies. These toolkits vary in their approach and level of detail, but it can be seen that more recent iterations provide not only a suite of green economy policy instruments that can be used by governments to green their economies. They also provide a range of diagnostic tools that can help with policy design, evaluation, and monitoring, as well as mechanisms that can help to mainstream green economy into national economic planning and decision making and build the necessary institutions and enabling conditions to support the implementation of national green economy plans. These are outlined in Boxes 4.1 to 4.4 below. As such, they provide a more useful product for governments and practitioners in applying green economy policies. They are also a step forward in delivering on Rio+20, which invited the development of green economy toolboxes and other resources.

Table 4.2: Review of policy measures identified in a selection of recent publications and initiatives

	Internalising		Incentives				Institutions				Investment						Information		Inclusion	
Country	Taxes, charges, fees, levies, pricing for 'bads' (i.e. pollution, resource use or proxy)	Cap-and-trade permit or certificate systems	Investment incentives–low-interest loans; micro-financing; exemptions, reducing import tariffs, insurance, rebates	Subsidies, feed-in tariffs and other direct support for 'goods'	Removing policy-induced distortions and perverse incentives (e.g. harmful subsidies)	Leveraging finance–PPPs, long-term guarantees, phased out support, removal of barriers to FDI, lower administrative burden, credit guarantees	Regulations–norms, standards, info disclosure, labelling, prohibitions, fines and enforcement, mandatory targets	Property right and access right laws, including IPR	Governance–accountability, transparency, enforcement, capabilities, anti-corruption, data & measurement, carbon inventories	Integrated planning, decision-making and resource management-EIA/SEA,IWRM,ICZM, LCA, disaster preparedness	Sustainable public procurement	Investment in natural capital–PES, protected areas, direct management and rehabilitation	Investment in sustainable agriculture	Investment in human capital–capacity building, training, skills	Investment in infrastructure–energy, water, transport, waste, ICT	Investment in Innovation–R&D, deployment, information sharing	Voluntary approaches–information provision, labelling, CSR, targets, agreements, educational initiatives	Measuring progress–green accounting, green targets and indicators, carbon inventories	Labour market policies–skills (re-)training, job search assistance, income support and benefits	Social protection–unemployment insurance and pensions, cash transfers, compensation for price increases, health care
UNEP (2011)	x	x	x	x	x	x	x	x	x	x	x	x	x	x	x	x	x	x	x	
Cosbey (2011)	x	x	x	x	x	x	x	x	x	x	x			x	x	x	x	x		
OECD/UN, World Bank (2012)	x	x	x	x	x	x	x	x	x	x	x	x	x	x	x	x	x		x	x
OECD (2011a)	x	x		x		x	x		x						x	x	x	x	x	
OECD (2011b)	x	x		x			x										x			
OECD (2011c)	x	x	x	x	x	x	x	x	x	x	x		x	x	x	x	x		x	x
UNESCAP (2012)	x	x	x	x		x	x		x	x		x	x		x					
World Bank (2011)	x	x	x				x				x					x	x			
Green Growth Leaders (2011)	x	x					x								x	x				
ILO,EU, ILLS (2011)	x	x					x				x	x	x	x	x	x				
Barbier (2011)	x	x	x	x	x	x	x	x	x	x	x	x	x	x	x	x	x	x		
World Bank (2012)	x	x	x	x	x		x			x		x		x	x	x			x	
UNEMG (2011)	x	x				x	x			x				x	x	x		x	x	x
Daly (2011)	x	x					x											x	x	
Green Economy Coalition (2011)	x	x	x	x			x		x	x		x	x	x		x			x	

In 2011 the OECD released a policy toolkit for green growth. *Tools for Delivering on Green Growth*—which provides a suite of policy options for addressing green growth constraints and challenges, focusing on green innovation, infrastructure investment, labour market transitions, consumer behaviour, and competitiveness (see Box 4.1). Many of the policy instruments outlined are economic instruments focused on internalising externalities and incentivising investment, and this is likely to be more useful for developed countries as it assumes that institutional capacity exists to effectively design and implement such policies.

Box 4.1. Tools for Delivering on Green Growth (OECD, 2011)

The following policy options are listed to address specific policy constraints as follows:

1. Inadequate infrastructure: Taxes, Tariffs, Transfers, Public-private partnerships
2. Low human and social capital and poor institutional quality: Taxes, Subsidy reform/removal
3. Incomplete property rights, subsidies: Review and reform
4. Regulatory uncertainty: Set targets, Create independent governance systems
5. Information externalities and split incentives: Labeling, Voluntary approaches, Subsidies, Technology and performance standards
6. Environmental externalities: Taxes, Tradable permits, Subsidies
7. Low returns on R&D: R&D subsidies and tax incentives, Focus on general-purpose technologies
8. Network effects: Strengthen competition in network industries, Subsidies or loan guarantees for new network projects
9. Barriers to competition: Reform regulation, Reduce monopoly power

Source: http://www.oecd.org/greengrowth/48012326.pdf

More recently, in 2012 the OECD published a preliminary draft of a policy framework specifically targeting developing countries which builds on previous analyses and sets out a policy framework based around three broad categories: (*i*) enabling conditions, (*ii*) mainstreaming mechanisms and (*iii*) policy instruments (see Box 4.2). This expands on the previous toolkit by incorporating not only a list of green economy policy measures, but also a number of mechanisms and tools to build institutional capacity and establish the evidence base for green economy policy design and implementation. This is an important addition to the policy toolkit for developing countries.

The G20 is also pursuing the development of a non-prescriptive good practice guide and toolkit for enabling national policy frameworks for inclusive green growth. As part of this process, two products are of particular note. The first is a report produced by the OECD, UN and World Bank on incorporating green growth policies into structural reform agendas which draws heavily on the green growth work of the OECD (see Box 4.3). The report was prepared primarily for G20 countries and focuses on improving framework policies, removing policy-induced distortions, and a suite of environmental policy instruments. As such, it does not incorporate tools for evaluating, selecting and designing policies.

The second is a preliminary toolkit of policy options to support inclusive green growth developed by the African Development Bank (AfDB), OECD, World Bank and UN6 (the "IGG Toolkit") (see Box 4.4). This toolkit was developed for G20 countries and focuses on a combination of policy instruments for incentivising investment and internalising externalities, as well as a number of tools for policy evaluation, integrated decision-making, monitoring of progress and attracting finance. As such, it advances beyond providing a short list of policy instruments to embrace a broader range of tools that governments can use to support more effective design, evaluation and implementation of green economy policies.

Box 4.2. Green Growth and Developing Countries (OECD, 2012)

This report proposes a green growth policy framework for developing countries which includes a suite of tools across 3 categories:

1. Six national enabling conditions for green growth created through a National Green Growth Plan:
   - Shift government expenditure;
   - More effective enforcement of legislation;
   - Education and training;
   - Resource and land rights regimes;
   - Creating enabling conditions for psychological and behaviour change;
   - Facilitating businesses to fully integrate sustainability and equity concerns.
2. Four green growth mainstreaming mechanisms:
   - Public Environmental Expenditure Review;
   - Strategic Environmental Assessment;
   - Councils for Sustainable Development;
   - Greening Accounting/Alternative Development Measures.
3. Eight green growth policy instruments:
   - Certification of Sustainable Production and Trade; o Subsidy Reform;
   - Payments for Ecosystem Services;
   - Environmental Fiscal Reform;
   - Green Energy Investment Frameworks and Incentives;
   - Inclusive Green Social Enterprise;
   - Sustainable Public Procurement;
   - Green Innovation.

Source: http://www.oecd.org/dac/environmentanddevelopment/greengrowthanddevelopment.htm

The participating organisations of the IGG Toolkit are working during the remainder of Mexico's G-20 presidency to finalize the toolkit, including integration of new or better tools on energy, water, and climate change. The authors are also discussing how best to disseminate the toolkit and transform it into a living document or 'wiki' platform, possibly to be hosted by the Green Growth Knowledge Platform or one of its affiliated programs.

There is substantial scope to refine further these toolkits so that they are intuitive, useful and easily accessible. The further development of green economy toolboxes could draw from recent experience in related fields, in particular low-carbon development, where process guidelines and toolkits are more advanced.

A number of UN agencies and other organisations have also developed methodologies for the analysis of green economy policies. Such methodologies are now beginning to form an integral part of the green economy toolkit, as can be seen in the IGG toolkit (Box 4.4) and OECD developing countries toolkit (Box 4.2). These toolkits incorporate a number of diagnostic tools and indices, such as long-term planning and modelling tools, labour market and income analysis, strategic environmental assessment, public expenditure reviews, and green accounting which can help in the selection, design and implementation of policies. There is scope to compile further information on these various methodologies, explore their utility in different country contexts for analysing green economy policies, and draw together examples and case studies.

Box 4.3: Incorporating Green Growth and Sustainable Development Policies into Structural Reform Agendas (OECD, World Bank, UN, 2012)

The report outlines a number of broad elements of structural packages to promote green growth and sustainable development as follows:

- Reforming the structure of taxes and charges to promote economic growth and make it greener
- Strengthening markets
- Inducing greener behaviour through regulatory and information policies
- Getting infrastructure "right"
- Targeting and leveraging investment by international corporate players
- Fostering innovation for greening growth
- Lowering barriers to the diffusion of green goods, services and technologies
- Encouraging job creation and equity for inclusive green growth

The report also includes a list of structural policy instruments across three broad categories as follows:

1. Improving framework policies: competition policies; tax reform; labour market policies; investment policies; network sector policies; and innovation.
2. Removing policy-induced distortions: removal of perverse subsidies; removal of barriers to trade and investment in green goods and services; ensuring well-targeted compensatory measures for the poor.
3. Implementing environmental policy instruments: cap-and-trade permit systems; taxes or charges on pollution or proxy; baseline-and-credit permit systems; subsidies and other direct support; deposit-refund systems; performance standards; technology standards; voluntary approaches; information and regulatory policies.

Source: http://www.oecd.org/eco/economicpoliciestofostergreengrowth/G20_report_on_GG_and_SD_final.pdf

Box 4.4: A Toolkit of Policy Options to Support Inclusive Green Growth (AfDB, OECD, UN and World Bank, 2012)

Outlines a practical and flexible green growth policy toolkit, which includes a step-by-step guide. It has four main categories of policy tools: 1. Incentivize: Tools for pricing pollution and natural resource use; Tools to complement pricing policies; and Tools to foster inclusiveness; 2. Design: Tools to manage uncertainty; 3. Finance: Financing and investment tools; and 4. Monitor: Monitoring tools. A number of policy tools are listed that fall into one or more of these categories:

- Environmental fiscal reform
- Public Environmental Expenditure Reviews
- Sustainable Public Procurement
- Strategic Environmental Assessments
- Social Protection Instruments
- Payments for Ecosystem Services
- Certification for Sustainable Production
- Green Innovation and Industrial Policies
- Project-Level Impact Assessment
- Integrated Water Resources Management
- Green Accounting.

Source: http://sustainabledevelopment.un.org/index.php?page=view&type=400&nr=695&menu=35

4.3.1 International Experience in Implementing National Strategies for Sustainable Development, Low-Emission Development, and Green Economy

Whilst most countries do not yet have an overall national strategy for a green economy, there is emerging international practice in the design and implementation of green economy policies in both developed and developing countries. This has been catalysed through international agreements which have promoted the development of national strategies for sustainable development, climate change and environmental management spanning back at least 20 years. In this context, the development of national low-carbon, green economy and green growth strategies can be seen as more recent iterations of national sustainable development strategies, which aim to build on past experience and more effectively integrate the three dimensions of sustainable development to address ongoing global crises and emerging priorities. National government efforts have also been facilitated by a number of international organisations, including UN organisations, that are committed to advancing the development of national strategies through advisory and technical assistance services and capacity building.

4.3.2 National Sustainable Development Strategies

Over the past two decades, a variety of strategies have been prepared on the basis of national needs and priorities. Many were either sponsored by multilateral financial institutions or were advocated by global conventions with financial support provided for their formulation. A variety of approaches have been adopted, including strategies targeting sustainable development, strategies dealing with economic recovery and growth, strategies providing a long-term national vision, strategies supporting social development/poverty reduction, and strategies focused on integrating environmental conservation into development.

Numerous international organisations have been involved in supporting the development of such strategies, including the World Bank, the International Monetary Fund (IMF), UNDESA, UNDP, the International Union for the Conservation of Nature (IUCN) and the OECD. The principal elements National Sustainable Development Strategy (NSDS) are tabulated in Table 4.3.

Overall, it would appear that the NSDS experience has had at best mixed results, with notable achievements in some areas but serious shortcomings in others. For green economy strategies to overcome the problems and challenges and become an important tool for implementing sustainable development, it is critical that we learn from the experience gained over 20 years of NSDS implementation. Indeed, more recent iterations of NSDS have attempted to address the challenges outlined above by adopting more integrated and participatory approaches, focusing on institutional reforms and enabling conditions, using more advanced diagnostic and measurement tools for policy evaluation, and applying economic instruments and market-based measures.

4.3.3 Low-Emission Development Strategies (LEDS)

Although the introduction of LEDS terminology emerged only recently in UNFCCC negotiations, planning for economic development that incorporates climate change mitigation and adaptation is not a new concept, and both developed and developing countries have been active in producing a variety of national climate change plans and strategies over the past two decades. A LEDS can perhaps best be described as a development paradigm that contributes to addressing the twin challenges of addressing climate change and advancing development. It seeks to promote economic growth or sustainable development while keeping greenhouse gas (GHG) emissions low, or lower than without interventions, and increasing resilience to climate change impacts.

The terminology used and the breadth of issues considered have evolved over time. Very recently, several countries have prepared strategies that are more clearly defined as LEDS, for example Guyana's Low-Carbon Development Strategy (2010), the UK's Low Carbon Transition Plan and Indonesia's Creating *Low Carbon Prosperity in Jambi* (2010). Boxes 4.5 to 4.7 below provide a quick reference guide to these strategies.

Table 4.3: Principles and good practices for developing national sustainable development strategies

Main Element	OECD (2006)	UNDESA (2002)	OECD (2001)
Policy integration	**Policy integration**—National strategies should give consideration to environmental, economic and social concerns in integrated approaches contained in national plans and reports.	Integrate economic, social and environmental objectives. Link different sectors	Integrate economic, social and environmental objectives. Ensure comprehensive and integrated strategy
Inter-generational timeframe	**Intergenerational timeframe**—National strategies should adopt long-term timeframes which enable inclusion of intergenerational principles and indicators.	Develop shared strategic and pragmatic vision. Link short term to medium/long term	Develop consensus on long-term vision.
Analysis and assessments	**Analysis and assessments**—Integrated assessment tools should be used in national reports to identify the environmental, economic and social costs and benefits of policy and strategy options.	Anchor strategy in sound technical and economic analysis. Build on existing mechanisms and strategies.	Base strategy on comprehensive and reliable analysis. Build on existing processes and strategies.
Coordination and institutions	**Co-ordination and institutions**—A wide range of government departments and agencies should be involved in the formulation and implementation of national strategies, with overall responsibility in the office of the Prime Minister or equivalent.	Ensure a strong institution or group of institutions spearheading the process	Embed strategy in high-level government commitment and influential lead institutions
Local and regional governance	**Local and regional governance**—local and regional authorities should be fully involved in the development of national strategies, with certain delivery aspects devolved to sub-national levels.	Link national, regional and global levels	Link national and local levels
Stakeholder participation	**Stakeholder participation**—stakeholders (e.g., business, unions, non-governmental organisations) should participate with government representatives in commissions responsible for developing and implementing national strategies.	Ensure access to information for all stakeholders, transparency and accountability. Develop partnerships among government, civil society, private sector and external institutions.	Ensure effective participation. Develop a people-centred strategy.
Indicators and targets	**Indicators and targets**—strategies should be based on structured indicator systems (enumerated in national plans and reports) to assist in monitoring progress and to serve as quantitative targets.	Base strategy on realistic, flexible targets	Include targets with clear budgetary priorities
Monitoring and evaluation	**Monitoring and evaluation**—independent bodies or processes should be established to act as watchdogs monitoring implementation of national strategies and providing recommendations for their improvement.	Include integrated mechanisms for assessment, follow-up, evaluation and feedback	Incorporate monitoring, learning and improvement

Box 4.5: Republic of Guyana: A Low-Carbon Development Strategy (LCDS)—Transforming Guyana's Economy While Combating Climate Change (2010)

The document sets out Guyana's strategy to forge a new low carbon economy over the coming decade. It identifies 8 priorities that will be the initial focus of LCDS implementation in 2010 and 2011, gives an outline of priorities for the period 2012–15, and sets out the framework for further consultation and strategy development on Guyana's long-term low carbon development. The strategy focuses on avoiding deforestation and using funding to enable low carbon economic development of new sectors. The strategy will lead to action in four areas:

- Investing in low-carbon economic infrastructure
- Facilitating investment and employment in low-carbon economic sectors
- Sustainably managing forest-based economic sectors, in particular forestry and mining
- Generally enhancing the nation's human capital and creating new opportunities for forest-dependent and other indigenous communities.

The development of the strategy was led by the Office of the President with the assistance of the Government of Norway and the International Institute for Environment and Development. The strategy augments the country's National Development Strategy and National Competitiveness Strategy, identifying areas where these broader development goals and targets can be achieved in a low- carbon manner.

Source: http://centralcontent.fco.gov.uk/central-content/campaigns/act-on-copenhagen/resources/en/pdf/DEC

Box 4.6: United Kingdom: The UK Low Carbon Transition Plan—National Strategy for Climate and Energy (2009)

This White Paper sets out the UK's transition plan for becoming a low carbon country: cutting emissions, maintaining secure energy supplies, maximizing economic opportunities, and protecting the most vulnerable. The plan includes a series of targets across five key areas within a 2020 timeframe:

- Protecting the public from immediate risk.
- Preparing for the future.
- Limiting the severity of future climate change through a new international climate agreement.
- Building a low carbon UK (aiming to cut UK emissions by 34% by 2020 and at least 80% by 2050 - compared with 1990 levels - through investment in energy efficiency and clean energy technologies).
- Supporting individuals, communities and businesses to play their part.

The plan was approved pursuant to the UK's Climate Change Act and the Department of Energy and Climate Change has overall responsibility for its delivery. An independent Committee on Climate Change was responsible for setting emissions reduction targets and has an ongoing review function. Under the plan, each government department is issued with a carbon budget and must publish a plan on how it will achieve this.

Source: http://centralcontent.fco.gov.uk/central-content/campaigns/act-on-copenhagen/resources/en/pdf/DECC-Low-Carbon-Transition-Plan

Box 4.7: Jambi Province, Indonesia:
Creating Low Carbon Prosperity in Jambi (2010)

This report evaluates the potential for low-carbon prosperity in Jambi Province, Indonesia. The report outlines a high-level a strategy for delivering economic growth while making deep cuts in carbon emissions. It focuses on three core elements:

1. CO_2 mitigation: estimating the size of current and future emissions; assessing the technical abatement potential and feasibility of abatement levers; developing an action plan to capture prioritised abatement opportunities.
2. Economic development: analysing existing competitive strengths and weaknesses; prioritising growth opportunities based on impact and feasibility; developing an action plan to capture prioritised growth opportunities.
3. Institutional enablers: developing a strategy for critical enablers that will underpin the success of the low-carbon growth strategy (institutions, monitoring and evaluation, financial mechanisms, spatial planning).

The report was led by the Office of the Governor of the Province of Jambi, the Dewan Nasional Perubahan Iklim and Indonesia's National Climate Change Council with the support of the Agence Francaise de Developpement, the ClimateWorks Foundation, the Norwegian Government and the Packard Foundation. Analytical support was provided by McKinsey and Company.

Source: http://photos.mongabav.com/10/ia mbi report 090810 english.pdf

Common elements of LEDS that have been formally communicated to the UNFCCC include: reducing energy demand, such as through investments in energy efficiency and infrastructure; ensuring energy security based on increasing renewable and low-carbon sources of energy as a proportion of total energy used; adopting supportive technologies and policies in non-energy sectors; and managing land sustainably.

Such measures are likely to bear significant financial costs and face significant challenges and barriers in implementation. Low carbon growth policy is a long-term process and there is often significant institutional inertia among local regulators and policy makers due to perceived institutional risks of adopting new technologies, existing vested interests, and a lack of capacity to assess economic benefits and costs.

A number of recent reviews of LEDS have been undertaken which identify challenges and lessons learned, as summarised in Table 4.4 below. Again, it will be important that governments and practitioners learn from this experience with LEDS in the development of green economy strategies.

4.4 GREEN ECONOMY AND GREEN GROWTH STRATEGIES

Building on international experience with NSDS and LEDS, a number of countries have begun to develop national green economy strategies. While these concepts are all closely related, green economy strategies could be seen as a further step in the evolution of integrated sustainable development strategies, promising new tools and a fresh approach for overcoming the gaps and challenges experienced over the past 20 years in the implementation of NSDS.

Table 4.4: Lessons learned for developing LEDS

Category	Lessons Learned
1. Fact base	A LEDS needs to build on a strong basis of high-quality and timely data on GHG emissions and socio-economic indicators, and the credibility of the research depends on the quality of and availability of data.
2. Capacity	Analytical capacity is needed for various tasks in the process, such as assessing the current situation and identifying alternative development pathways. Collaboration with international experts may improve the analysis, but national capacity is essential to ensure that the strategy is sufficiently rooted in the reality of the specific country.
3. Awareness and leadership	Government, the private sector and civil society stakeholders need to be aware of how low-carbon development can affect them. This awareness is essential to create buy-in for the strategy and its implementation. Strong, senior leadership from government is a key success factor for developing a LEDS that is integrated across all policy areas.
4. Government coordination	Clear roles and policy mandates need to be established. Governments on the subnational level needs to be engaged in the strategy as early as possible, since they are typically crucial for the implementation. To establish momentum for implementation, a LEDS ideally needs to be integrated into the mainstream national decision-making process.
5. Stakeholder Involvement	Engage stakeholders from the start of the process to provide and improve input, and to create support for the strategy. Lack of time, resources and commitment may lead to late involvement and a narrow base for participation, which in turn may create a gap between the strategy and on-the-ground realities.

The central focus on the economy could assist with engaging central planning and finance ministries, integrating with national expenditure and fiscal priority setting, and the utilisation of economic instruments and fiscal reform to implement sustainable development. Building on recent experience with LEDS, the shift to green economy strategies could move beyond the integration of climate change with development to address a wider range of environmental, social and economic issues in a more comprehensive and integrated manner.

4.4.1 Leadership, Engagement, Institutions and Integration with National Budgetary Processes

It can also be seen that countries are using a range of different institutional support structures and arrangements for whole-of-government engagement in the development and implementation of the strategy. Such arrangements are particularly important where a line agency is the lead to ensure engagement across government and the involvement of central economic planning and finance ministries and a broader range of stakeholders.

It would also seem that most countries are utilising whole-of-government coordinating bodies to develop their national strategies, some of which remain in place to support strategy implementation. Building institutional capacity for strategy implementation was also a key priority targeted in some of the developing country strategies reviewed, in particular Guyana, Rwanda and Ethiopia. This includes the establishment of new coordination offices, technical committees, research centres, and funds (see Box 4.8 below). Indonesia's provincial plan for Jambi also includes some useful recommendations regarding the proposed establishment of a new coordination and delivery unit for low carbon development (see Box 4.9 below).

Box 4.8: Proposed institutional Arrangements for Strategy Implementation in Guyana, Rwanda and Ethiopia

Guyana (Republic of Guyana, 2010)

To ensure successful execution of its strategy, Guyana is developing five new enhanced institutional capabilities:

1. An Office of Climate Change (OCC) established in the Office of the President to consolidate and streamline existing Government efforts to encompass, among other things, the coordination of engagement with multilateral processes and negotiations including UNFCCC. It has the overall coordinating responsibility for the strategy.
2. A strategy Project Management Office established to drive key projects under the strategy and report directly to the President.
3. A Guyana REDD Investment Fund (GRIF) will be established to manage forest payments, to reduce the cost of capital on other essential investments, and over the long-term to act as a permanent investment fund for low carbon investments.
4. A strengthened EPA will ensure that social and environmental safeguards are applied to the appropriate internationally recognised standards for all GRIF investments.
5. The REDD Secretariat at the Guyana Forestry Commission will be the implementing agency for REDD readiness activities, including a MRV system.

Rwanda (Republic of Rwanda, 2011)

The strategy acknowledges that it will require large amounts of finance and human capacity to be implemented, requiring significant support from development partners, civil society and the private sector, and leveraging of foreign direct investment. Capacity building is underway in government but needs to be scaled up to meet the needs of the strategy. Initially, this will require technical assistance from the international community, and local staff will need to study and gain experience abroad until the technical courses are running in Rwanda.

Moving forward, the government is finalising legislation for a National Fund for Climate and the Environment within the Ministry of Finance and Economic Planning as a 'basket fund' that will play a key role in managing climate funds that flow into Rwanda. Two additional new organisational structures will also be established - a Centre for Climate Knowledge for Development (CCKD) and a Technical Coordinating Committee (TCC) to lead and facilitate the flow of knowledge among the different organisations.

The CCKD will be a focus organisation for interpreting climate information in the form required for each sector. It will be a multi-disciplinary organisation with expertise in climate change and sectors. It will work with the TCC to provide information to relevant Ministries and stakeholders. The TCC would comprise department heads from the Revenue Authority, Rwanda Natural Resources Authority, the Energy Water and Sanitation Authority, the Rwanda Housing Authority, the Rwanda

Transport Development Agency, the Private Sector Federation, the Rwanda Development Board and other agencies and representatives from civil society, academia, development partners and the private sector.

Ethiopia (Federal Democratic Republic of Ethiopia, 2011)

The Government is establishing a permanent platform for implementation of the strategy. Overall responsibility will lie with Ethiopia's Environmental Council, which is chaired by the Prime Minister and comprises members drawn from Federal Ministries, Presidents of National Regional States, and representatives of non-governmental bodies, the private sector and trade unions. A Ministerial Steering Committee and a Technical Committee sit under the Environmental Council, as well as eight sub-technical committees focusing on specific sectors. The Government will install a subsidiary body to govern the Climate-Resilient Green Economy initiative under the co-responsibility of the Ministry of Finance and Economic Development and the EPA.

4.4.2 Costs and Financing

It is important to undertake sound analysis of policy measures, identify the underlying trade-offs and synergies among economic, environmental and social objectives and use this information to set policy priorities. An important component of this is cost-benefit analysis which requires adequate information on the costs associated with implementing policies. It is unlikely that policies will be implemented unless the costs are calculated and a clear pathway to funding is provided, particularly where these costs are high. The lack of adequate information on costs and sources of financing has been another challenge experienced in the implementation of NSDS over the years.

Box 4.9: Recommendations for Jambi's (Indonesia) Delivery Support Unit

This strategy from Indonesia is at the provincial level for Jambi Province, but nevertheless includes some useful recommendations for establishing a delivery unit for implementing low carbon strategies as follows:

1. It must have a direct relationship with and a clear mandate from the highest levels of government;
2. It needs to include representatives from different levels of government;
3. Relationships and decision-making rights must be clearly defined between the new unit, ministries, and other stakeholders;
4. Employee compensation and value proposition must be competitive with the commercial sector to attract top talent; and
5. A rigorous performance management framework around a few priority outcomes must be developed.

These indicative costs are outlined in Table 4.5 below. This highlights the significant funding that will be needed to support green economy as well as the need to mobilise capital investment in the early years of development. However, it is also recognised that not all of this expenditure is necessarily additional to business as usual investment.

Table 4.5: Indicative costs of implementing green economy strategies

Country	Costs indicated in strategy
Ethiopia	The strategy will require US$150 billion over 20 years, including 80 billion in capital investment and 70 billion in operating and programme expenses, with high initial capital expenditure of US$22 billion required by 2015. This averages out at US$7.5 billion per annum, which is a considerable 25% of Ethiopia's annual GDP of US$29.7 billion (as at 2010).
Guyana	The strategy identifies a requirement for US$1billion in essential capital projects fully or partially funded through private investment assisted by REDD+ payments, which represents almost 45% of annual GDP (US$2.23 billion in 2010). A further US$1billion is also predicted to be required for climate change adaptation, bringing total costs to almost 90% of GDP. However, there is no set timeframe for these investments and this may be expended over several years.
South Africa	Total funding identified in South Africa's green economy accord amounts to US$28.46billion over approximately 5 years1, or approximately US$5.7 billion per annum. This represents approximately 1.56% of annual GDP of US$363.9 billion (as at 2010).
RoK	The government has earmarked US$97 billion of public investment for supporting green growth from 2009 to 2013, which is about 2% of the country's annual GDP.
UK	The plan estimates that total net costs of policies set out in the 10 year plan including both public and private costs are between US$39 billion and US$49 billion. Averaged at US$3.9 to 4.9 billion per annum over 10 years, this represents 0.17 to 0.21% of annual GDP (i.e. of US$2.25 trillion as at 2010).
Jambi, Indonesia	In the first year, between US$ 19 million and 39 million will be required to establish basic readiness functions to support low-carbon growth. From 2011–2030, ongoing running costs to support implementation of carbon abatement and sustainable livelihood opportunities will gradually increase and reach between US$373 million and US$676 million in 2030, assuming capture of the full 55 Mt CO_2e in potential abatement.

The necessary investment varies significantly between countries and it is evident that, for developing countries in particular, the required investment as a percentage of annual GDP is substantial.

In addition to the countries outlined in the above table, France's national strategy provides some financing targets such as the allocation of 3% of GDP to research and development from 2010 onwards (to be financed by both government and business) as well as mobilising 1 billion additional Euros for research and development on sustainable development by 2012. Grenada's roadmap also includes some costing estimates for potential energy-related projects.

Linked to the issue of costs associated with strategy implementation is the need to identify clearly sources of funding and establish the appropriate institutional arrangements to attract external funding. As mentioned above, a clear objective in some of the strategies reviewed for developing countries is to facilitate access to emerging climate finance. For example, the strategies for Guyana, Rwanda and Ethiopia all prioritise the establishment of appropriate financial frameworks necessary for attracting and managing climate finance (see Table 4.6). Rwanda's strategy highlights that "a financial framework is just as important as an institutional framework and needs to be implemented as soon as possible to allow climate finance to flow into the country."

Table 4.6: Financial Frameworks for attracting climate finance

Country	Financial Frameworks
Guyana	A Guyana REDD Investment Fund (GRIF) will be established to manage forest payments, to reduce the cost of capital on other essential investments, and over the long-term to act as a permanent investment fund for low carbon investments.
Rwanda	A National Fund for Climate and the Environment (FONERWA) is being established and will play a key role in managing climate funds that flow into Rwanda. The Fund will seek to employ a wide range of public financing mechanisms, such as performance-based grants, loan guarantees, lines of credit, and public venture capital to create an attractive investment environment for low-carbon activities. The Government is in the process of finalising a Bill to operationalise this basket fund which will be housed in the Ministry of Finance and Economic Planning.
Ethiopia	The Government plans to govern the Climate Resilient Green Economy initiative under the co-responsibility of the EPA and the Ministry of Finance and Economic Development (MoFED). The UNDP has offered its support in establishing a Multi-Donor Trust Fund within MoFED through which international funds can be channelled.

These countries also identify particular sources of climate finance which best align with their natural resource endowments and national circumstances.

Box 4.10 below provides a summary of sources of finance identified by Rwanda and Ethiopia. Whilst this is identifying a clear pathway to potential funding, it remains uncertain whether or not funding will be forthcoming and, if so, whether or not it will be sufficient to implement their strategies. In the case of Ethiopia, given the significant funding target of $20 billion annually, this seems unlikely. As such, additional sources of financing would be critical.

4.5 USE OF A MIX OF GREEN ECONOMY POLICY INSTRUMENTS

Governments have a mix of policy instruments at their disposal to green their economies, including economic instruments for internalising externalities, measures for incentivising green or sustainable practices, measures to enhance institutional capacity and regulatory capabilities, green investment across key sectors, information-based and voluntary measures, and complementary social policies to drive inclusion.

Experience with the implementation of NSDS highlights that while a mix of policy measures has been pursued by countries, economic instruments appear to be under-utilized. This is problematic because countries increasingly rely on the market to allocate resources efficiently and market-based instruments can thus provide the policy levers necessary to change behaviours and deliver desired policy outcomes, in principle at least cost. Markets comprise a complex array of actors such as consumers, producers, investors, and lenders, each of whom have the potential to impact on the environmental and social systems that support our well-being. Establishing appropriate market conditions through economic policies, incentives and marketbased measures is critical for establishing appropriate 'rules of play' and sending the right market signals that ensure that resources are allocated as efficiently as possible (*e.g.* by taxing 'bads' and in some instances subsidising 'goods').

Box 4.10: Identified Funding Sources and Financing Priorities for Developing Countries

Rwanda's priorities for strategy financing (Republic of Rwanda, 2011)

The 5 priorities for the finance pillar are:

- Operationalise the National Fund for Climate and the Environment
- Secure grants from the Green Climate Fund, the Adaptation Fund and other climate funds targeted at LDCs
- Promote CDM and voluntary carbon projects and push for simplified baseline calculations and monitoring methods at UNFCCC
- Investigate and employ environmental fiscal reforms, a feed-in tariff, a green investment index and public financing mechanisms to encourage green consumerism and investment
- Encourage conversation through PES schemes.

Ethiopia's proposed funding sources (Federal Democratic Republic of Ethiopia, 2011)

A funding pool of at least $20 billion annually should be obtained from various climate finance schemes set up to foster the green economy initiatives of developing countries. In the short term this could take the following forms:

- Bi-/multilateral grants primarily for project setup, capacity building, technology development and dissemination
- Bi-/multilateral pay-for-performance deals, *i.e.*, payments linked to verified GHG abatement
- Trading schemes or offset markets, *i.e.*, emission reduction, for example resulting from the CDM, sold to companies (ETS) or committed countries (cap and trade) or via voluntary carbon markets.

The policy instruments utilised and priority sectors targeted vary from country to country based on national circumstances.

It can be seen that only the UK, RoK and France adopt economic instruments in their strategies for internalising externalities, such as taxes, charges and levies on pollution and resource use, or cap-and-trade permit systems. However, almost all developed and developing countries adopt economic instruments that provide incentives for green investment, including low-interest loans, micro financing and tax or tariff exemptions. For developing countries, these tended to be small-scale investment incentives such as micro-financing, micro-insurance, small grants and government-backed loans, rebates and cash-back schemes. Again, this highlights potential gaps associated with institutional capacities in developing countries (or could also be associated with a lack of political willingness, resistance from stakeholders or other factor) to implement more complex fiscal reforms and market-based instruments.

Other areas of policy focus for both developed and developing countries appear to be government investment in infrastructure, in particular in the energy and transport sectors, as well as the use of regulations such as technical and performance standards, building codes, environmental regulation, and mandatory targets.

For developing countries, there was also a tendency to focus on governance and institutional reforms to build capacities, including in analysis, implementation, and enforcement. Other common policy measures included investment in human capital through education and training schemes, investment in natural capital (including through REDD+, payments for ecosystem services), and investment in sustainable agriculture.

With regard to developed and emerging economies, their strategies tended also to incorporate measures for investment in innovation (including R&D), voluntary approaches targeting awareness raising (*e.g.* focusing on consumption), as well as measuring progress through targets and indicators and establishing institutional arrangements to monitor progress and coordinate followup actions as required.

■ ■ ■

5

Guidelines to Green Agriculture

5.1 INTRODUCTION

This chapter makes a case for investing in greening the agriculture sector, emphasising the potential global benefits of making this transition. It provides evidence to inspire policymakers to support increased green investment and guidance on how to enable this transformation, which aims to enhance food security, reduce poverty, improve nutrition and health, create rural jobs, and reduce pressure on the environment, including reducing GHG emissions.

5.2 GENERAL BACKGROUND

Agriculture is a major occupational sector in many developing countries and is an important source of income for the poor. World Bank statistics (2010) show agricultural value-added as a percentage of GDP to be 3 per cent for the world as a whole, and 25 per cent for low income countries (LICs), 14 per cent for lower middle income countries (LMICs), 6 per cent for upper middle income countries (UMICs) and 1 per cent for high income countries (HICs). Approximately 2.6 billion people rely on agricultural production systems - farming, pastoralism, forestry or fisheries—for their livelihoods.

To date, global agricultural productivity has more than kept up with population growth. However, agricultural productivity per worker and per land unit varies a great deal across countries. Agricultural productivity per worker in 2003–05 was 95 times higher in HICs than in LICs, and this difference increased compared with 1990–1992, when it was 72 times higher. Industrial agriculture mostly practiced in developed countries, continues to generate high levels of production—more than 50 per cent of the world value added in agriculture and food processing—but it also accounts for proportionally more adverse environmental impacts than lower-yield traditional farming. Agriculture in developing countries is becoming more productive. Over the above period, aggregate agricultural productivity per worker in developed countries increased by 21 per cent, albeit from a very low base.

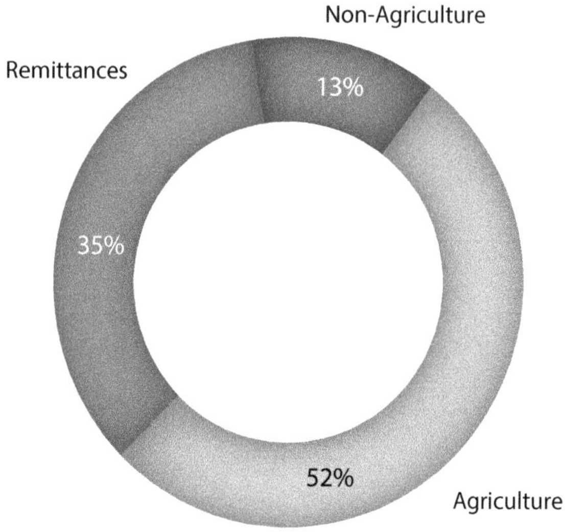

Fig. 5.1: Total average contribution to poverty reduction from growth of agricultural, remittance and non-farm incomes in selected countries.

Despite the increasing productivity of agriculture, nearly 1 billion people remain malnourished. Between 2000 and 2007, over a quarter (27.8 per cent) of children under the age of five in LICs were malnourished. Moreover, over half of food-insecure families are rural households, often in countries such as India that have food surpluses. A transition in the agricultural paradigm must also assist in meeting this challenge.

Agriculture also has tremendous potential to alleviate poverty. A large proportion of the rural population and labour force in developing countries is employed in agriculture. On average, the contribution of agriculture to raising the incomes of the poorest is estimated to be at least 2.5 times higher than that of non-agriculture sectors in developing countries. Underscoring the relationship between increasing yields and return on labour with poverty, estimate that for every 10 per cent increase in farm yields, there was a 7 per cent reduction in poverty in Africa and more than a 5 per cent poverty- reduction effect for Asia. Growth in manufacturing and services do not show a comparable impact on poverty reduction. The World Bank reported that an increase in overall GDP derived from agricultural labour productivity was, on average, 2.9 times more effective in raising the incomes of the poorest quintile in developing countries than an equivalent increase in GDP derived from non-agricultural labour productivity. Using crosscountry regressions per region, found greater effects from agricultural growth on poverty (defined as less than US$ 2 per day per person) reduction in sub-Saharan Africa and South Asia. (This trend was not seen in East Asia and Latin America where there were greater poverty-reducing effects of growth originating in non-agriculture sectors).

Despite the potential contribution of agriculture to poverty alleviation, mainly owing to the urban bias of many national government policies, rural sectors in most developing countries have not received the levels of public investment required to support the development of a thriving agriculture sector. Government expenditure on agriculture in developing countries dropped from 11 per cent in the 1980s to 5.5 per cent in 2005, with the same downward trend observed in official development assistance going to the agriculture sector, which fell from 13 per cent in the early 1980s to 2.9 per cent in 2005. In Africa, governments publicly committed in the Maputo Declaration of 2000 to spending 10 per cent of their GDP on agriculture, including rural infrastructure spending. However, only eight countries had reached the agreed level by 2009.

Between 1980 and 2000, an inverse association was noted between the contribution of agriculture to GDP and public spending on agriculture as a percentage of agricultural GDP as shown in Fig. 5.2, which distinguishes between agriculture-based, transforming and urbanised countries.

The result of this long-term neglect of the agriculture sector in developing countries is that rural poverty rates consistently exceed those in urban areas, with more than 75 per cent of the world's most impoverished people living in rural areas, and many seeking ways to migrate to cities. We note that in this scenario, poverty can result in environment-related economic consequences if crop production is based upon unsustainable land use, which in turn results in the depletion of soil nutrients and cultivation of unsuitable, marginal land that can lead to soil erosion, degradation of ecosystems and the reduction of natural habitats for biodiversity.

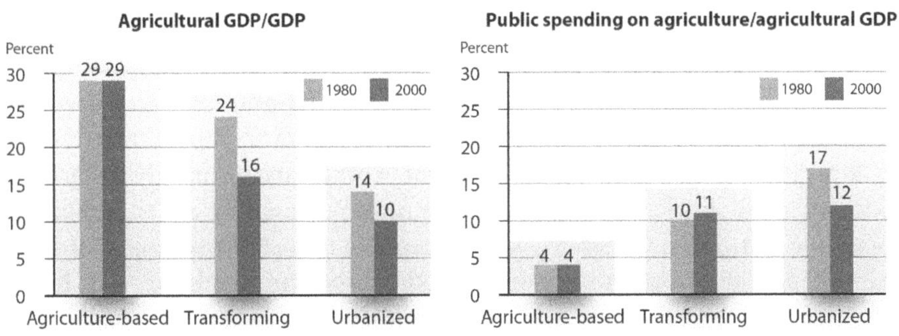

Fig. 5.2: Contribution of agriculture to GDP and public expenditure on agriculture as a proportion of agricultural GDP.

5.3 CONVENTIONAL/INDUSTRIAL AGRICULTURE

Conventional (industrial) agriculture is characterised by farming practices that rely on use of external farming inputs. Most of the large scale industrial farming is considered energy-intensive (using 10 calories of energy for every calorie of food produced), whose high productivity (kg/ha) relies on the extensive use of chemical fertilisers, herbicides, pesticides, fuel, water, and continuous new investment (*e.g.* in advanced seed varieties and machinery).

The impressive productivity gains of the Green Revolution of the last few decades took place mainly in conventional agriculture. These productivity gains were triggered by investment in agricultural research and expansion in public-sector extension services. The productivity increases of the Green Revolution relied primarily on the development of higher-yield varieties of major cereal crops (*i.e.* wheat, rice and corn/maize), a significant increase in the use of irrigation, inorganic fertilisers, pesticide/herbicide use and fossil fuel-based farm machinery.

Despite substantial gains in total crop production, the consequences of the revolution have not been entirely positive. Production gains have been highly correlated with increased use of non-renewable resource inputs, and have often entailed significant environmental costs due to their overuse (Fig. 5.3). Industrial agriculture consumes on average 10 exosomatic energy calories (derived from fossil fuel energy resources) for every food endosomatic energy calorie (derived from human metabolism of food) that is produced and delivered to the consumer. This energy-intensity, in many cases, is encouraged by subsidising inorganic fertiliser, fuel and electric power used on farms. In addition, biodiversity losses have resulted from production subsidies targeted at a limited number of crops. Industrial agriculture has also resulted in shrinking the agricultural labour force even as farm outputs have dramatically increased, a trend intensified to some extent by subsidies for farm mechanisation.

5.3.1 Traditional/Small Farm/Subsistence Agriculture

Traditional (subsistence) smallholder agriculture typically relies on indigenous and traditional knowledge that is based on farming practices used for several generations, has limited or no use of off-farm inputs, and results in low-productivity, low value added per worker and primarily reliant on extracting soil nutrients with insufficient replenishment by either organic or inorganic fertilisers. Generally, it is susceptible to yield losses due to erratic rainfall, pest and weed infestations and other production- related risks. It can trap already poor farmers in a downward spiral of growing poverty and social marginalisation.

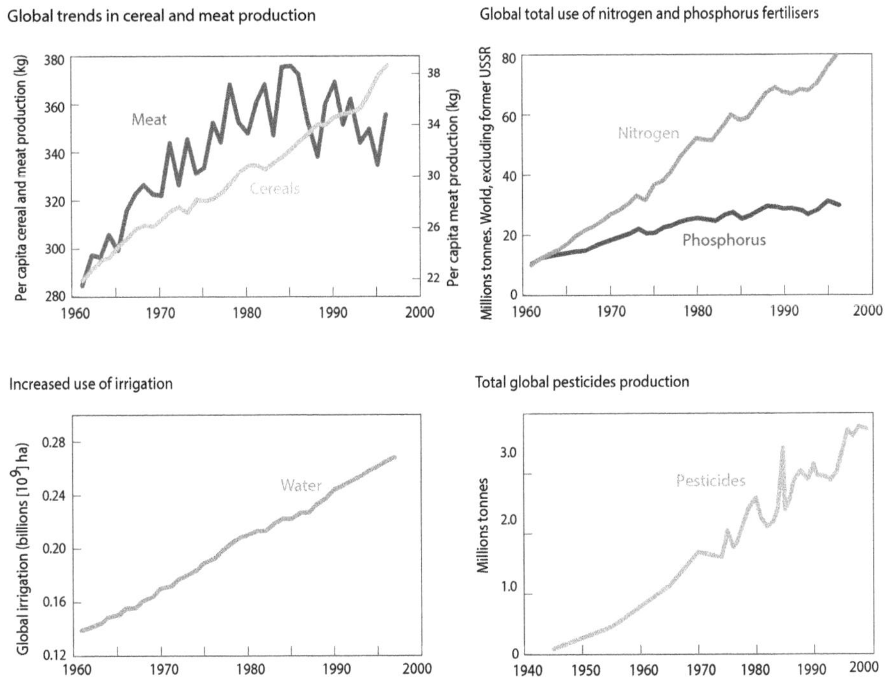

Fig. 5.3: Global trends in cereal and meat production, nitrogen and phosphorus fertiliser use, irrigation and pesticide production.

Traditional agriculture has limited scope for capital intensive farm mechanisation and intensive use of external agrochemical inputs. Many smallholders' plots, overarchingly located in developing countries, are too small to realise the economies of scale required for most of the available commercial farm machinery. In addition, the high cost of purchased inputs, such as chemical fertilisers, pesticides and seeds, generally require that at least some portion of the crops produced must be sold to recover costs. Failure to modernise land tenure systems, which can facilitate distribution, consolidation, and the use of land as security for bank loans are important barriers to the commercialisation of small-scale agriculture in many developing countries. Commercialisation is further limited by inadequate road transportation linking food-producing areas to large urban centres. For these reasons, value added per worker in developing countries is far below that of industrialised economies. Whereas the average value added per agricultural worker in OECD countries in 2003 was US$ 23,081 (which grew at 4.4 per cent per year between 1992 and 2003, in Africa, the figures were only US$ 327 and 1.4 per cent, respectively.

Worldwide, there are 525 million small farms, 404 million of which operate on less than two hectares of land. These small farmers in the developing world produce the majority of staple crops needed to feed the planet's population. Their highest share is in Africa where about 90 per cent of all agricultural production is estimated to be derived from small farms. In many instances their contribution is growing at the national level. While the issue is contested, there is substantial evidence that smaller farms have higher yields than large farms. In Kenya, the share of national agricultural production contributed by smallholders increased from 4 per cent in 1965 to 49 per cent in 1985. In India, smallholders contributed over 40 per cent of food grain production in 1990–91, compared with only a third of the total in 1980. As of the late 1990s, they also owned the majority of livestock and dominated the dairy sector.

Box 5.1: Agriculture at a Crossroads

The key message of the Assessment of Agricultural Knowledge, Science and Technology for Development, published in 2009 is: "The way the world grows its food will have to change radically to better serve the poor and hungry if the world is to cope with a growing population and climate change while avoiding social breakdown and environmental collapse." The Assessment calls for a fundamental shift in agricultural knowledge, science and technology (AKST) to successfully meet development and sustainability objectives. Such a shift should emphasise the importance of the multifunctionality of agriculture, accounting for the complexity of agricultural systems within diverse social and ecological contexts and recognising farming communities, farm households, and farmers as producers and managers of ecosystems. Innovative institutional and organisational arrangements to promote an integrated approach to the development and deployment of AKST are required as well. Incentives along the value chain should internalise as many negative externalities as possible, to account for the full cost of agricultural production to society. Policy and institutional changes should focus on those least served in the current AKST approaches, including resource-poor farmers, women and ethnic minorities. It emphasises that small-scale farms across diverse ecosystems need realistic opportunities to increase productivity and access markets.

Source: IAASTD.

Despite their higher output per hectare and the significant contribution they make to food production, however, small farmers are often very poor. In a survey of smallholder households, 55 per cent in Kenya and 75 per cent in Ethiopia, respectively, fell below the poverty line. Low prices, unfair business practices and lack of transportation, storage and processing infrastructure contribute to this situation. Half of all undernourished people, three-quarters of malnourished African children and the majority of people living in absolute poverty are found on small farms. In the majority of countries, poor rural people are both sellers of food commodities and buyers of foodstuffs, at different times of the year. Typically, they sell immediately after harvest, usually at very low prices, to meet their immediate cash requirements, and buy food in the months prior to the following harvest, usually at higher prices, to meet their food needs.

It is expected that expanding smallholder production through green agricultural practices and greater commercialisation and integrating them into supply chains will create more better rewarding jobs in rural areas. As farmers get wealthier, they are likely to withdraw from occasional labour. Wealthier farmers are also likely to spend more on locally-produced goods and services leading to multiplier effects. Rural linkage models in Africa have estimated multiplier effects ranging from 1.31 to 4.62 for Burkina Faso, Niger, Senegal and Zambia.

5.4 THE GREENING OF AGRICULTURE

The greening of agriculture refers to the increasing use of farming practices and technologies that simultaneously:

- maintain and increase farm productivity and profitability while ensuring the provision of food and ecosystem services on a sustainable basis;
- reduce negative externalities and gradually lead to positive ones; and
- rebuild ecological resources (*i.e.* soil, water, air and biodiversity natural capital assets) by reducing pollution and using resources more efficiently.

A diverse, locally adaptable set of agricultural techniques, practices and market branding certifications such as Good Agricultural Practices (GAP), Organic/Biodynamic Agriculture, Fair Trade, Ecological Agriculture, Conservation Agriculture and related techniques and food supply protocols exemplify the varying shades of green agriculture.

Farming practices and technologies that are instrumental in greening agriculture include:

- restoring and enhancing soil fertility through the increased use of naturally and sustainably produced nutrient inputs; diversified crop rotations; and livestock and crop integration;
- reducing soil erosion and improving the efficiency of water use by applying minimum tillage and cover crop cultivation techniques;
- reducing chemical pesticide and herbicide use by implementing integrated and other environmental friendly biological pest and weed management practices; and
- reducing food spoilage and loss by expanding the use of post-harvest storage and processing facilities.

The greening of agriculture does not imply ruling out technologies or practices on ideological grounds. If a technology works to improve productivity for farmers, and does not cause undue harm to society and the environment, then it is very much part of the efforts for greening of agriculture. Although natural methods of pest and weed management and organic sources of fertiliser and seed are at one end of a green agriculture spectrum, the highly efficient and precise use of inorganic fertilisers, pest controls and technological solutions may also be included in the broad spectrum of sustainable farming practices. The Foresight Report presents resembling ideas given the need for the global food system to deliver much more than just food, and food security in the future. So greening of high input dependent agriculture, which has a high ecological footprint, could start by making the use of inputs most precise and efficient, gradually moving toward farming practices that have low or no ecological footprint.

To be able to measure success in moving towards the objectives of greening agriculture, two categories of indicators are proposed in Table 5.1.

Table 5.1: Potential indicators for measuring progress towards green agriculture

Action indicators	Outcome indicators
Number of enacted and implemented policy measures and officially approved plans that promote sustainable agriculture (including trade and export policy measures, payment for ecosystem services through agriculture, etc.)	Percentage and amount of land under different forms of green agriculture (organic, GAP-good agriculture practices, conservation, etc.)
Level of governmental support to encourage farmers to invest in conversion to green agriculture and get the farm and the product certified	Decline in use of agro-chemicals as a result of conversion to green agriculture; and the number and percentage of farmers converting to green agriculture
Percentage of agricultural budget that is earmarked for environmental objectives	Increasing proportion of Payments for Environmental Services as a percentage of total farm income
Proportion of available producer support utilised for environmental objectives as a percentage of total agricultural producer support	Number of agriculture extension officers trained in green agriculture practices
Approved measures that reduce or eliminate barriers to trade in technologies and services needed for a transition to a green agriculture.	Number of enterprises set up in rural areas, especially those that produce local natural agricultural inputs, to offer off-farm employment opportunities.

5.5 CHALLENGES AND OPPORTUNITIES

Today, agriculture stands at a crossroads. There are calls for changing the way food is produced and distributed if the poor and hungry are to be served better and if the world is to cope with a growing population and climate change. This section presents some major challenges and opportunities in transitioning to a green agriculture.

5.5.1 Challenges

Agriculture is facing a multitude of challenges on both the demand and supply side. On the demand side, these include food security, population growth, changing pattern of demand driven by increased income, and the growing pressure from biofuels. On the supply side, limited availability of land, water, mineral inputs and rural labour as well as the increasing vulnerability of agriculture to climate change and pre-harvest and postharvest losses are the main challenges.

1. Increasing Demand for Food

The most significant factors contributing to the increasing demand for food are the continued growth of the global population, especially in developing countries (Fig. 5.3), and a rise in income levels in emerging economies. Demand for meat and processed food is rising with growing affluence. The current global population of more than 6 billion, of which 925 million are undernourished, is forecast to reach 8.5–9 billion by 2050, and per capita incomes are expected to rise by as much as a factor of 20 in India and 14 in China, respectively. Figure 5.4 shows that rural populations are increasingly migrating to urban and peri-urban areas in developing countries. This has consequences for food demand and field-to-table supply chains because the diets of urban dwellers show an increased proportion of processed foods. The prospect of the human population expanding by almost a third by 2050, combined with an expected rise in per capita demand for meat, dairy and vegetable products, requires geographically-focused efforts and a change in agricultural production patterns.

2. Competing Demand from Biofuels

Growing interest in producing first-generation liquid biofuels to augment and replace petroleum-based transportation fuels is adding to the demand for starch, sugar and oilseed food commodities. For example, the production of ethanol and bio-diesel fuels are predominantly based on food commodity feed stocks such as corn, sugarcane, soy, canola, sunflower and palm oil. Despite growing ethical, environmental, and economic concerns surrounding the use of food staples for producing these biofuels, there is continued public—and private-sector interest in their development. No matter where these crops are grown, they will inevitably compete with food crops for land, water and nutrients. Figure 5.4 shows food prices tracking fuel prices. At present, this alignment of food and energy prices may primarily result from the cost of fossil fuels used as an input in food production. But it is expected that the pattern will become more marked because of the competition for food crops that are used to produce biofuels.

As a result, significant efforts are being made to develop technologies for second-generation biofuels, which can be produced from non-food biomass feedstock such as ligno-cellulosic wood and crop-residue wastes, perennially-grown switch grass and algae. Such technologies can potentially enable the production of biofuels to be scaled up with fewer adverse impacts on global food security. However, much more analysis is needed regarding the degree to which converting large quantities of cellulosic feedstock to biofuels would displace the recycling of organic nutrients from crop residues to arable land, pastures and forests.

3. Limited arable Land and Scarce Water

Approximately 1.56 billion hectares or 12 per cent of the earth's total land surface area is arable land being used to produce crops for human and livestock consumption.

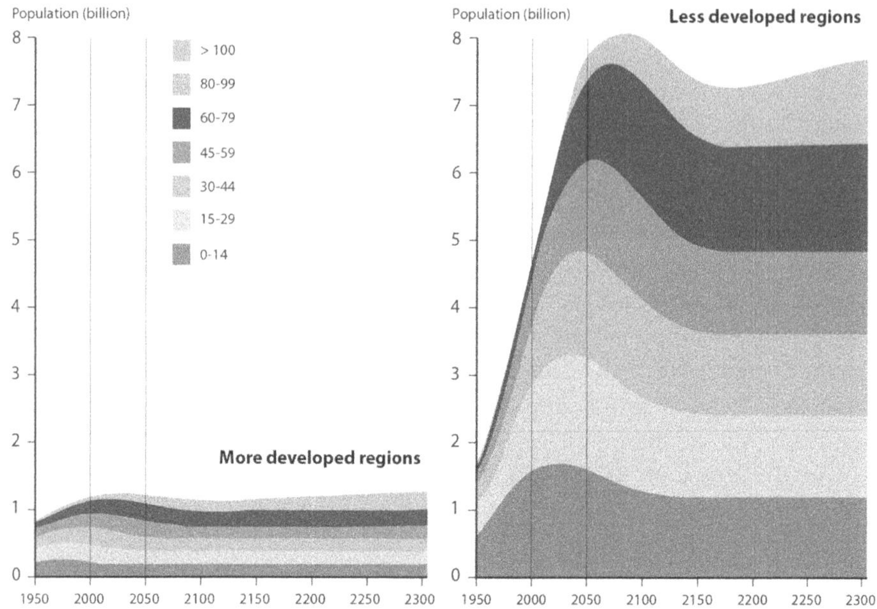

Fig. 5.4: Distribution of population by age in more developed and less developed regions (1950-2300).

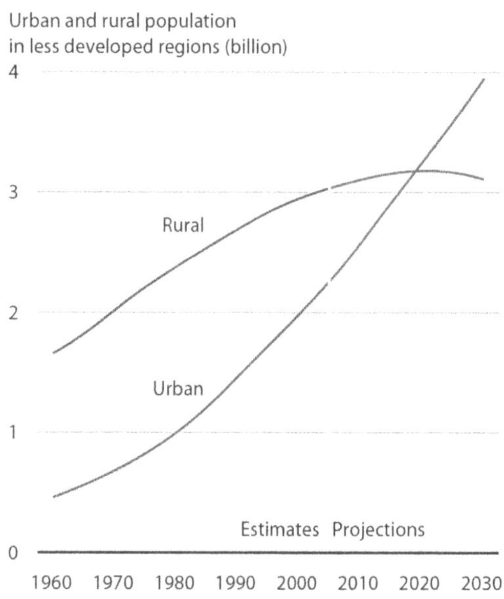

Fig. 5.5: Urban and rural population trends in developing regions.

In addition, some 3.4 billion hectares of pasture and woodland are now used for livestock production. The agricultural productivity of the available arable land is extremely varied. Crop yields in developed countries are generally far greater than the yields realised in most developing countries. These productivity differences result from different levels of natural soil fertility;

fertiliser, pesticide and herbicide use; quality of cultivated plant species and seeds; availability and access to water; farmers' education and access to information, credit and risk insurance and the degree of agricultural mechanisation.

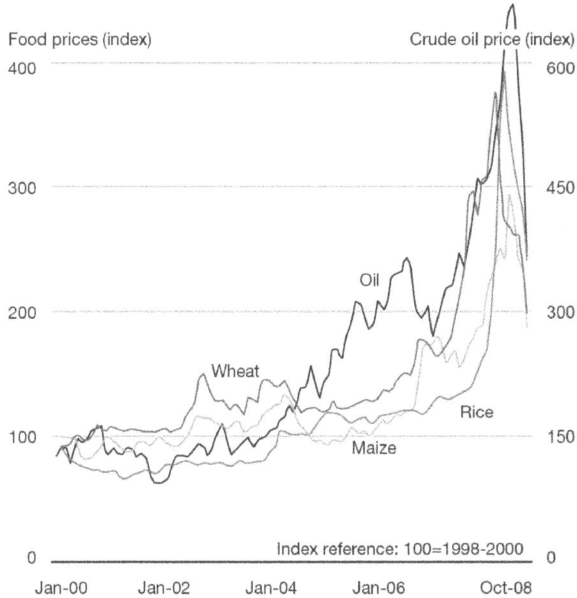

Fig. 5.6: Trends in food commodity prices, compared with trends in crude oil prices.

Only limited additional land can be readily brought into agricultural production through conversion or rehabilitation. Moreover, the often highly fertile arable land surrounding cities is rapidly being converted into residential and commercial development as urbanisation gathers pace. Expanding cultivated areas is no longer the obvious way to increase production (exceptions are parts of sub-Saharan Africa and Latin America where some savanna areas could be brought into production). Furthermore, over-grazing by livestock and extended drought conditions are accelerating the desertification of fragile arid and semi-arid regions. Agriculture has contributed to land degradation in all regions, but is most severe in input-intensive production systems (notably in East Asia, Latin America, North America and Europe). Agricultural activities account for around 35 per cent of severely degraded land worldwide. Given the high risk of further deforestation, developing countries will need to meet food-supply gaps by simultaneously increasing productivity and greening their agricultural practices, rather than seeking widespread expansion of arable land.

The agriculture sector is the largest consumer of fresh water, accounting for 70 per cent of global use, including rainfall run-off. A majority of crop lands are exclusively rain-fed, and only 24 per cent of arable land is cultivated with the help of irrigation from flowing surface waters or groundwater aquifers. This distinction is important because irrigated fields are much more productive and produce nearly a third of all agricultural output.

The increasing disruption of historical rainfall patterns experienced in many areas of the world is a cause for great concern since rain-fed farming is the dominant form of agriculture. The Intergovernmental Panel on Climate Change (IPCC) Fourth Assessment Report concluded that many observed changes in extremes, such as more frequent, heavy precipitation events and longer, more intense droughts, are consistent with warming of the climate system. While affecting rain-fed agriculture, precipitation changes also adversely affect the recharge rates of aquifers and watersheds. The continued worsening of water-stress conditions suggests that efforts to increase the use of irrigation will gradually increase agricultural production costs. Clearly, practices that increase water-use efficiencies are required to alleviate this trend.

Figure 5.7 shows projections for global water stress in the future. The figure also underscores the need for increased coordination in water use nationally and across borders. In this context, the Mekong River Commission, which coordinates the watershed development plans of member states, is one of several promising supra-national river basin initiatives.

Box 5.2: Opportunities for Improved Sanitation Systems and Organic Nutrient Recycling

There is a critical need to recover and recycle nutrients from organic waste streams and use them as productive inputs of organic fertiliser. Enormous quantities of valuable organic nutrients could be recovered from intensive livestock farming; food processing sites; municipal green wastes; and human sewage wastes in both rural and urban communities. It is particularly important to maximise the recovery of phosphorous nutrients from organic wastes; as a mineral, phosphate is essential to agricultural productivity and it has been estimated that economically recoverable global reserves may be depleted in 100 years. Technologies are under development that would eliminate pathogens and other toxic elements from these waste streams and recover commercial quantities of phosphorus. It is expected that the rising costs of inorganic fertilisers will help accelerate research and commercialisation of such organic nutrient-recovery technologies.

4. Limited Availability of Mineral Inputs

Industrial farming practices are dependent on inorganic fertilisers. In turn, the production and prices of these depend on the availability of fossil fuels, minerals and petro-chemicals. In this context, the demand for two major minerals—potassium and phosphorous—used in fertiliser production, has been increasing. But known supplies of readily accessible, high-grade stocks, especially phosphate rock, are falling. Estimates of the longevity of these stocks vary dramatically. Nevertheless, only one-fifth of the phosphorus mined for food production actually contributes to the food we consume, while the remainder is either polluting the world's water or accumulating in soils or urban landfills. Although it is expected that the increasing prices of phosphates and other minerals will lead to increases in supplies, including recovery of phosphate from wastewater treatment facilities, these prices are likely to continue to put upward pressure on the cost of fertilisers and food prices, which affects the poor's access to food disproportionately.

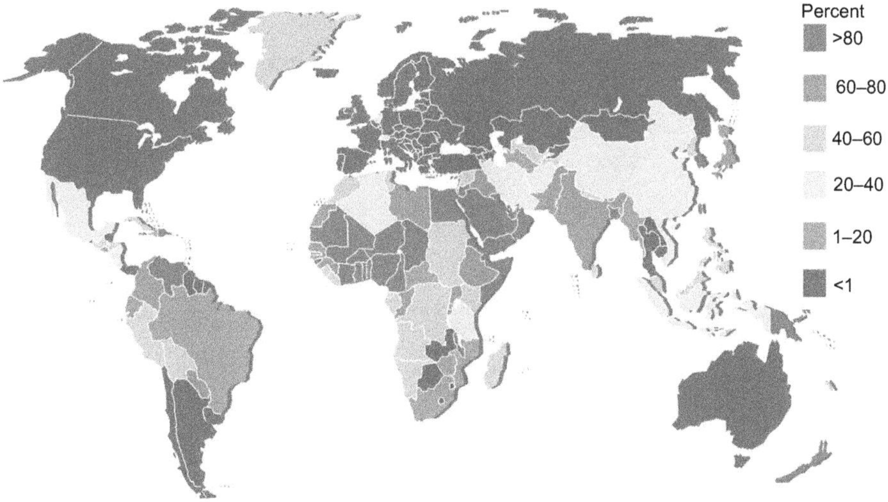

Fig. 5.7: Percentage of country populations that will be water stressed in the future.

5. Post-harvest Spoilage

Today, the volume of food produced globally is more than sufficient to feed a healthy population. But significant amounts of food produced around the world are lost or wasted after harvesting. As Fig. 5.8(b) shows, in developed countries this primarily occurs in the retail, home and municipal food-handling stages. For example in the USA, around 40 per cent of all food produced is wasted, resulting in losses of all embedded inputs such as energy (equivalent to wasting 350 million barrels of oil per year), water (equivalent to about 40 trillion litres of water every year) and huge volumes of fertilisers and pesticides. Losses in developed countries are often caused by factors such as retailers' rejection of produce due to poor appearance or supersized packages leading to post-retail spoilage. The latter can account for up to 30 per cent of the food bought by retail distributors. Post-retail food losses tend to be lower in developing countries. There, they mainly result from a lack of storage facilities, on-farm pest infestations, poor food-handling and inadequate transport infrastructure. For example, rice losses in developing countries may be as high as 16 per cent of the total harvest. Thus, there is ample scope for increasing food supplies and food security in developing countries through simple targeted investments in post-harvest supply chains.

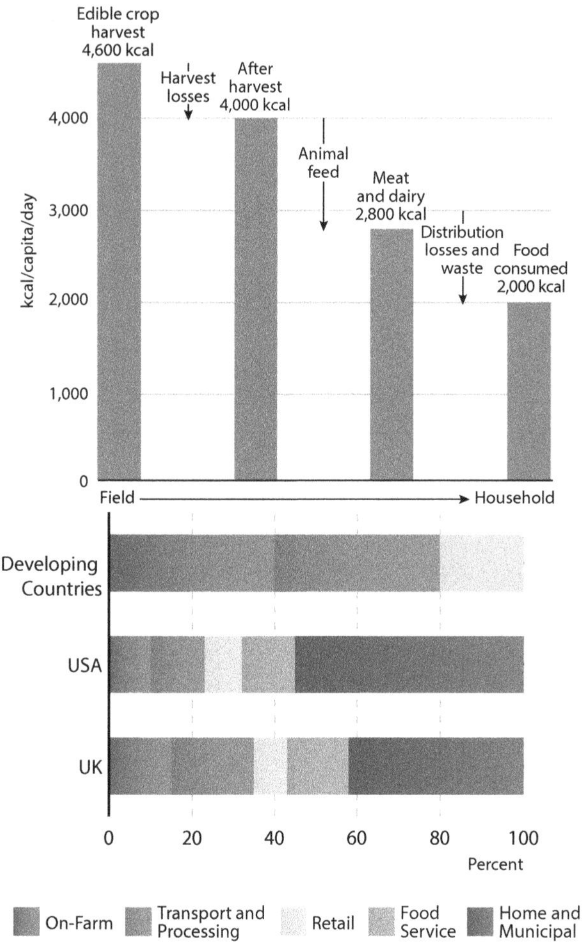

Figure 5.8a–b: The makeup of total food waste.

6. Rural Labour

The accelerating migration of rural populations to urban and peri-urban areas in developing regions of the world (Fig. 5.4) has resulted in significant demographic changes in rural populations. Working-age men are likely to relocate to cities in search of employment, reducing the pool of men available for agricultural work. This rural out-migration of men has also resulted in a dominant role for women as smallholders in these regions; more than 70 per cent of smallholders in sub-Saharan Africa are women. These demographic changes, while offering economic opportunities, have placed additional responsibilities on women, who invariably also have to care for their children and the elderly.

7. Increased Vulnerability of Agriculture Due to Climate Change

Modelling by the IPCC suggests that crop productivity could increase slightly at mid- to high-latitudes for mean temperature increases of up to 1–3°C (depending on the crop). However, at lower latitudes, especially in the seasonally dry and tropical regions, crop productivity could decrease as a result of even small local temperature increases (1–2°C).

Further warming could have increasingly negative impacts in all regions. Climate change scenarios suggest that by 2080 the number of undernourished people will increase, mostly in developing countries, by up to 170 million above the current level. Intergovernmental Panel on Climate Change modelling indicates that an increased frequency of crop losses due to extreme climate events may overcome any positive effects of moderate temperature increases in temperate regions.

Box 5.3: Innovations in the agricultural Supply Chain Increase Shareholder and Societal Value

For investors, water risk exposure is increasingly becoming material for mitigating investment risk in companies. For example, Robeco Asset Management invests in mainstream companies and encourages them, through active dialogue, to implement policies and innovative practices that mitigate risks resulting from water scarcity to their operations and reputations. In doing so, it also encourages companies to find solutions that can enhance their performance, increase shareholder value and therefore contribute in the long-term to building and sustaining a green economy.

Cotton, one of the most water-intensive crops, is the focus of a dialogue with companies in the textile industry to develop water-efficiency targets and adopt sustainable supply-chain practices. Through Better Cotton Initiative (BCI), a platform has been created for exchange of experiences on the use of efficient irrigation technologies, farmer education programmes and reduction in the use of pesticides and acceptance of transparent sourcing efforts.

Source: Based on the information from Robeco Asset Management received through Lara Yacob, Senior Engagement Specialist (2010)

In South Asia and sub-Saharan Africa, where some of the poorest people live and farm, the scenarios of climate change's impacts on agriculture present a dire picture. Recent studies confirm that Africa is the most vulnerable continent to climate change because of multiple abiotic and biotic stresses and the continent's low adaptive capacities. Yields in Central and South Asia could decrease up to 30 per cent by the mid-21st century (IPCC 2007a). In drier areas of Latin America, climate change is expected to lead to salinity and desertification of some agricultural land, reducing the productivity of some important crops and animal husbandry.

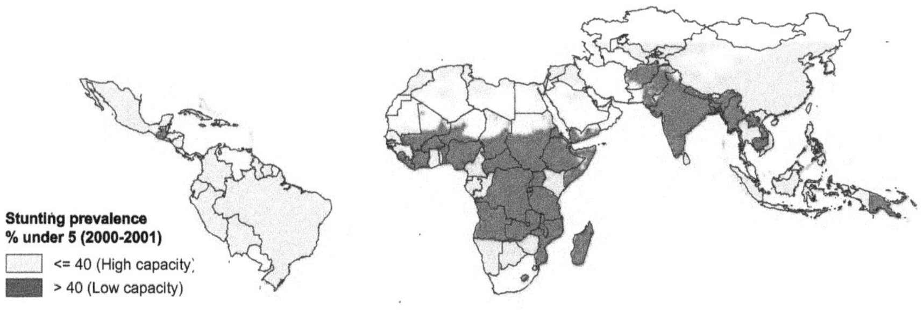

Fig. 5.9: Expected future food insecurity.

5.6 OPPORTUNITIES

Many opportunities exist for promoting green agriculture. They include increased awareness by governments, donor interest in supporting agriculture development in low income countries, growing interest of private investors in sustainable agriculture and increasing consumer demand for sustainably produced food.

1. Government Awareness

Governments, particularly in developed countries, have become increasingly aware of the need to promote more environmentally sustainable agriculture. Since the mid-1980s, OECD countries have introduced a large number of policy measures addressing environmental issues in agriculture. Some of these are specific to the agriculture sector, including the practice of linking general support to environmental conditions; others are included in broader national environmental programmes. The result is that the environmental performance of agriculture has begun to improve in OECD countries.

The proportion of global arable land dedicated to organic crops has increased from a negligible amount in 1990 to around to 2 per cent in 2010, and as much as 6 per cent in some countries. The extent of soil erosion and the intensity of air pollution have fallen; the amount of land assigned to agriculture has decreased even as production has increased, and there have been improvements in the efficiency of input use (fertilisers, pesticides, energy, and water) since 1990. However, subsidies for farm-fuel have continued to be a disincentive to greater energy efficiency.

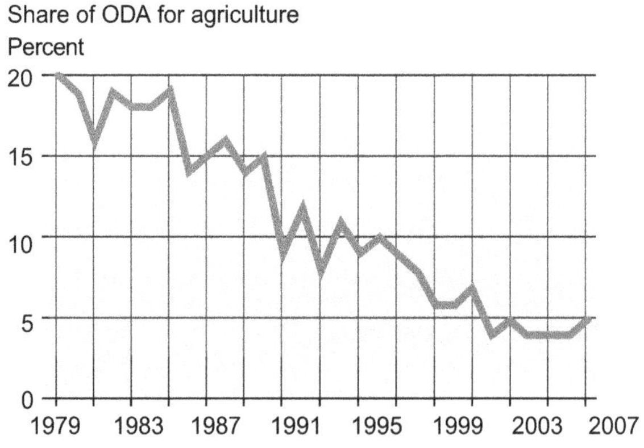

Fig. 5.10: Share of overseas development assistance for agriculture (1979–2007).

2. Donor Support for Agriculture Development

Agriculture-related Overseas Development Assistance (ODA), which has fallen steadily over the past 30 years, began to pick up in 2006 as the current food crisis escalated. In 2009, at the G8 summit in Italy, wealthy nations pledged US$ 20 billion for developing-country agriculture. However, there is a pressing need to ensure that these investments, as Ban Ki-moon put it, "breathe new life into agriculture, one which permits sustainable yield improvements with minimal environmental damage and contributes to sustainable development goals". Recently, the Food and Agriculture Organisation (FAO), World Bank, the United Nations Conference of Trade and Development (UNCTAD) and the International Fund for Agricultural Development (IFAD) have jointly proposed Principals for Responsible Agricultural Investments.

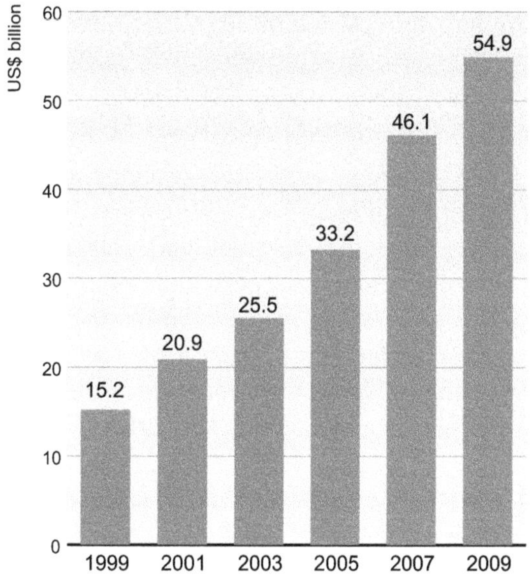

Fig. 5.11: Global trade in organic food and drinks (1999-2009).

3. Private Funding Interest

Preferential access to credit and investment capital is one of the most important incentives to catalyse a transition to greener agriculture. The number, volume and rate of return of sovereign wealth funds (SWFs), pension funds, private equities and hedge funds with investment in agriculture, are increasing. Major financial institutions are expanding their green portfolios to offer investment credit to companies that manufacture and market products that enable more efficient use of agricultural inputs and introduce innovative private enterprises (see Box 5.3). The public sector, especially in developing countries, should support finance mechanisms (*e.g.* loan-guarantee funds) that can leverage larger multiples of private capital loans to smallholders who need working capital to undertake sustainable agriculture practices.

4. Increasing Consumer Demand for Sustainable Food

Over the last few years, consumer demand for sustainably produced food has increased rapidly. Purchasing patterns of fairtrade products have remained strong despite the global economic downturn. In 2008, global sales of fair-trade products exceeded US$ 3.5 billion. Data collected by the International Trade Centre (ITC) and the Forschungsinstitut fur biologischen Landbau (FiBL) shows that the major markets for organic food and beverages expanded on average by 10 to 20 per cent per year between 2000 and 2007 and reached US$ 54.9 billion in 2009. This figure does not include markets for organic fibre, cosmetics and other luxury products. This demand has driven a similar increase in organically managed farmland. Approximately 32.2 million hectares worldwide are now farmed organically. In addition, as of 2007, organic wild products were harvested on approximately 30 million hectares.

5.7 THE CASE FOR GREENING AGRICULTURE

Both conventional and traditional agriculture generate substantial pressure on the environment, albeit in different ways. With very different starting positions, the pathways to green agriculture will vary substantially and will have to be sensitive to local environmental, social and economic conditions. Industrial agriculture needs to lessen its reliance on fossil fuels, water and other inputs. Both large and small farms can benefit from more on-farm recycling of nutrients by reintegrating livestock, which provide manure, and the cultivation of green manures to improve and maintain soil fertility.

1. The Cost of Environmental Degradation Resulting from Agriculture

Several studies have estimated the cost of externalities caused by current agricultural practices, which include those from use of inputs such as pesticides and fertilisers leading, for example, to the pollution of waterways and emissions from farm machinery and food-related transport.

Agricultural operations, excluding land use changes, produce approximately 13 per cent of anthropogenic global GHG emissions. This includes GHGs emitted by the use of inorganic fertilisers agro-chemical pesticides and herbicides; (GHG emissions resulting from production of these inputs are included in industrial emissions); and fossil fuel-energy inputs. Agriculture also produces about 58 per cent of global nitrous oxide emissions and about 47 per cent of global methane emissions. Both of these gases have a far greater global warming potential per tonne than CO_2 (298 times and 25 times respectively). Moreover, methane emissions from global livestock are projected to increase by 60 per cent by 2030 under current practices and consumption patterns. The expansion of agricultural land at the expense of forests has been estimated to represent an additional 18 per cent of total global anthropogenic GHG emissions.

A study by Jules Pretty *et. al.* estimated the annual costs of agricultural externalities to be US$ 2 billion in Germany and US$ 34.7 billion in the USA. This amounts to between US$ 81 and US$ 343 per hectare per year of grassland or arable land. In the UK, agriculture's total environmental externality costs, including transporting food from the farm to market and then to consumers, have been calculated to be £ 5.1 billion per year for 1999/2000, a cost greater than annual net farm income. In China, the externalities of pesticides used only in rice systems have been estimated to amount to US$ 1.4 billion per year in health costs to people, and adverse effects on both on- and off-farm biodiversity. The national pollution census in China revealed that agriculture was a larger source of water pollution than industry, discharging 13.2 MT of pollutants. In Ecuador, annual mortality in the remote highlands due to pesticides is among the highest reported anywhere in the world at 21 people per 100,000 people. The economic benefits of Integrated Pest Management (IPM) based systems that eliminate these effects are increasingly beneficial. Land degradation is costing ten Asian countries an economic loss of about US$ 10 billion, equivalent to 7 per cent of their combined agricultural GDP.

At the same time, as a result of the poor management of fertiliser usage during the last half-century, the phosphorus content in freshwater systems has increased by at least 75 per cent, and the flow of phosphorus to the oceans has risen to approximately 10 million tonnes annually. The combined effects of phosphate and nitrogen water pollution, much of it linked to the use of inorganic fertilisers is the main cause of eutrophication, the human-induced augmentation of natural fertilisation processes which spurs algae growth that absorbs the dissolved oxygen required to sustain fish stocks. The estimated costs of the eutrophication in the USA alone run as high as US$ 2.2 billion annually.

Not all agricultural externalities are quantified and thus the calculations above probably underestimate the total cost to society. Conventional agriculture, for example, causes millions of cases of pesticide poisoning per year, resulting in over 40,000 deaths. It is important to note that most such cases remain unreported.

Farmers who use chemical/synthetic farm inputs are significantly more indebted, especially in developing countries. For example, in Central India, cotton farmers bought inputs with loans at annual interest rates between 10–15 per cent (from cooperative societies) to over 30 per cent (from private money lenders). By contrast, those engaged in organic agriculture were far less likely to take loans owing to lower production costs and greater use of on-farm inputs.

Although there is a difference of opinion on the issue, Jalees has argued that the main cause for the extremely high rate of suicide among Indian farmers is the debt-servicing obligations for working capital (*e.g.* fertilisers, pesticides and GM seeds) costs.

The following section presents some on- and off-farm investment strategies that will help minimise, eliminate and gradually reverse the environmental and economic costs resulting from currently predominant forms of agriculture.

5.8 INVESTMENT PRIORITIES FOR GREENING AGRICULTURE

1. Investments in R&D and Agribusinesses

One of the major reasons for the wide spread adoption of the Green Revolution that greatly increased agricultural productivity was the level of first public, then private-sector investment in R&D and the subsequent dissemination and commercial implementation of the results. These gains were also achieved with the introduction of irrigation and greater application of inorganic agrochemical inputs. A new wave of investment is needed to develop, deploy and diffuse resource-efficient technologies and agricultural inputs, farming practices, and seed and livestock varieties that would counter the environmental externalities that are often associated with the green revolution.

The International Assessment of Agricultural Knowledge, Science and Technology for Development noted that ROI in AKST across commodities, countries and regions on average are high (40–50 per cent) and have not declined over time. They are higher than the rate at which most governments can borrow money. The commercial rate of return, however, should not be the only determinant of the decision to invest in R&D for greening agriculture. The social rate of return would be considerably higher if rural communities could adequately monetise the ecosystem, livelihood and socio-cultural benefits that would accrue with their adoption of greener agriculture practices and land stewardship.

Research to improve the performance of biological nitrogen fixation processes, breeding plant, livestock and aquatic species for improved yields and adaptive resilience and developing perennial cereal crops would enable significant reductions in the energy, water and fertiliser inputs needed to cultivate commodity grains. Such research may require several decades to produce commercially viable crop varieties with these beneficial attributes. However, the impacts would be significant in terms of providing options for future generations' dependency on expensive fossil fuel-based fertilisers and adapting to expected climate change.

2. Plant and Animal Health Management (PAHM)

Field trials of improved PAHM practices have resulted in increased profitability of farms. Various intercropping strategies utilise selected plant species' biochemical emissions to either attract or repel different insects, nematodes and other pests. One of the most effective such techniques is known as "push-pull," which involves intercropping, for example, certain species of legumes and grasses with maize. Aromas produced by legumes planted on the perimeter of a field repel (push) maize pests, while scents produced by the grasses attract (pull) insects to lay their eggs on them rather than the maize.

The implementation of push-pull in eastern Africa has significantly increased maize yields and the combined cultivation of N-fixing forage crops has enriched the soil and has also provided farmers with feed for livestock. With increased livestock operations, the farmers are able to produce meat, milk and other dairy products and they use the manure as organic fertiliser that returns nutrients to the fields. In small-holder farming operations, the ability to support livestock for meat, milk and draft animal power is an important added benefit of this strategy. An economic analysis of a push-pull field trial in East Africa with 21,300 farmers revealed a benefit-cost ratio of 2.5 to 1. The income returns for labour were US$ 3.7 per person/day with push-pull as opposed to US$ 1 person/day with their previous maize mono-cropping practice. The gross revenue ranges between US$ 424 and US$ 880 per hectare under push-pull and US$ 81.9 to US$ 132 per hectare in maize mono crop. Similar systems are being field-trialed for other cropping systems and it is likely that comparable rates of return will be realised.

Box 5.4: Cost of Training Smallholder Farmers in Green Agriculture Practices

In a recent report on organic agriculture, the ADB concluded that the cost of transition for farmers to move from conventional agricultural practices to organic practices, including the cost of certification, was approximately US$ 77–170 per farmer for an average farm size of 1 hectare (ADB 2010). Training costs were estimated at US$ 6–14/farmer. These are fairly modest compared to the overall investment required for extricating farmers from poverty (an approximate investment of US$ 554–880, according to the World Bank. Yet there remain additional costs. These are the costs of enabling policies that allow research and development, market linkages and creating incentive systems on the demand and supply side. These costs cannot be understated and obviously require multilateral and bilateral support in the international arena.

Another example of PAHM practices is seen in Cameroon. In this case study, cocoa farmers were trained in pruning, shade adjustment and phytosanitary harvesting methods that effectively maintained yields comparable to conventional practices that used multiple applications of fungicides. The farmers who practiced these techniques used 39 per cent fewer fungicides. Although labour costs increased by 14 per cent, total production costs decreased by 11 per cent relative to conventional practices. By introducing green farming, methods that relied on more knowledgeable labour inputs, a much larger share of the total costs of cocoa production was paid to workers within the local community. Imports of fungicide chemicals were also reduced, saving valuable foreign exchange. Additional benefits included reduced health costs and less environmental pollution.

Investments in PAHM should focus on research, training and investments in natural pest- management processes that defend, defeat and manage the many organisms that threaten agricultural production. While there are a wide range of low-cost natural bio-control practices that improve the ability of plants and livestock to resist and suppress biotic stresses and combat pests, during the past few decades there has been a substantial increase of private and, to a much lesser degree, publicly-funded efforts to develop genetically modified (GM) crops to overcome pest and weed problems. After initial success, there is growing evidence of an evolving resistance to GMO crops by many pests and weeds. The IAASTD report (2009) recommended that research on the ecological, economic and social questions concerning the widespread application of GM crops should be increased, particularly in the public R&D sector, whose scientific advances could be more broadly and equitably available for use in developing countries.

Table 5.2 presents selected evidence on the costs and benefits of plant and animal health management strategies (PAHM). Plant and animal health management practices reduce farmers' input costs and their exposure to hazardous chemicals while effectively supporting productive crop yields. Plant and animal health management practices also reduce or replace the use of chemical insecticides that often kill non-targeted insects. Many insect species killed as collateral damage from such insecticides have beneficial environmental and agricultural roles as pollinators and as predators of other pests, and are part of the natural food chain.

Evidence presented in Table 5.2 show that all PAHM interventions are highly profitable. Intercropping is a particularly useful strategy with high benefit to cost ratios of 2.5 to 1. Compared with mono-cropping strategies push pull strategies and intercropping both imply an increased use of labour, but demonstrated returns are more than 200 per cent.

Similarly, pest management strategies that include introducing new predator species in Africa to combat losses caused by the mealy bug have proven to be extremely effective. Most significant costs are associated with research development and extension but the resulting increase in effective produce and diminished post-harvest losses contribute to more than an order of magnitude increase in returns. Unlike push-pull, these types of strategies are usually managed at a country or inter-country level and thus benefit from scale, while providing benefits to all farmers, regardless of their size and their possibility to invest in pest control.

Table 5.2: Selected evidence on benefits and costs of plant and animal health management

Strategy	Crop and country	Costs	Benefits	Trends in revenues and profits after including additional costs of greening
Intercropping	Maize intercropped with Desmodium uncinatum, East Africa.	Most costs are associated with additional labour costs.	Maize grain yield increases ranged from double to five times in plots using push-pull strategies compared to monocropped plots. Levels of pests reduced significantly and were completely eliminated in some. (Reductions ranged from 75% to 99%).	Benefit to cost ratio is 2.5 to 1 using the push-pull strategy. Gross revenues with push-pull were US$ 424–880/ha compared to 82-132/ha using a monomaize cultivation strategy.
Pest Management	The wasp predator to fight the cassava bug in Africa. Cocoa in Cameroon.	The cost of introducing the wasp across cassava growing countries in Africa (19782003) is estimated at US$ 14.8 million. This includes research and distribution costs. For cocoa, IPM meant that labour costs increased by 14%. But total production costs decreased by 11% due to reduced use of fungicides.	Introducing the wasp predator introduction helped avoid 60% of the losses caused by the cassava mealy bug. In cocoa plantation, IPM reduced cost of fungicides by 39%.	Benefit cost ratio of 149 to 1 for the wasp predator strategy, across all cassava growing countries in Africa, 1978–2003. Reduced costs of fungicides in the context of obtaining similar yields can lead to increase in profitability for the farmers.
Pest Management	The wasp predator to fight the cassava bug in Africa. Cocoa in Cameroon.	The cost of introducing the wasp across cassava growing countries in Africa (19782003) is estimated at US$ 14.8 million. This includes research and distribution costs. For cocoa, IPM meant that labour costs increased by 14%. But total production costs decreased by 11% due to reduced use of fungicides.	Introducing the wasp predator introduction helped avoid 60% of the losses caused by the cassava mealy bug. In cocoa plantation, IPM reduced cost of fungicides by 39%.	Benefit cost ratio of 149 to 1 for the wasp predator strategy, across all cassava growing countries in Africa, 1978–2003. Reduced costs of fungicides in the context of obtaining similar yields can lead to increase in profitability for the farmers.
Bio-pesticides	Fungal spores in fighting grasshopper in Benin, maize and cassava, cowpea and groundnuts crops.	Estimated cost for effective intervention was US$ 4/ha.	Cumulative mortality of grasshoppers after 20 days of spraying was over 90%.	Bio-pesticides have small costs and major benefits of avoided damage. Yield losses due to grasshoppers can reach 90% in cowpea and 33% in maize.

3. Scaling up Adoption of Green Agriculture by Partnering with Leading Agribusinesses

A small number of corporations control a large share of the global agribusiness. The four biggest seed companies control more than half of the commercial seed market , the biggest ten corporations (four of them are among the top 10 seed companies) together control 82 per cent of the world pesticides business. The share of the top-ten corporations in the global market for food processing is 28 per cent, and the top 15 supermarket companies represent more than 30 per cent of global food sales. Investment decisions of these approximately 40 companies have the power to determine, to a large extent, how the global agriculture sector could endorse and encourage green and sustainable farming practices.

By greening the core business operations and supply chains, these corporations can play a major role in supporting a transition to greener agriculture. In addition, they can provide investments to develop and implement viable strategies for ensuring global food security based on optimal use of inorganic inputs and building capacity to recycle on-farm nutrients. Investing in building consumer awareness about benefits of sustainable agrifood products is another area that offers benefits for the environment and these businesses. One of the promising developments in the area of agribusiness and NGO partnerships to promote green agriculture is the Sustainable Food Laboratory.

4. Strengthening the Supply Chains for Green Products and Farm Inputs

Demand for sustainably produced products is increasing but it is concentrated in developed countries. Investments in developing new markets in developing countries and expanding existing markets in developed countries could (*i*) create new and high return employment opportunities for on- and off-farm sectors (*e.g.* certification auditors); (*ii*) shorten the field-to-market supply chains, and thus offer better prices to farmers in these countries; and (*iii*) help maintain the price premiums, which can range from 10 per cent to more than 100 per cent over a variety of conventionally-produced foods. A major challenge in this regard is consumer demand for less expensive food and high demand elasticities associated with premium prices for organic food and other products. As incomes rise and consumers learn more about lifestyle diseases, and in the absence of good food safety regulations or lack of their implementation, the negative health effects of some cheaper, conventionally produced foods, we expect to see in upper and middle income consumers an increasing willingness to pay for more environmentally sustainable and ethically produced (*e.g.* fairtrade, etc.) foods at prices that would cover their higher costs.

The limited availability of substantial quantities of natural fertiliser and pesticides in many countries is a major constraint to the growth of sustainable farming practices. Large-scale composting of organic matter and recovery of livestock manures for commercial organic fertiliser products will be required in most farming regions. Investments in the production, supply and marketing of non-synthetic, natural inputs for farming will not only offer competitive returns but will also help in setting up new small-scale businesses in rural areas. The bulk and volume of organic fertilisers that are required for equivalent applications of inorganic fertilisers make them not very cost-effective for long distance transport, thus necessitating relatively localised or regional compost-production capacities.

5. Farm Mechanisation and Post-harvest Storage

Appropriate mechanisation of small and medium farms can significantly increase agricultural productivity and help green the farming practices. The degree to which there is access to farm mechanisation equipment (both draft animal and modern fuel-powered technology) will substantially determine achievable levels of productivity per unit of labour and of land. Use of (*i*) more energy-efficient cultivating machines that incorporate plant residues into the soil to increase fertility, (*ii*) zero-tillage and minimal-tillage direct seeders for optimum planting uniformity and minimal topsoil disturbance, (*iii*) precision application systems for more efficient use of agrochemicals, (*iv*) drip and sparkling irrigation, and (*v*) harvest and postharvest operations that include village-level processing of farm products and by-products are central to the green mechanisation of farms.

Since most farm mechanisation technologies require modern fuels or electric power to operate and fossil fuel price increases are seen as inevitable, it is important that non-conventional energy sources such as biodiesel fuels and biogas power generation and process heat be developed and used in mechanised farming systems in developing countries. While there are examples of rural bioenergy production technologies operating throughout the world, in most cases these technologies remain uncompetitive mainly due to subsidies and policy support for fossil fuels and related farm machinery.

Coupled with farm mechanisation, which may negatively affect on-farm employment opportunities, investment in off-farm employment opportunities is needed. Food packaging and processing in rural areas would enable new non-farm jobs and could improve market access for agricultural produce. However, the feasibility of added value processing would be substantially determined by the quality of rural road infrastructure that connect to urban centres, ports and airports and the availability of skilled labour capable of operating food-handling facilities. In those cases where rural food processing is implemented, the residues from food processing should be composted or processed into organic fertilisers in order to avoid waste and to return needed organic nutrients to the nearby farm land.

With regard to post-harvest storage, simple technologies with small investments can make a big difference. Small holder farmers with limited access to dry and sanitary storage and cold chain facilities often suffer post harvest food losses that can range from 20 per cent to more than 30 per cent of their crop yields. Furthermore, without crop storage systems, farmers are usually compelled to sell their entire crop immediately at the time of harvest when market prices are much lower than levels possible several months after harvest (Kader and Rolle 2004). Investments in post-harvest storage can bring multiple economic and development benefits (Box 5.5).

Box 5.5: Simple Storage: Low Investment, High Returns

An FAO programme that supported the production and use of household and community- scaled metal silos for grain storage estimated that farmers who invested in silos were able to earn nearly three times the price for maize sold four months following harvest as opposed to the price paid at harvest (US$ 38/100 kg of maize compared with US$ 13/100 kg). The production costs for these metal silos ranged between US$ 20 for a 120 kg small- capacity unit to US$ 70-US$ 100 for an 1800 kg large-capacity silo in a variety of countries. Most farmers realised a full return on their investment within the first year of use. The FAO reports that although reducing post-harvest losses could be relatively quickly achieved, less than 5 per cent of worldwide agricultural research and extension funding currently targets this problem.

Similar improvements in reducing postharvest losses are possible with cost-effective hermetically sealed packaging materials and handling processes that protect grains and pulses from insect and mold contamination. A notable example of such technologies is the Purdue Improved Cowpea Storage (PICS) system, which is composed of two polyethylene bags and a third outer bag of woven polypropylene. The PICS materials are made by several West African manufacturers and have proven to offer safe and inexpensive storage of cowpea and other grains for 4–6 months and longer.

6. Improving Soil and Water Management and Diversifying Crops and Livestock

One of the most significant consequences of conventional agriculture is the rapid depletion of soil organic matter (SOM). Repeated cultivation degrades soils and lowers crop yields hence increases production costs. Strategies for better soil management have been experimented in Colombia, England, Mexico, Morocco and the USA. Results show yield increases ranging from 30 per cent to 140 per cent. Some of these strategies include, growing and integrating back in soil nitrogen fixing fodder and green manure crops such as pea, ferns and cloves or rice straw, no-tillage and planting new seeds in crop residues, using waste biomass or biochar (still needs research to fully understand its true potential), and organic and mineral fertilisers. Table 5.3 presents evidence from field trials and plots in Colombia, England, Morocco, Mexico and the USA that show yield increases ranging from 30 per cent to 140 per cent resulting from better soil management strategies. Nonetheless, each strategy does require some additional investments. Strategies such as nitrogen-fixing fodder or green manure mainly involve additional labour costs: additional labour is required to distribute fodder over land and for sowing and growing green manure plants. In addition, in some countries, the cost of fodder can be substantial since it can be used alternatively for feeding animals. Nevertheless, crop yield increases as high as 40 per cent are capable of making the investments profitable for farmers.

Table 5.3: Selected evidence on benefits and costs of soil management strategies

Strategy	Crop and country	Costs	Benefits	Trends in revenues and profits after including additional costs of greening
Use of nitrogen-fixing fodder and cultivating green manure	Cultivation of maize in Spain and rice in India, Indonesia and Philippines.	Costs varied depending on methods and country. Rice straw use (for green manure) costs ranged from US$ 18/ha in Indonesia and Philippines, to US$ 40/ha in India. Azolla (type of fern) for nitrogen fixing and green manure meant additional costs ranging from US$ 34/ha in India, to US$ 48/ha in the Philippines.	Maize crop yields increased approximately 40% in the first year, 5% in second year and 20% in year three. No significant increases in yields were observed in rice crops compared to the use of inorganic fertilisers but result in long term soil improvements. Maize crop yields increased after the first year, by 28%, 30% and 140% in the last 3 years of the study. No impact was seen on soybean crop yields.	Revenues increased even though there was no difference in the costs of using green manure over inorganic fertiliser for rice crops.
No-tillage practices	Maize in Mexico, wheat in Morocco and cereal grain crop in England. Sorghum and maize in Botswana, Maize, sorghum and cowpea in Nigeria. Soybean in Australia.	The capital costs for a small scale No-tillage planting system are estimated to be US$ 25,000 to 50,000 (ICARDA). No tillage system was cheaper by US$ 156/ha when rented from a contractor in England, compared to renting tilling systems. In Botswana, cost per household of tractor was US$ 218.	Maize yields increased by 29 per cent; wheat yields by 44 per cent. No impact on total cultivated areas, crop yields and total crop output in traditional tillage systems vs. animal power or manual usage (Botswana and Nigeria). An average yield increase in soybean yields of 27% over 14 years in no-tillage vs. till systems.	No-tillage systems are economically profitable, even after incorporating the costs of no-till systems. (Baker 2007).
Biochar use	Cultivation of maize intercropped with soybean (Colombia) and wheat (USA).	Biochar production costs range between US$ 87–350/tonne depending on source of inputs and mode of production.	Maize crop yields increased after the first year, by 28%, 30% and 140% in the last 3 years of the study. No impact was seen on soybean crop yields.	In the US, wheat production increased sufficiently to generate a profit of US$ 414/acre, but only while using low-price biochar. Higher-cost biochar reduces profits.

The use of a no-tillage system strategy mainly requires additional capital outlays, which can be significant. In countries with developed markets for agricultural equipment no-tillage systems can be cheaper than using tilling machinery, in developing countries the investment in farm equipment may represent a significant barrier. Farmer cooperatives and extension services can help defray these costs.

Biochar usage represents a costly investment, mainly because of the high cost of production for biochar (US$ 87–350 per tonne depending on the source of inputs and mode of production). Although it can bring significant increases in crop yields, biochar profitability is still highly dependent on the cost of production.

Similarly, the use of water for irrigation is rapidly exceeding the natural hydrological rate of recharge in many river basins. Practices such as flooding fields, poor drainage and excessive pumping imply that there are many opportunities for using ground and rainwater in more efficient and sustainable ways. Some sustainable water-use strategies include drip irrigation systems, pressurised water pipe and sprinkler systems and use of manual treadle pumps. According to some studies , drip irrigation has resulted in yield gains of up to 100 per cent, and water savings of 40 to 80 per cent.

Using leaf and straw mulch reduces surface evaporation and helps to retain moisture near plant roots, thus increasing water-use efficiency. Landscape contouring and vegetative barriers are an effective means of minimising rainfall runoff and retaining moisture in fields. Using drought-resistant varieties of crops can also help conserve water. For example, System Rice Intensive (SRI) practices substantially reduce the amount of water and other external inputs through decreased planting densities, which require less seed and fewer workers. The approach generally achieves between 40 per cent and 200 per cent greater crop yields compared with conventional flooded rice cultivation. Table 5.4 demonstrates that most water-saving technologies can bring about increased profits despite additional infrastructure and operating costs. Most water-saving techniques require additional equipment and increased working capital to cover the costs of increased labour use. Additional labour is required for strategies such as the use of mulching fields, raising plant beds and aligning furrows, and in other land contouring strategies. Such labour costs are nevertheless easily recovered through increased crop yields, and the reduced risk of losses during drought or dry years.

Table 5.4 shows that investment costs in drip irrigation systems and in manual treadle pumps are recovered more quickly; returns to investments have on average been more than 10-fold. These technologies have demonstrated their effectiveness in reducing income vulnerability and uncertainty for small-holder farmers across the continent. Drip irrigation systems also allow the more efficient use of water and are particularly useful for multiple cropping; in Nepal women farmers have been able to earn additional incomes by growing high value crops on otherwise barren land. Strategies such as the use of drought-resistant varieties of crops mainly involve investment in research and distribution of new seeds. In this context, estimated returns on investment are an order of magnitude higher, especially as witnessed in water-starved regions of Africa.

Table 5.4: Selected evidence on benefits and costs of water management strategies

Strategy	Crop and country	Costs	Benefits	Trends in revenues and profits after including additional costs of greening
Cover mulch	Grain in India; Groundnut in India	In groundnut cultivation the cost of wheat straw mulch was US$ 58/ha. Cultivation required 5 tonnes of mulch per hectare. Black plastic covers cost much more (US$ 1.8 /kg, vs. straw at US$ 0.01/kg).	Average yields for grain and straw were the highest in fields that received cover mulch of 6 tonnes/ ha: Yields increased by 130–149% over 3 years. Using wheat straw mulch cover increased pod yield of groundnut by 17–24%. Using both- wheat straw mulch and black plastic covers led to yield increases of 30 to 86% across test fields.	For groundnut crops, analysis of profitability showed that both systems (wheat straw and wheat straw with plastic cover) have positive income returns of US$ 92/ha and US$ 42/ha respectively. For grain crops, long-term profitability is possible with the use of mulch depending on the costs of mulch.
Furrow contouring	Corn in China .	Technique used plastic covers and constructed furrows. Costs of plastic and labour are not provided.	Corn yields increased by 60-95% during drought years, 70–90% in wet years and 20–30% in very wet years.	Revenues and profits are likely to be positive and increase, except during very wet year.
Manual treadle pump	Major staples including cassava, maize, rice and yam in Ghana and 2009) and a variety of crops, Zambia.	Depending on region the cost of a manual treadle pump in Ghana was US$ 89. Users had to pay additionally for labour. Total production costs increased by US$ 162/farm on average. In Zambia the cost of suction pumps ranged from US$ 60–77 and cost of pressure pumps was US$100–120.	In Ghana, treadle pump users were able to grow multiple crops. In Zambia Treadle Pump users of were able to grow three crops a year.	Incomes for Treadle Pump users increased by more than 28 per cent in Ghana. On average users earned almost US$ 343/ farmer over non-users in Ghana. In Zambia, incomes rose more than six- fold. Farmers earned US$ 125 with bucket irrigation on 0.25 ha of land to US$ 850–1,700.
Drip irrigation	Vegetables in Nepal Maize and vegetables in Zimbabwe.	On average farmers had to pay US$ 12/ farmer in Nepal for drip irrigation system (perforated tubing and a suspended water container).	Barren land became more productive in Nepal. In Zimbabwe no significant differences in yield were observed. Water use reduced by 35%.	In Nepal, women farmers earned an additional US$ 70 annually by selling surplus vegetables.
Using low-water varieties of crops	Maize varieties in 13 countries of eastern, southern and West Africa.	US$ 76 million was invested in cultivating low-water varieties of crops over 10 years in these countries.	Average yield increases estimated to be between 3–20%.	Maize yield increases translate into uS$ 0.53 billion. The ratio of returns to investment is estimated to be between 7 and 11 times.

The success of these strategies also implies that agronomic research and development on improving water management practices in rain-fed agriculture and on tilling practices has been successful although much more is required. A strategy that remains relatively untapped is community-led watershed management. Watershed management has conventionally meant large hydraulic engineering efforts that are applied to local streams or river basins to establish a network of water reservoirs, catchment areas and other water impoundment and storage infrastructures. However, community-led watershed management strategies that protect and improve soil, water and plant resources in a catchment area are rapidly gaining traction and are rapidly becoming a lucrative opportunity for farmers who can benefit from *Payment for Ecosystem Schemes* (PES). These community led watershed management strategies offer important opportunities for increased efficiencies in irrigation.

As far as crop and livestock diversification is concerned, genetic resources for plant and animal breeding are the basis for food production. Genetically diverse crops can combine the best traits of local varieties of crops derived from indigenous species and other higher yielding varieties. Similarly, selecting and mating local animal breeds with high-performance breeds increases diversity and can bring significant biological, social and economic benefits. Replenishing soil nutrients with biological nitrogen fixation and crop-residue recycling, reducing thermal stress and water evaporation rates, and attracting beneficial insects for pollination and pest predation, and deterring pests are all important benefits of crop diversification. Combining the horticultural production of higher-value vegetables and fruits with the cultivation of cereals and cash commodity crops can raise farm income, along with grass-fed livestock, which also enables people to acquire protein and calories derived from otherwise inedible biomass resources. Recycling of livestock manures as organic nutrients for soil is an essential element of greening agriculture. In addition, there are numerous opportunities for combining a wide variety of trees and shrubs with the cultivation of crops, horticulture and specialty crops (*e.g.* coffee, tea, vanilla, etc.) to maximise the output of a farm.

Diversification strategies are not useful to ensure diminished vulnerability but also to increase profitability and yields of existing farming systems. Table 5 presents selected evidence for costs and benefits of agricultural diversification strategies in Asia and Africa. Diversifying across crops has demonstrated increased yields in India and Bangladesh and shows potential for recovering research and extension costs. In both Africa and Asia, diversifying into animal husbandry has meant increased profits. The main on-farm costs for all these strategies is usually the cost of increased labour, but also the cost of training and learning new practices. In addition, diversification into animal husbandries may involve important capital costs in farm equipment. In countries where employment opportunities are few, diversification represents a potent poverty alleviation strategy for both the farmer and the labourer. After the analysis of costs of current agriculture and some strategies for a managed transition away from BAU, the following section lays out the benefit expected from greening the agriculture sector.

5.9 THE BENEFITS OF GREENING AGRICULTURE

The greening of the agriculture sector is expected to generate a range of benefits including increased profits and income for farmers, gains at the macroeconomic level, enabling the sector to adapt to climate change and benefits for ecosystem services.

1. Profitability and Productivity of Green Agriculture

No business is sustainable unless it is also profitable. Many studies have documented the profitability and productivity of sustainable farms, both in developed and developing countries. An FAO study that analysed 50 farms, mostly in the USA, reported: "The overwhelming majority of cases show that organic farms are more economically profitable".

Box 5.6: Investment in Sustainable Agriculture–Case Study

Current trends of population growth, climate change and resource scarcity make sustainable agriculture a compelling investment opportunity. Sustainable Asset Management AG (SAM) taps into this potential through its sustainable theme funds, investing in companies that offer cost- effective, eco-friendly technologies that enable more efficient use of water or more sustainable food production.

SAM has pursued water investments because the need for adequate water supplies is one of today's major challenges. Advanced micro or drip irrigation systems can halve farmers' water requirements and limit the need for chemicals while boosting yields by up to 150 per cent. Countries affected by water shortages are adopting these technologies at rapid rates (see chart).

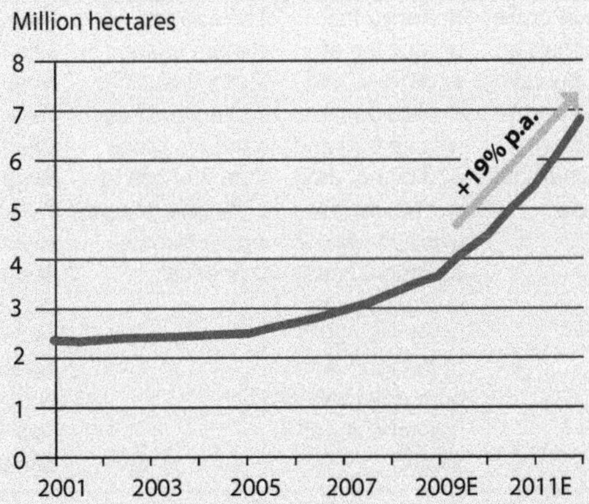

There are various examples of higher productivity and profitability in developing countries. Another study showed an average yield-increase of nearly 80 per cent as a result of farmers in 57 poor countries adopting 286

recent best practice initiatives, including integrated pest and nutrient management, conservation tillage, agroforestry, aquaculture, water harvesting and livestock integration. The study covered 12.6 million farms, encompassing over 37 million hectares (3 per cent of the cultivated area in developing countries). All crops showed water use efficiency gains, with the highest improvement occurring in rain-fed crops. Carbon sequestration potential averaged 0.35tC/ ha/year. Of projects with pesticide data, 77 resulted in a decline in pesticide use by 71 per cent, while yields grew by 42 per cent. In another example, bio-dynamic farms recorded a 100 per cent increase in productivity per hectare due to the use of soil-fertility techniques such as compost application and the introduction of leguminous plants into the crop sequence.

Table 5.5: Selected evidence on benefits and costs of agricultural diversification

Strategy	Crop and country	Costs	Benefits	Trends in revenues and profits after including additional costs of greening
Crop diversification	Rice with pigeon pea, groundnut and blackgram in India. Variety of crops in Bangladesh.	US$ 41.8 million allocated to promoting crop diversification for a 5-year plan in Bangladesh. Empirical study shows reduced variable cost for diversified farmers of US$ 40/ farm (Jan. 1997 exchange rate).	In India, intercropping of rice with pigeon pea, groundnut and blackgram approximately tripled the yield of crops (rice and alternative crops) vs. rice alone.	In Bangladesh, similar net profits were earned by diversified and non diversified farmers; but positive environmental benefits accrued to the diversified farms.
Diversification into animal husbandry and horticulture	Variety of crops and animals in Africa. Survey of crops and countries in Africa and South East Asia.	In Kenya the production of snowpeas and French beans, require 600 and 500 labour days per ha, respectively. In Mexico, the horticultural sector required more than 20% of the total labour days within the agricultural sector.	The impacts of climate change on farms diversified into animal husbandries range from 9% loss to 27% gain depending on climate scenarios.	Profits of farmers diversified into horticulture were consistently higher compared to non-diversified farmers (29% in Bangladesh to 497% in Kenya). Estimates show that integrated or diversified farms have the potential to become more profitable compared to non-integrated farms 50 years from now, in the context of climate change.

For small farms in Africa, where the use of synthetic inputs is low, converting to sustainable farming methods has increased yields and raised incomes. In a project involving 1,000 farmers in South Nyanza, Kenya, who were cultivating, on average, two hectares each, crop yields rose by 2–4 tonnes per hectare after an initial conversion period. In yet another case, the incomes of some 30,000 smallholders in Thika, Kenya rose by 50 per cent within three years after they switched to organic production.

A significant part of a farm's production costs is linked to its energy inputs and organic agriculture tends to be more energy-efficient. Growing organic rice can, for example, be four times more energy-efficient than the conventional method. The study also shows that organic farmers required 36 per cent of the energy inputs per hectare compared with conventional rice farmers. Niggli *et. al.* found that organic agriculture reduces production systems' energy requirements by 25 to 50 per cent compared with conventional chemical-based agriculture. Energy consumption in organic farming systems is reduced by 10 to 70 per cent in European countries and by 28 to 32 per cent in the USA compared with high-input systems, with the exception of certain crops including potatoes and apples, where energy-use is equal or even higher.

Market price premiums often exist for certified sustainably produced products, however this incentive may not be adequate in the long run unless there is a commensurate increase in global consumer demand for sustainable agricultural products (*e.g.* in countries other than primarily the EU and USA). Premium price incentives are likely to relatively decrease in response to supply and demand elasticities. However, if prices of conventionally grown food (crops and animals) included the costs of their externalities, sustainable products may become relatively less expensive than conventional products. Furthermore, if the positive ecosystem service benefits of sustainable practices were valued and monetised as incremental payments to green farmers, greener agriculture products would become more competitive with conventional products.

2. Macroeconomic Benefits from Greening Agriculture

Significant secondary macro-economic and poverty reduction benefits are expected from greening agriculture. Investments aimed at increasing the productivity of the agriculture sector have proved to be more than twice as effective in reducing rural poverty than investment in any other sector. The greatest success stories in terms of reducing hunger and poverty are from China, Ghana, India, Vietnam and several Latin American nations, all of which have relatively higher net investment rates in agriculture per agricultural worker than most developing countries. The World Bank has estimated that the cost of achieving the first Millennium Development Goal (MDG 1) amounts to between US$ 554 and US$ 880 per head (based on growth in income in general), while a study published by the Asian Development Bank Institute has concluded that the cost of moving a household out of poverty through engaging farmers in organic agriculture could be only US$ 32 to US$ 38 per head.

Box 5.7: Innovative Sustainable and Social Capital Investment Initiatives

Institutional investments for greening agriculture are emerging. For example, Rabobank Group is supporting sustainable agriculture through the launch of the Rabo Sustainable Agriculture Guarantee Fund and supporting initiatives such as the Dutch Sustainable Trade Initiative (IDH), the Schokland Fund and Round Table of Sustainable Palm Oil (RSPO), the Round Table on Responsible Soy (RTRS), and the Better Sugar Initiative (BSI). In addition, it has launched programmes to improve the financial strength and resilience of small farmers in developing countries via the Rabobank Foundation and Rabo Development. It has also introduced new financial services such as the Sustainable Agricultural Fund to try out innovative financing models such as the Xingu River Basin Project in Brazil, under which 83 hectares have been replanted in the last two years. Rabobank has invested nearly US$ 50 million to purchase carbon emission reduction credits that are created by the Amazon reforestation by farmers.

Another example of social capital investment institutions is the Acumen Fund, which has channelled investment worth millions of US dollars to private entrepreneurs in developing countries, enabling businesses and other initiatives to flourish, from those that provide drip-irrigation products to those operating village-scale biogas power-generation services. Acumen provides both patient capital investments and business management capacity-building support to the private businesses in their portfolio.

Box 5.8: Organic Versus Conventional Cotton Production

An Indo-Swiss research team compared agronomic data of 60 organic and 60 conventional farms over two years and concluded that cotton-based organic farming is more profitable. Organic farming's variable production costs were 13–20 per cent lower and inputs were 40 per cent lower. But yields and profits margins were 4–6 per cent and 30–43 per cent higher respectively during the two years. Although crops grown in rotation with cotton were sold without a price premium, organic farms achieved 10–20 per cent higher incomes compared with conventional agriculture. Similarly, an impact assessment study for organic cotton farmers in Kutch and Surendranagar in eastern India, concluded that farmers who participated in the project enjoyed a net profit gain of 14 to 20 per cent resulting from higher revenues and lower costs. The updated version of the study surveying 125 organic cotton farmers concluded that 95 per cent of respondents found their agricultural income had risen since adopting organic agriculture, on average by 17 per cent. Most farmers attributed this largely to the reduced cost of production and an increase in output price. Also found in Andhra Pradesh that organic cotton was much more profitable.

In addition, green agriculture directs a greater share of total farming input expenditures towards the purchase of locally-sourced inputs (*e.g.* labour and organic fertilisers) and a local multiplier effect is expected to kick in. Overall, green farming practices tend to require more labour inputs than conventional farming (*e.g.* from comparable levels to as much as 30 per cent more), creating jobs in rural areas and a higher return on labour inputs. This is especially important for developing countries, where large numbers of poor people continuously leave rural areas in search of jobs in cities and growing proportions of young people are imposing enormous pressures for job creation. In addition, most developing countries run substantial trade deficits with the lack of foreign exchange representing a key resource constraint. Greening agriculture can relax the foreign-exchange constraint by reducing the need for imported inputs and by increasing exports of sustainable agrifood products. Reducing deficits would enable these countries to purchase technology and other critical inputs for their economies.

Table 5.6: Incremental annual agricultural investment figures by region needed to counteract climatechange impacts on child malnutrition

Scenario	South Asia	East Asia and the Pacific	Europe and Central Asia	Latin America and the Caribbe-an	Middle East and North Africa	Sub-Sa-haran Africa	Devel-oping coun-tries
NCAR with developing-country investments							
Agricultural research	172	151	84	426	169	314	1,316
Irrigation expansion	344	15	6	31	-26	537	907
Irrigation efficiency	999	686	99	129	59	187	2,158
Rural roads (area expan-sion)	8	73	0	573	37	1,980	2,671
Rural roads (yield in-crease)	9	9	10	3	1	35	66
Total	1,531	934	198	1,162	241	3,053	7,118
CSIRO with developing-country investments							
Agricultural research	185	172	110	392	190	326	1,373
Irrigation expansion	344	1	1	30	-22	529	882
Irrigation efficiency	1,006	648	101	128	58	186	2,128
Rural roads (area expan-sion)	16	147	0	763	44	1,911	2,881
Rural roads (yield in-crease)	13	9	11	3	1	36	74
Total	1,565	977	222	1,315	271	2,987	7,338

3. Climate Adaptation and Mitigation Benefits, and Ecosystem Services

Making agriculture more resilient to drought, heavy rainfall events, and temperature changes is closely linked to building greater farm biodiversity and improved soil organic matter. Practices that enhance biodiversity allow farms to mimic natural ecological processes, enabling them to better respond to change and reduce risk. The use of intra and inter-species diversity serves as an insurance against future environmental changes by increasing the system's adaptive capabilities. Improved soil organic matter from the use of green manures, mulching, and recycling of crop residues and animal manure increases the water holding capacity of soils and their ability to absorb water during torrential rains.

The International Food Policy Research Institute (IFPRI) estimates that an additional US$ 7.1–7.3 billion per year are needed in agricultural investments to offset the negative impact of climate change on nutrition for children by 2050 (Table 5.6). The International Food Policy Research Institute recommended investments were needed primarily for basic infrastructure such as rural roads in Africa and expanded irrigation, and for agricultural research. However, assessments of green investment options that would include agro-ecological soil fertility enhancement; water-use efficiency improvements for rain-fed farming; breeding for drought and flood tolerance; integrated pest management; and post harvest handling infrastructures, still remain to be done.

The IPCC estimates that the global technical mitigation potential from agriculture by 2030 is approximately 5,500–6,000 Mt CO_2-eq/yr. Soil carbon sequestration would be the mechanism responsible for most of this mitigation, contributing 89 per cent of the technical potential. Therefore, agriculture has the potential to significantly reduce its GHG emissions, and possibly to function as a net carbon sink within the next 50 years. The most important opportunity for GHG mitigation is the application of carbon-rich organic matter (humus) into the soil. This would significantly reduce the need for fossil fuel-based and energyintensive mineral fertilisers and be a cost-effective means of sequestering atmospheric carbon. Further GHG mitigation gains could be achieved by improving yields on currently farmed lands and reducing deforestation pressures and by adopting no/low tillage practices that reduce fuel usage.

The environmental services provided by greening farms are substantial. The Rodale Institute, for example, has estimated that conversion to organic agriculture could sequester additional 3 tonnes of carbon per hectare per year. The carbon sequestration efficiency of organic systems in temperate climates is almost double (575–700 kg carbon per ha per year) that of conventional treatment of soils, mainly owing to the use of grass clovers for feed and of cover crops in organic rotations. German organic farms annually sequester 402 kg carbon/ha, while conventional farms experience losses of 637 kg. From such studies, it is possible to approximate that if only all the small farms on the planet employed sustainable practices, they might sequester a total of 2.5 billion tonnes of carbon annually. Such verifiable carbon sequestration levels could be equivalent to US$ 49 billion in carbon credits per year, assuming a carbon price of US$ 20/tonne. The FAO has documented that a widespread conversion to organic farming could mitigate 40 per cent (2.4 Gt CO_2-eq/yr) of the world's agriculture greenhouse gas emissions in a minimum implementation scenario; and up to 65 per cent (4 Gt CO_2-eq/yr) of agriculture GHG emissions in a maximum carbon sequestration scenario.

Furthermore, emissions of nitrous oxides and methane could be reduced if farmers use nitrogen and other fertilisers more efficiently, including through precision applications and introducing improved crop varieties that more effectively access and use available nitrogen in the soil. Greening agriculture also has the potential to eventually become self-sufficient in producing nitrogen through the recycling of manures from livestock and crop residues via composting; and by increased intercropping rotations with leguminous, nitrogen-fixing crops.

Table 5.7: Results from the simulation model (a more detailed table can be found in the Modelling chapter)

	Year	2011	2030		2050	
	Scenario	Baseline	Green	BAU2	Green	BAU2
Agricultural sector variables	**Unit**					
Agricultural production	Bn US$/Yr	1,921	2,421	2,268	2,852	2,559
Crop	Bn US$/Yr	629	836	795	996	913
Livestock	Bn US$/Yr	439	590	588	726	715
Fishery	Bn US$/Yr	106	76	83	91	61
Employment	M people	1,075	1393	1,371	1,703	1,656
(b) Soil quality	Dmnl	0.92	0.97	0.80	1.03	0.73
(c) Agriculture water use	KM3/Yr	3,389	3,526	4276	3,207	4,878
Harvested land	Bn Ha	1.20	1.25	1.27	1.26	1.31
Deforestation	M Ha/Yr	16	7	15	7	15
Calories per capita per day (available for supply)	Kcal/P/D	2,787	3,093	3,050	3,382	3,273
Calories per capita per day (available for household consumption)	Kcal/P/D	2,081	2,305	2,315	2,524	2,476

Additional ecosystem benefits resulting from greening of agriculture include better soil quality with more organic matter, increased water supply, better nutrient recycling, wildlife and storm protection and flood control. Systems that use natural predators for pest control also promote on-farm and off-farm biodiversity and pollination services.

5.10 MODELLING: FUTURE SCENARIOS FOR GREEN AGRICULTURE

The UNEP in its report has made an assessment of a scenario in which an additional 0.16 per cent of the global GDP is invested in green agriculture per year (equalling US$ 198 billion) between 2011 and 2050. This is as part of a green investment scenario in which an additional 2 per cent of global GDP is allocated to a range of key sectors. In the part of the modelling exercise, which focused on agriculture sector, these additional green investments are undertaken equally in the following four activities:

- **Agricultural management practices:** one-fourth of the investment is assumed to be invested in environmentally sound practices;
- **Pre-harvest losses:** another one-fourth of the additional budget is invested in preventing pre-harvest losses, training activities and pest control activities;
- **Food processing:** one-fourth of the investment is assumed to be spent on preventing post-harvest losses, better storage and improved processing in rural areas.
- **Research and Development:** the remaining one- fourth amount is assumed to be spent on research and development especially in the areas of photosynthesis efficiencies, soil microbial productivity, climate adaptation biological processes, and improvements of energy and water-use efficiency.

The green scenario is compared with a BAU2 scenario, where the same amount of additional investment is made in conventional and traditional agriculture over the 40-year period.

The results are stark. Overall, the green investments lead to improved soil quality, increased agricultural yield and reduced land and water requirements. They also increase GDP growth and employment, improve nutrition and reduce energy consumption and CO_2 emissions.

- *Agricultural production and value added:* In the green scenario, total agricultural production (including agricultural products, livestock, fishery and forestry) increases significantly compared to other scenarios. This change is driven by increased crop production, which is able to satisfy a growing population that is projected to reach 9 billion by 2050. Similarly value added in agricultural production increases by 9 per cent compared with the BAU2 scenario. It is important to note that despite an increase in agricultural production and value added, there is no increase in area harvested. This suggests positive synergies between ecological agriculture investments and forest management. Similarly, improved water-efficiency reduces water demand by almost one-third by 2050, compared with the BAU2 scenario. On the other hand, energy consumption increases by 19 per cent in 2050 compared with BAU2, due to higher production volumes.
- *Livestock production, nutrition and livelihoods:* Additional investment in green agriculture also leads to increased levels of livestock production, rural livelihoods and improved nutritional status. An increase in investment in green agriculture is projected to lead to growth in

employment of about 60 per cent compared with current levels and an increase of about 3 per cent compared with the BAU2 scenario. The modelling also suggests that green agriculture investments could create 47 million additional jobs compared with BAU2 over the next 40 years. The additional investment in green agriculture also leads to improved nutrition with enhanced production patterns. Meat production increases by 66 per cent as a result of additional investment between 2010–2050 while fish production is 15 per cent below 2011 levels and yet 48 per cent higher than the BAU2 scenario by 2050. Most of this growth is caused by increased outlays for organic fertilisers instead of chemical fertilisers and reduced losses because of better pest management and biological control.

- *GHG Emissions and biofuels:* Total CO_2 emissions to increase by 11 per cent relative to 2011 but will be 2 per cent below BAU2. While energy-related emissions (mostly from fossil fuels) are projected to grow, it is worth noting that emissions from (chemical) fertiliser use, deforestation and harvested land decline relative to BAU2. When accounting for carbon sequestration in the soil, under ecological practices, and for synergies with interventions in the forestry sector, net emissions decline considerably.

The UN Report also specifically analyses the generation of agricultural waste, residues and biofuels in these models. In the green economy case, we assume that investment is allocated to second-generation biofuels, which use agricultural residues, non-food crops and are primarily grown on marginal land. On average it is found that the total amount of fresh residues from agricultural and forestry production for second- generation biofuel production amounts to 3.8 billion tonnes per year between 2011 and 2050 (with an average annual growth rate of 11 per cent throughout the period analysed, accounting for higher growth during early years, 48 per cent for 2011–2020 and an average 2 per cent annual expansion after 2020). Using the IEA's conversion efficiency standards (214 litres of gasoline equivalent (lge) per tonne of residue) we project that additional green investments lift the production of second-generation biofuels to 844 billion lge, contributing to 16.6 per cent of world liquid fuel production by 2050 (21.6 per cent when first-generation biofuels are considered). This would cost US$ 327 billion (at constant US$ 2010 prices) per year on average and would require 37 per cent agricultural and forestry residues. The IEA estimates that up to 25 per cent of total agricultural and forestry residues may be readily available, and economically viable (IEA Renewable Energy Division 2010), for second-generation biofuel production. Residues not used for second-generation biofuels are expected to be returned to the land as fertilisers, and in other cases may be used as livestock feed. More details on the projections on first- and second-generation biofuels production are available in the Modelling and Energy chapters.

Overall, combining these results with research from other sources following results were found:

- Return on investments in BAU agriculture will continue to decrease in the long run, mainly owing to the increasing costs of inputs (especially water and energy) and stagnated/decreased yields;
- The cost of the externalities associated with brown agriculture will continue to increase gradually, initially neutralising and eventually exceeding the economic and development gains; and ? By greening agriculture and food distribution, more calories per person per day, more jobs and business opportunities especially in rural areas, and market-access opportunities, especially for developing countries, will be available.

While any of the proposed measures contributes to the shift towards a green agriculture sector, the combination of all these interacting actions together will yield positive synergies. For instance, the investment in more sustainable farming practices leads to soil conservation, which increases agricultural yield in the medium to longer term. This allows more land for reforestation, which in turn reduces land degradation and improves soil quality. The higher yield and land availability also benefits the promotion of second-generation biofuels, which may help mitigate the effects of climate change.

5.11 GETTING THERE: ENABLING CONDITIONS

Despite the clear logic and economic rationale for moving more rapidly towards greener agriculture, the transition will require a supportive policy environment and enabling conditions that could help level the playing field between conventional and green agricultural practices.

Environmental and economic performance in agriculture is most likely to be improved by employing a mix of policies. There needs to be a greater use of regulations and taxes that impose penalties for pollution in order to include externality costs into market prices for these inputs, as well as economic incentives that reward green practices. There are also opportunities for applying market solutions as alternatives to direct regulation, for example, by using tradable permits and quotas to reduce pollution from greenhouse gases and water-borne nutrients. In general, governmental subsidies for farmer (producer) support should be increasingly decoupled from crop production and alternatively be retargeted to encourage farmers' efforts and investments in adopting green agriculture practices.

In the absence of good governance, collusion and excessive profit taking are constant dangers for incentive programmes. Instilling greater levels of transparency could help reduce such abuses of public-support programmes. In this section we present some of the key conditions that will facilitate a transition to a green agriculture.

5.11.1 Global policies

At the global level, the enabling conditions are synonymous with improvements to the international trading system and economic development cooperation for promoting sustainable agriculture. An enabling environment for greening agriculture should include a range of interventions at various points along the entire agri-food supply chain:

1. Elimination of Export Subsidies and Liberalising Trade in Agricultural Products

Current multilateral trade policies at the global level have primarily focused on the gradual reduction and removal of national tariff barriers. While such policies aim at facilitating trade, many developing nations are concerned that they are not well positioned to benefit from such trade policies as are the more developed nations.

These concerns are particularly relevant while domestic subsidies and other producer-support programmes remain in many developed countries. These measures effectively distort and diminish any competitive advantages that developing nations might have. In addition, subsidies have effectively reduced global commodity prices, making it frequently unprofitable to produce certain products in many developing countries, especially for smallholder farmers. This combination of international trade laws and national subsidies can impede development of commercial agriculture in many developing countries, negatively affecting their efforts to achieve economic growth and poverty reduction.

Such trade and subsidy policies need to be reformed to liberalise trade in environmentally-friendly products and services while allowing developing countries to protect some domestic food crops (special products) from international competition when they are particularly important to food security and rural livelihoods. TheWorld Trade Organisation (WTO) already makes a dispensation for countries with a per capita GDP of less US$ 1,000 (Amsden 2005). Furthermore, agricultural subsidies need to be redirected to encourage more diverse crop production with long-term soil health and improved environmental impacts. A major shift of subsidy priorities is needed in which governments would help reduce the initial costs and risks of farmers' transition efforts to implement sustainable farming practices.

2. Market Power Asymmetry

Asymmetric market power in trade is an important issue for WTO competition policy. Leading firms are predominantly located in industrialised countries and maintain significant control over the food system standards and regulatory processes at all stages of the supply chain. In such market conditions, primary producers generally capture only a fraction of the international price of the commodity. Thus, the degree of poverty reduction and rural development benefits of supplying global trade have been limited. A recent study shows that even in a resource-rich country like the United States, despite rapid increase in prices for food commodities since 2006, "small-to-mid-scale family farmers had lower farm incomes in 2009 than they did earlier in the decade when prices were lower". A green agriculture system would require trade policies that redress these chronic asymmetries.

3. Food Safety Standards

The already stringent food safety standards and verifiable logistics management systems that are applied in international markets are likely to become more sophisticated over the next few decades. Currently, most domestic food supply chains in developing countries have relatively low levels of food safety and handling practices. Improving capacity to develop and implement sanitary and food safety standards that can ensure compliance with international requirements can increase prospects for small farmer communities to supply international markets. Furthermore, it is particularly important to support international efforts to harmonise the variety of sustainable and organic certification protocols and standards. Today's fragmented certification procedures impose high transaction and reporting costs on farmers and limit their access to international markets.

Another important issue is that the cost of certification and reporting is to be borne only by sustainable producers while polluters can market their products freely. The burden of proof must be shifted to the polluter through introduction of certification protocols and labeling schemes which, at a minimum, show the quantities of different agrochemical inputs used in the production and processing of a product, and whether the product contains GMOs or not.

4. Intellectual Property

The application of Intellectual Property (IP) regimes has, in some cases, restricted the results of agricultural research and development being made available as public goods. Private-sector and often public-sector IP rights restrict the access of many in developing countries to research, technologies and genetic materials. Supporting the implementation of the World Intellectual Property Organisation's (WIPO) Development Agenda and providing improved access to and reasonable use of IP that involves traditional knowledge, ecological agriculture techniques and genetic resources in international IP regimes would help advance development and sustainability goals.

5.11.2 National Policies

At the domestic public policy level, the key challenge is creating the conditions that would encourage more farmers to adopt environmentally sound agriculture practices instead of continuing to practice unsustainable conventional farming methods.

1. Support for Improved Land Tenure Rights of Smallholder Farmers

In order for farmers to invest capital and more labour into the transition from brown to green agriculture, major land reforms will have to be implemented, particularly in developing countries. In the absence of more secure rights to specific plots of land for many years into the future, many poor farmers are unlikely to take on additional risks and efforts to gradually build up the natural capital of their farms beyond a one or two-year horizon.

2. Targeting Programmes for Women Small-holder Farmers

Small-farm diversification often requires a division oflabour at the household level that may result in gender-based distribution of management roles and responsibilities for both on and off-farm tasks. This has resulted in the majority of small-holder farms, especially in Africa, being run by women. Securing collective and individual legal rights to land and productive resources (*e.g.* water, capital), especially for women, indigenous people and minorities is important. Improving women's access to working capital through microfinance is an option that would allow much greater numbers of small-scale producers to procure green inputs and related mechanisation technologies.

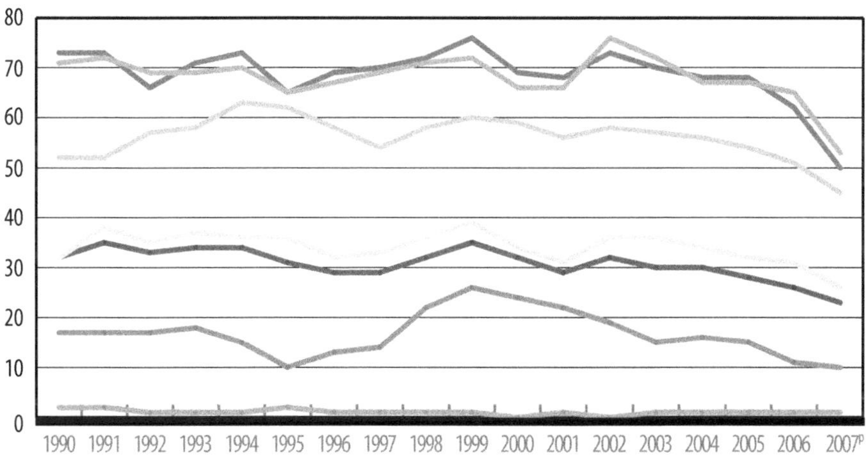

Switzerland
Norway
Japan
European Union*
OECD**
USA
New Zealand

* EU-12 for 1990-94, including ex-GDR; EU-15 for 1995-2003; EU-25 for 2004–06; then EU-27.
** Austria, Finland and Sweden are included in the OECD total for all years and in the EU from 1995. The Czech Republic, Hungary, Poland and the Slovak Republic are included in the OECD total for all years and in the EU from 2004. The OECD total does not include the six non-OECD EU member States.

Fig. 5.12: Estimated producer support by country (as a percentage of total farmer income).

3. Public Procurement of Sustainably Produced Food

Government-sponsored food programmes for schools and public institutions and public procurement policies should be encouraged to source foods that are sustainably produced. The Strategic Paper on Public Procurement, prepared by the UK Department for Environment, Food and Rural Affairs (DEFRA) in January 2008, provides a good example of how organic and sustainable products can be supported through public procurement policies.

5.11.3 Economic Instruments

Agriculture's environmentally damaging externalities could be reduced by imposing taxes on fossil fuel inputs and pesticide and herbicide use; and establishing specific penalties for air emissions and water pollution caused by harmful farming practices. Alternatively, tax exemptions for investments in bio-control integrated pest management products; and incentives that value the multi-functional uses of agricultural land have proven effective in improving the after tax revenues for farmers that practice sustainable land management. The OECD countries have developed a wide range of policy measures to address environmental issues in agriculture, which include economic instruments (payments, taxes and charges, market creation, *e.g.*, tradable permits), community based measures, regulatory measures, and advisory and institutional measures (research and development, technical assistance and environmental labelling).

In OECD countries, the partial shift away from production-linked support has enabled the agricultural sector to be more responsive to markets, thus improving growth. Importantly, some support measures have been linked to specific environmental objectives, research and development, information, and technical assistance, food inspection services, biodiversity, flood and drought control, and sinks for greenhouse gases and carbon storage. There is a need to strengthen these recent trends in developed countries and replicate them in those developing countries that offer farm subsidies in order to target these funds to specific objectives for greater and sustainable economic and environmental performance.

Payment for Ecosystem Service (PES) can further incentivise efforts to green the agriculture sector. This is an approach that verifies values and rewards the benefits of ecosystem services provided by green agricultural practices. A key objective of PES schemes is to generate stable revenue flows that help compensate farmers for their efforts and opportunity costs incurred in reducing environmental pollution and other externality costs that adversely impact the shared commons of the local, national and global environment. Such PES arrangements should be structured so that small-scale farmers and communities, not just large landowners, are able to benefit. Innovative PES measures could include reforestation payments made by cities to upstream communities in rural areas of shared watersheds for improved quantities and quality of fresh water for municipal users. Ecoservice payments by farmers to upstream forest stewards for properly managing the flow of soil nutrients, and methods to monetise the carbon sequestration and emission reduction credit benefits of green agriculture practices in order to compensate farmers for their efforts to restore and build SOM and employ other practices described in this chapter are important elements of PES programmes that have been implemented to date.

5.11.4 Capacity Building and Awareness-Raising

The availability and qualitative capabilities of rural labour are critical resources needed for implementing green agriculture practices. Green agricultural practices emphasise crop and livestock diversification; local production of natural fertiliser and other more labourintensive farm operations. The seasonal variability of crop-specific farming tasks affects temporal labour surpluses and shortages, which must be managed throughout the year. Whether rural labour provides an advantage or a constraint for the adoption of green agriculture practices is highly contextual with specific regional and national conditions. The relative age and gender distribution of rural populations, their health, literacy and family stability, gender equity with respect to access to training and financial services, and other factors will determine the degree to which rural farming communities respond to public and private encouragement of their adoption of green agriculture.

1. Supply Chains, Extension Services and NGOs

Green farming practices in developing countries must be promoted and supported by information outreach and training programmes that are delivered to farmers and their supply-chain partners. These enhanced and expanded training programmes should build upon established agriculture extension service programmes in those countries where they are now functioning. However, in order to effectively use existing agriculture extension services, it should be recognised that some extension services over the past 50 years have failed due to a pervasive attitude that small farmers need to be "taught". The green agriculture paradigm requires participatory learning in which farmers and professionals in agro-ecological sciences work together to determine how to best integrate traditional practices and new agro-ecological scientific discoveries. Efforts should also be made to partner with NGOs that support farmers, field schools, demonstration farms and other such initiatives. It is also important to support small and medium business enterprises that are involved in supplying agriculture inputs; particularly those firms that offer green agriculture products and services such as organic certification auditing and reporting.

2. Integrating Information and Communications Technologies with Knowledge Extension

Support is needed to improve farmers' access to market information including through IT in order to enhance their knowledge of real market prices so that they can better negotiate the sale of their crops to distributors and end customers. There are also opportunities to support the construction of meteorological monitoring telemetry stations that could support national and regional weather forecasting capabilities that would help farmers determine best times for planting, fertiliser applications, harvesting and other critical weather-sensitive activities. Such networks could help support the introduction of innovative financial services such as weather-indexed crop insurance that would help reduce risks associated with adopting new technologies and shifting to green practices and marketing methods.

3. Better Food Choices

In an era where global human health is undermined by malnourishment and obesity, there is an opportunity to guide and influence people's food consumption into a greater balance with sustainably produced and more nutritious foods. Raising awareness about better food and its availability at affordable prices can reduce and reshape food demand trends. In this regard, there is a need to invest in public education and marketing that would encourage consumers to adopt more sustainable dietary habits.

Large-scale industrial farming practices, in many cases, pose enormous public health risks due to the overuse of inputs such as antibiotics, pesticides and synthetic growth hormones. There are neither policies nor any labels that transparently display the level of use and residues of these inputs. Introducing labelling schemes that can help consumers to make informed choices will dramatically shift the consumer behaviour towards safe and healthy food.

■ ■ ■

6

Guidelines to Green Fisheries

6.1 INTRODUCTION

Fish are one of the planet's most important renewable resources. Beyond their crucial role in marine and freshwater ecosystems, fish make a vital contribution to the survival and health of a significant portion of the world's population. Marine fisheries provide nutrition and livelihoods for millions of people in coastal communities, notably in South and South-East Asia, West Africa and Pacific Island states. As coastal populations continue to grow, the future benefits these resources can provide will depend on how well fisheries can be greened.

Often, fisheries managers and policy-makers are under pressure to sacrifice the long-term health of marine fish resources in favour of perceived short-term economic benefits to the fishing industry and consumers. Gaining a better understanding of the potential contribution and impact of marine fish populations on the global economy will provide broader, longer-term, economic and social perspectives.

The total catch from the world's marine capture fisheries rose from 16.7 million tonnes in 1950 to 80.2 million tonnes in 2005. It reached a peak of 85.3 million tonnes in 1994. For these 56 years, fish comprised about 86 per cent of the total landings, with crustaceans and molluscs accounting for 6 per cent, and 8 per cent respectively. The total landed value (gross output value) of the world's marine capture fisheries was about US$ 20 billion in 1950. It increased steadily to about US$ 100 billion in the late 1970s and remained at that level throughout the 1980s despite further increases in the total landings.

Box 6.1: Inland Capture Fisheries

Around the world, inland fisheries are an increasingly important factor for communities because of increasing consumption per capita and the inability of people to purchase other animal protein. In a recent *State of World Fisheries and Aquaculture* report, the UN Food and Agriculture Organisation (FAO) estimates that inland fisheries generate 10 million tonnes in landings annually; this amounts to about 11 per cent of the total capture fisheries catch from both inland and marine sources (FAO 2009). South-East Asia's Mekong river system, which is home to more than 850 freshwater species including many economically important species of catfish and carp, is estimated to provide fisheries landings worth around US$ 2 billion per year.

The top ten countries captured about a third of the global annual catch in 2005, with an estimated landed value of nearly 50 per cent of the global total. This implies that for the world to succeed in greening the fishing sector, the ten countries listed in Table 6.1 will have to be committed participants.

Table 6.1: Top ten marine fishing countries by fleet capacity

	Fishing Effort (million kW sea days)	Landings (million t)[2]	Landed value (2005 real US$ billion)
Russia	432	3	3.2
Japan	398	4	14.4
China	301	10	15.2
Taiwan	261	1	2.7
USA	225	4.8	4.2
Spain	147	0.9	1.3
Korea Republic	138	1.6	2.5
France	116	0.6	1
New Zealand	115	0.5	1.1
Italy	100	0.3	1

6.2 CHALLENGES IN GLOBAL FISHERIES

1. Overfishing

In the early 1970s, fishing activity expanded, particularly in Asia, but also along the Chilean coast, where large quantities of anchoveta were taken, and along the coast of West Africa. By 2005, there was a contraction of high-value areas. However, there has been a considerable expansion of fisheries into the high seas, most notably in the North Atlantic and South Pacific.

The spatial expansion of marine fisheries around the world partially masks the extent to which fisheries have been overfished. In fact, the FAO believes that only about 25 per cent of the commercial stocks, mostly of low-priced species, are currently underexploited, 52 per cent are fully exploited with no further room for expansion, 19 per cent overexploited and 8 per cent depleted.

Studies have estimated that by 2003, some 29 per cent of the world's marine fisheries had collapsed in the sense that their current catch level was less than one-tenth of the maximum registered catch. In the business-as-usual (BAU) scenario, half the amount of fish available in 1970 would be available by 2015 and only one-third in 2050. Practices such as "fishing down marine food web," where species are targeted and fished to depletion from largest to smallest species, can bring about significant changes to the balance of species in the ecosystem.

Halting the fishing of vulnerable, overexploited species and establishing conditions so that stocks can recover are clearly major challenges that have to be achieved despite demand for fish. Explaining the scale of the issue is a challenge in developed and developing countries and catalysing policy reform is particularly difficult when there are legitimate fears that fish stocks might not recover even if complete bans on fishing in certain areas are enforced.

2. Subsidies

Fisheries subsidies are defined here as financial transfers, direct or indirect, from public entities to the fishing sector, which help the sector make more profit than it would otherwise. Such transfers are often designed to either reduce the costs of fishing or increase revenues. In addition, they may also include indirect payments that benefit fishers, such as management and decommissioning programmes. Subsidies have gained worldwide attention because of their complex role in trade, ecological sustainability and socio-economic development.

It is widely acknowledged that global fisheries are over-capitalised, resulting in the depletion of fishery resources. There are many reasons for the decline of fishery resources, but the contribution of subsidies to the expansion of capacity and overfishing cannot be over-emphasised.

The challenge is that once subsidies are provided they become entitlements, which makes them politically difficult to remove. Only concerted action by groups such as civil society organisations, international bodies and governments can bring about the removal of such subsidies. Also, one strategy that may help is to keep the amount of the subsidy within the fishing community but divert it from increasing overfishing to enhancing fish stocks.

3. Small-scale Fisheries

A key issue along any coast is that of the local small-scale fisheries (SSF), which often provide crucial food supplies, sustain regional economies and support the social and cultural values of the areas, but are threatened as pressures on coastal areas are growing. This poses what is undoubtedly a major socioeconomic challenge: how to balance current and future needs for fishery resources.

While all fisheries face a range of challenges, for SSF many of the challenges are related to factors that are external to the fisheries *per se* but within the broader social-ecological system. These include (1) negative impacts of industrial and foreign fleets, depleting coastal fish stocks, and in some cases destroying coastal fishing gear; (2) degradation of coastal environments and fish habitat, through land-based sources of marine pollution, development of urban areas, shrimp farming, tourism, mangrove extraction, etc., leading in each case to reduced fish stocks; (3) infrastructure challenges, such as limitations on transportation of fish products; and (4) global forces, such as climate change and globalisation of fish markets, that can negatively affect the small-scale fisheries. In addition, over-fishing by SSF contributes to the problem in many cases. It is important to recognise that given the above external factors, solving the sustainability challenge for SSF requires coordinated, multi-faceted approaches that aim to improve fishery governance at a local level—so that coastal fishers are involved in developing, and thereby support fishery management measures—while simultaneously dealing with other fleets, and market and infrastructure issues to improve coastal environmental quality. An integrated approach is thus unavoidable.

4. Greening Aquaculture

According to FAO, aquaculture supplies around 50 per cent of the world's seafood. There are many challenges to aquaculture as a source of animal protein in a green economy. Many farms still rely on wild caught fish as feedmeal and oil. The potential for disease from fish farms impacting wild populations is also an issue. Finally, there is the potential that fish farms can pollute the environment because of the waste they produce. Given these challenges, it is clear that current aquaculture practices need to be modified to make fish farming green.

The sector needs to (1) be organised to ensure minimal environmental degradation; (2) stop the farming of carnivorous fish such as salmon, bluefin tuna and seabass until non-wild fish sources of fish meal are developed; (3) adopt integrated technologies that would make fish farming as self-contained as possible; and (4) develop reliable management systems for green aquaculture practices.

5. Climate Change and Greenhouse Gas Emissions in Fisheries

Climate change has begun to alter marine conditions, particularly water temperature, ocean currents, upwelling and biogeochemistry, leading to productivity shocks for fisheries. Shifts in species distribution that appear to be caused by changes in sea temperature are well documented, as are variations in growth rates. Climate change may also alter the phonology of marine organisms, creating mismatches between the availability of prey and predator requirements and leading to coral bleaching and habitat loss for reef-associated fish species. These changes would affect the distribution and volume of catch worldwide, thereby affecting global fisheries socially and economically. For instance, recent studies estimate that climate change may lead to significant losses in revenues, profits and/or household incomes, although estimates are considered preliminary.

Box 6.2: Subsidies and Small-scale Fisheries

Moves to shift to a green economy can provide opportunities to invest in SSF in a manner that enhances sustainability of the resource base as well as the coastal economy and society. The key lies in using the investments to build institutional strength and suitable incentives at a local scale. Measures such as subsidies and investment strategies can be used as incentives to change human behaviour positively, supporting long-term objectives in moving the fishery toward sustainability, without serious negative impacts. For example, this could involve providing funds to encourage certain actions such as conversion of fishing gear to less damaging choices, or a shift from fuel-intensive to more labour-intensive fishing methods.

In the context of SSF, this implies a careful examination of which subsidies are truly sustainable, equitable and moving in the direction of conservation. For example, a fuel subsidy is common in fisheries, but this tends to promote more fuel-intensive and capital-intensive fleets, which leads not only to overfishing, but also to inequitable expansion of catching power for some (those who can take advantage of the subsidy) at the expense of others (with less capital). On the other hand, a subsidy that is used to provide more secure livelihoods for coastal fishers, and one that leads to a shift of SSF, where necessary, to more ecologically suitable methods, may be very helpful. The subsidy issue also relates to the balance of small- scale and industrial fishing. Past subsidies on vessel construction and on fuel led to a favouring of industrial fleets that are too capital—and fuel-intensive. A better policy would be to orient subsidies as incentives to balance industrial and small-scale fisheries, thereby generating both human and ecological benefits.

6.3 OPPORTUNITIES IN GLOBAL FISHERIES

Greening the world's fisheries will help restore damaged marine ecosystems. When managed intelligently, fisheries will sustain a greater number of communities and enterprises, generating employment and raising household income, particularly for those engaged in artisanal fishing.

1. Jobs Supported by Global Fisheries

The world's fisheries provide livelihoods to millions of people in coastal regions and contribute significantly to national economies. They are relied upon as a safety net by some of the world's poorest, providing cash income and nutrition.

When considering post-catch activities and workers' dependants, the number of people directly or indirectly supported by marine fisheries is about 520 million or nearly 8 per cent of the world's population.

There has been a steady increase in fisheries employment in most low- and middle-income countries, while in most industrialised countries, the trend has been towards a decrease in the number of people employed in capture fisheries. For example, since 1970, the number of fishers has fallen by 61 per cent and 42 per cent in Japan and Norway, respectively.

2. Recreation and Tourism

Marine recreational activities (MRAs) such as recreational fishing, whale watching and diving have grown in popularity in recent years and they have consequently come to the forefront of discussion and research on the ecological, economic and social impacts of more benign forms of interacting with the sea.

To estimate the value of MRAs, in a recent study Cisneros-Montemayor and Sumaila found that currently, recreational fishing occurs in 118 maritime countries and that country-level data on expenditure, participation and employment are available in 38 of these countries (32 per cent of total). The authors estimated that in 2003, nearly 60 million recreational anglers around the world generated a total of about US$ 40 billion in expenditure, supporting over 950,000 jobs. In their analysis, countries with data account for almost 95 per cent of estimated total expenditure and 87 per cent of participation, so the authors argue that this estimate likely provided a close approximation to actual recreational fishing effort and expenditure.

3. Marine Protected Areas

Marine protected areas (MPAs) have been implemented in many countries and are regarded as a very important management instrument for fisheries. The assumption underlying the MPAs is that they can conserve the resources and increase the biomass therein, and consequently benefit surrounding areas through species migration and enhanced recruitment. To fully utilise MPAs as a management tool, the Johannesburg Plan of Implementation adopted at the World Summit on Sustainable Development in 2002 aims to establish a global network of MPAs covering 10–30 per cent of marine habitats by 2012. This deadline was extended to 2020 and the target lowered to 10 per cent at the CBD meeting in Nagoya, Japan in late 2010.

4. Consumer Awareness

In recent years, we have seen a relative explosion in the number of programmes that seek to help consumers make informed decisions in terms of sustainability about their consumption of fish products. Although such programmes are not without criticism, it is clear that consumer awareness of marine fishery issues, if properly designed and implemented, would be an important driver of greening world fisheries as such awareness programmes expand into more and more places around the world.

6.4 THE COST OF GREENING GLOBAL FISHERIES

A key element of greening the fisheries sector involves moving from the current situation where we are not fishing the resource in a sustainable manner to one where the fish we catch each year is equal to or less than the growth of wild stocks. To make the change from the current state of affairs would require some investment into adjusting fishing capacity, managing transitions in labour markets, management programmes and scientific research.

1. Identifying Greening Efforts

There is widespread agreement that the world's fisheries are currently operating at overcapacity. Advances in technology have made it possible for a much smaller global fleet to catch the maximum sustainable yield, but the global fishing capacity keeps on growing owing to the common property nature of fisheries and the provision of fishing subsidies by many maritime countries of the world. Also, the use of sometimes damaging fishing methods such as bottom-trawling, unselective fishing, pollution and human-induced variations in climate has changed the productivity of many aquatic environments.

The issue of overcapacity can be addressed by investigating some of the common sources of excess fishing capacity. In several places, fishing is considered employment of last resort, attracting people with few other job options. Investing in re-training and education programmes for fishers and creating alternative employment has been successful in reducing fishing pressure, especially in places that are known for artisanal fishing.

2. The Cost of Fishing Fleet Adjustment

The world's current fishing capacity is widely estimated to be 2.5 times more than what is needed to land the maximum sustainable yield (MSY). This implies that in order to shift the fishing industry to MSY levels, we would need to trim excess fishing capacity. However, the cumulative power of the global fleet is presently increasing at a rapid rate, notably in Asia.

It is estimated that some 4 million boats are actively engaged in marine fisheries. If we assume that current fishing capacity is between 1.5 and 2.5 times the level needed to maximise sustainable catch, fishing effort would need to be reduced by between 40 and 60 per cent. This means that the active fishing fleet may need to be reduced by up to 2.4 million vessels.

To achieve this goal, the UNEP has recommended the following plan of action. Many large-scale vessels are concentrated in developed countries while small-scale vessels are mostly found in developing countries. Although the same green economy result could potentially be achieved by making cuts to just large-scale vessels, this would be ineffective in areas dominated by small-scale fishing that are currently overfished, such as in India and Senegal. In this scenario, it should be possible to put three-quarters of the responsibility for cutting fishing effort on large-scale vessels, with the remaining quarter filled by small-scale vessels.

By focusing effort reductions on large-scale vessels, the total cost of adjustment to green world fisheries in this scenario is much less costly than the first scenario, requiring a one-time total investment of between US$ 190 billion to US$ 280 billion with a mean of US$ 240 billion to decommission vessels and provide for workers as they transition to other forms of employment. It would also be necessary to increase management expenditure by 25 per cent to US$ 2 billion on an annual basis.

3. Cost-benefit Analysis of Greening Fisheries

Greening the fisheries sector would lead to an increase in value added from fishing, globally, from US$ 17 billion to US$ 67 billion a year. This is a net increase of US$ 50 billion a year. Given that the cost of restructuring the global fishing fleet under scenario three is a one-time investment of about US$ 240 billion, benefits would be realised very quickly if fish stocks recover fast. Discounting the flow of US$ 50 billion per year over the next 50 years at 3 per cent and 5 per cent, real discount rates represent a present value from greening ocean fisheries of US$ 960 and US$ 1,325 billion, which is between 4 and 5.5 times the mean estimate of the cost of greening global fisheries. This signals that there is potentially a huge green advantage.

4. Managing Fisheries

Effective management is crucial for ensuring a green marine fisheries sector, although this has so far proved difficult to achieve. There are several areas of management where increased investment can be extremely beneficial. These include:

- Stock-assessment programmes;
- Monitoring and control programmes; and
- Establishment of marine protected areas (MPA).

Stock assessment programmes are basic for fishery managers who require reliable statistics to inform them of the state of fish stocks so that they may keep a careful eye on whether fishing effort is appropriate for the sustainable use of the stock.

Monitoring and control programmes are those that allow fisheries managers to determine whether fishers are acting in compliance with catch quotas or not. Such programmes are also necessary in terms of mitigating the impact of illegal and unreported fishing activities.

Historically, MPAs have not been used as a major tool in the management of the world's fisheries. However, their role as a management tool has become more popular in recent years. Marine Protected Areas attempt to maintain the health of fish stocks by setting aside an area of the ocean that is free from fishing activity—allowing mature fish in these areas to escape into unfished areas, thereby ensuring the future resilience of the fishery.

6.5 ENABLING CONDITIONS: INSTITUTIONS, PLANNING, POLICY AND REGULATORY REFORM AND FINANCING

6.5.1 Building Effective National, Regional and International Institutions

We need effective institutions at all levels of government, from the local to the provincial/state to the national, regional and international because of the migratory nature of many fish stocks. Many fish stocks spend their lives completely in the Exclusive Economic Zones (EEZ) of countries—they do not migrate across EEZs of other countries or straddle into the high seas. For these fish stocks, effective national institutions are all that is needed. Then we have fish stocks that are shared by two or more countries, the so-called transboundary fish stocks that live completely within the EEZs of more than one country. For these fish stocks, participants in the fishery must agree on the management of the stock in order to make it effective. Then there are fish stocks that are partly or wholly located in what is left of the high seas. It has for a long time been a concern that the regulation of these fisheries is ineffective and that regulation of stocks that are governed by one or more coastal states but which straddle periodically into the high seas is undermined by the open access to the high seas. This prompted a conference on high seas fisheries in the 1990s under the auspices of the UN. This resulted in what is usually called the *UN Fish Stocks Agreement,* which vests the authority to regulate high seas fisheries in *Regional Fisheries Management Organisations* (RFMOs) , whose functioning was recently reviewed by Cullis-Suzuki and Pauly and generally found wanting.

6.5.2 Regulatory Reform

The basic requirement for a successful management of a fish stock is limiting the rate of exploitation to some sensible level. This necessitates (1) a mechanism to set such a target catch level and (2) a mechanism to monitor and to enforce it. The basic question to ask is whether the scientific, administrative and law-enforcing capability is in place to make this happen. The presence of strong social norms and cultural institution are great tools for enforcement where they work.

In practice, effective management institutions would have in place mechanisms for providing scientific advice, as well as a mechanism to set the rate of exploitation on the basis of that advice and in such a way that it maximises long-term benefits in the form of food supplies or fishing rent (difference between revenues and costs adjusted for subsidies). The latter requires an efficient and uncorrupted administration that strives for the best possible economic (or food supply) situation of the country in question.

6.5.3 The Economics of Fishery Management Tools

The basic fishery management tools can be grouped into (1) output controls; (2) input controls; and (3) auxiliary measures. Both (1) and (2) control the rate of exploitation, which is the fundamental factor that needs to be controlled, as stated earlier.

Output controls mean limiting the total amount of fish that can be caught. We do not know what this means in terms of rate of exploitation unless we know what the size of the fish stock is. This can only be estimated with a considerable and possibly high degree of imprecision. However, where feasible, the target output should be set on the basis of maximizing either food supply or fishing rent, depending on what is deemed most appropriate.

Under some circumstances, *effort controls* could be better than quota controls. This can happen if quotas are difficult to monitor, or if the size of the fish stock cannot be estimated while we can be reasonably certain that it is always evenly distributed in a given area so that a unit of effort produces a given rate of exploitation.

Then there are several measures which are termed *auxiliary*, as they do not primarily address the basic problem of controlling the rate of exploitation but promote greater yields from fish stocks in various ways. The rationale for such measures is to discourage people from seeking fish that they are not authorised to take. While this is indeed desirable, such regulations are economically wasteful and one should look for ways to achieve the desired outcome in less wasteful ways.

6.5.4 Managing the Transition Process

This would be most challenging when we are dealing with depleted fish stocks that need to be rebuilt. This situation arises because the capacity of the fishing fleet has outgrown the available resource, and so the fleet would have to be downsized. Both of these necessitate a cutback in fishing activity. Fish quotas that are lower than contemporary and recent catches which have depleted the fish stock are necessary to rebuild the stock. Such small quotas mean that some of the fishing capacity is redundant, and even with rebuilt stocks it is highly likely to remain redundant if a repeated depletion of the stock is to be avoided. All this implies investment in the fish stocks as it were, through foregone earnings in the short-term for the purpose of obtaining higher benefits in the future.

■ ■ ■

7

Guidelines for Green Water Sector

7.1 INTRODUCTION

In green economies, the role of water in both maintaining biodiversity and ecosystem services and in providing water is recognised, valued and paid for. The use of technologies that encourage efficient forms of recycling and reuse is encouraged. In a green economy there is emphasis on the pursuit of opportunities to invest in sectors that rely upon and use natural resources and ecosystem services. Investing in green sectors, including the water sector, more jobs and greater prosperity can be created. Arguably, these opportunities are strongest in areas where people still do not have access to clean water and adequate sanitation services. Early investment in the provision of these services appears to be a precondition for progress. Once made, the rate of progress will be faster and more sustainable, thus making transition to a green economy possible.

Water in the green economy focuses on the socio-economic opportunities that proper water management provides to social and economic development, whilst also safeguarding freshwater ecosystems. In green economies, the role of water in both maintaining biodiversity and ecosystem services and in providing water is recognised, valued and paid for.

7.2 THE WORLD'S WATER RESOURCES

Access to the world's water resources is heavily dependent upon the nature of the water cycle. While a massive amount of water reaches the earth's land surface, much less, around 40 per cent, makes its way into creeks, rivers, aquifers, wetlands, lakes and reservoirs, before cycling back into the atmosphere (see Fig. 7.1). Of the water that is extracted for human purposes, on average, approximately:

- 70 per cent is used for agricultural purposes;
- 20 per cent is used by industry (including power generation); and
- 10 per cent is used for direct human consumption.

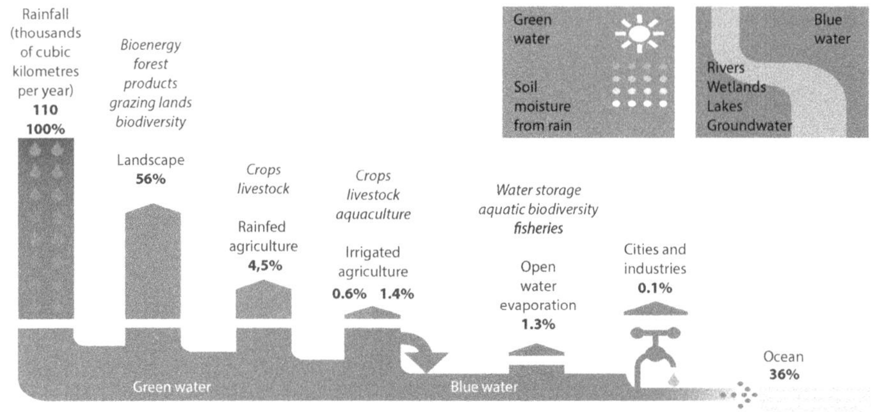

Fig. 7.1: Green water refers to rainwater stored in the soil or on vegetation, which cannot be diverted to a different use. Blue water is surface and groundwater, which can be stored and diverted for a specific purpose.

Given that the vast majority of usable fresh water is channelled towards agriculture, any global consideration of water allocation must consider the factors that determine the efficiency of water use in the sector. Irrigated land produces around 40 per cent of the world's food. One of the biggest challenges facing water managers is to find a way to significantly increase the productivity of irrigated agriculture so that water can be transferred to other sectors without adversely affecting the environment or food security. In many parts of the world there are few opportunities to enhance supplies at reasonable cost.

But general observations can be misleading. No two water bodies are the same. Managing large, complex, transboundary water systems typically requires a different approach to overseeing smaller water systems, where local issues are often all that need to be considered. In developing countries, water management and investment are typically geared towards reducing poverty and enabling economic development; the priority for developed nations tends to be maintaining infrastructure and supplying access to water at reasonable cost. In both cases, there is a need to focus more on long-term sustainability of the systems and services provided. Demand and supply also vary greatly. In Singapore, for example, almost all water is extracted for urban and industrial purposes, while in many other parts of the world, the majority of water is extracted for agricultural or mining purposes.

7.3 WATER: A UNIQUE NATURAL RESOURCE

Unlike most other natural resources, water flows readily across and through landscapes in complex ways that affect its availability and opportunities to manage it. Understanding these water flows is critical to the design of investment programmes and policies necessary to support a transition to a green economy.

7.3.1 Services from Natural Infrastructure

Water makes an irreplaceable contribution to ecosystem services that stem from the earth's natural capital and vice versa. Protecting the natural ecosystems of river basins and restoring degraded catchment areas is crucial to securing the world's water supplies, maintaining their quality, regulating floods and mitigating climate change. The role of other ecosystems, such as forests, wetlands and floodplains in providing access to water also needs to be recognised and quantified. Gauging the true value that these ecosystems provide is a key part of charting a course to a green economy.

Recent analysis is showing a close global correlation between the threats to biodiversity and threats to water security. Regions where threats to human water security is high, but the threat to biodiversity is low, are rare. When the threat to human water security is high, usually the threat to biodiversity is high. This suggests that there may be considerable opportunities for governments to improve biodiversity outcomes by investing in water security. Water- dependant ecosystems also play an important role in the provision of cultural benefits.

7.3.2 Water Accounting

As water flows through and across land, it is used and reused. This makes information about water difficult to assemble and use for management. When, for example, a policy promotes a more efficient irrigation system, it is critical to decide whether or not the savings are to be used to expand irrigation or returned back to the river or aquifer from which the water was taken. Gains in one area can be associated with losses in another area. When the savings are not returned back to the river or aquifer, the result can be a significant reduction in the quantity of water available to the environment and to other users.

Another common water accounting error is to assume that ground and surface water systems are not connected to one another and to administer them separately. Many rivers play an important role in replenishing aquifers, while aquifers can provide much of a river's base flow. Failing to account for these interactions can result in the serious problems of over-use and degradation. One administrative solution is to reverse the onus of proof and require managers to assume that ground and surface water resources are linked and manage them as a single connected resource until such time as disconnection can be shown.

Land-use changes can have similar effects on the volume of water available for use. For example, whenever a plantation forest is established, a hillside is terraced, or a farm dam is constructed, run-off is usually reduced. As a result, the quantity of water available for extraction from a river or aquifer is less than it otherwise would be. Accounting for water in a way that is consistent with the hydrological cycle and that avoids double counting of its potential is critical to developing the robust allocation and management systems that underpin a green economy.

7.3.3 Water and Energy

The interdependence of water and energy demands also needs careful attention as arrangements are put in place for a transition to a green economy. There are at least two dimensions to this relationship.

First, water plays an important role in energy generation, notably as a coolant in power stations. In the USA, for example, 40 per cent of industrial water-use is for power-station cooling, although water-use efficiency varies with the technology used. By 2030, it is expected that 31 per cent of all industrial water-use in China will be for cooling power plants. Generally, as countries become wealthier and more populous, industrial demand for water is expected to increase. In China, more than half of the increase in demand for water over the next 25 years is expected to result from a significant expansion in its industrial sector, which will need to be accommodated through a simultaneous reduction in the amount of water used for irrigation in the agricultural sector.

Second, the water supply and sanitation sector is a large consumer of energy. Relative to its value, water is heavy and in energy terms expensive both to pump over long distances and to lift. In California, USA, where large volumes of water are transported over long distances, the water sector consumes 19 per cent of the state's electricity and 30 per cent of its natural gas.

In developed countries, the relatively high energy costs of pumping and treating water for household, industrial or mining purposes are broadly accepted. In developing countries, great care must be taken to ensure that water treatment and distribution systems remain affordable. The relatively modest financial returns from food production in both developed and developing countries means it rarely pays to pump water over long distances for agricultural purposes. In recognition of this, Saudi Arabia has recently shifted its food security policy from one that subsidises water use at home to one that invests in the development of agriculture in other countries where water supplies are more abundant. This is enabling Saudi Arabia to access food at more affordable prices and use the revenue saved for other, more sustainable, purposes.

7.4 CHALLENGES AND OPPORTUNITIES

Together, Cambodia, Indonesia, the Philippines and Vietnam lose an estimated US$ 9 billion a year because of poor sanitation (based on 2005 prices). This amounts to around 2 per cent of their combined GDP, varying from 1.3 per cent in Vietnam, 1.5 per cent in the Philippines, 2.3 per cent in Indonesia and 7.2 per cent in Cambodia.

The annual economic impact of inadequate sanitation is approximately US$ 6.3 billion in Indonesia, US$ 1.4 billion in the Philippines, US$ 780 million in Vietnam and US$ 450 million in Cambodia. In these four countries, the total value of this impact is US$ 8.9 billion per year.

In 1991, a cholera epidemic swept through most of Peru and cost US$ 1 billion to control. If one tenth of this amount (US$ 100 million) had been spent on the provision of sanitation services, the epidemic would not have occurred.

7.4.1 Challenges

1. Poverty, Access to Clean Water and Adequate Sanitation Services

Nearly 1 billion people lack access to clean drinking water and 2.6 billion lack access to improved sanitation services. As a direct result, every year, 1.4 million children under the age of five die due to a lack of access to clean water and adequate sanitation services. In east Nigeria and north Cameroon, every 1 per cent increase in use of unprotected water sources for drinking purposes is directly associated with a 0.16 per cent increase in child mortality.

Gleick argues that failure to provide people with affordable and reliable access to water and sanitation services is one of humankind's greatest failings. Lack of sanitation makes people sick. When water is unclean, water-borne diseases such as diarrhoea and water-washed diseases including scabies and trachoma are common. Diarrhoea is the third most common cause of child mortality in West Africa after malaria and respiratory infections. New water-borne diseases such as the Whipple disease are still emerging.

The adverse impacts of water-borne disease on an economy can be large. When people are sick, they cannot work and, among other costs, considerable expenditure on medical treatment is needed.

The adverse impacts of inadequate access to clean water, however, do not stop with water-borne disease. When water is not on tap, people (mainly women and children) must either spend a large amount of time fetching water or pay high prices for it to be carted to them. In Western Jakarta, Indonesia, the cost of water purchased from a water cart is ten to fifty times the full cost to a water utility of establishing a reliable mains water supply. In certain circumstances, it is challenging to find a way to convince governments and private investors to go ahead despite a widespread perception that poor people are not able to pay for water (services) and that it is not cost-efficient to supply water to informal settlements. A lack of easy access to clean water also erodes the capacity of the poorest to engage in other activities.

From a government perspective, when water supply and sanitation services are inadequate, large amounts of revenue are spent dealing with the impacts of disease, rather than generating wealth.

2. Water Scarcity

Exploring opportunities to invest in the construction of dams, the International Water Management Institute (IWMI) has identified two types of water scarcity: physical scarcity and economic scarcity. In regions where there is physical scarcity, the sustainable supply limit has been reached and little opportunity to construct more dams remains. In regions where the scarcity is economic, however, it is possible to increase supplies if the financial resources necessary to build a new dam can be found.

The OECD has estimated that by 2030 nearly half the world's population (3.9 billion people) will be living under conditions of severe water stress (Fig. 7.2). The reasons for the emergence of this scarcity include:

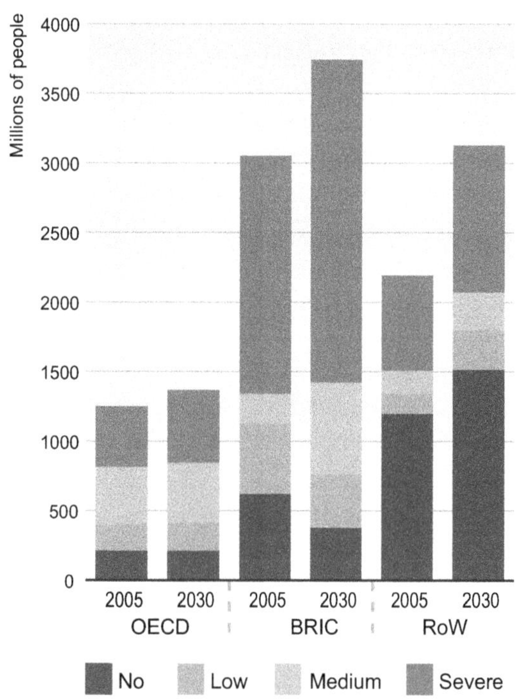

Fig. 7.2: Number of people living in waterstressed areas in 2030 by country type.

- *Population increase*—by 2030 the world's population will have increased by 2.4 billion people. All of these people will to demand access to water for basic needs, to supply industrial goods and grow food;
- *Increased living standards*—as countries develop and people become wealthier, they tend to consume more water and more water-intensive products such as meat;
- *Over-exploitation*—around the world a considerable proportion of aquifers and river systems are overused. It has been estimated that 15 per cent of India's total agricultural production is being delivered via groundwater depletion—the situation that occurs when extraction exceeds replenishment;
- *Water pollution*—an increasing number of water supplies are becoming contaminated by pollutants, with the consequence that less water is available for use or it costs much more to make it usable;
- *Ecosystem degradation*—over the last 50 years ecosystems have been degraded faster than ever before. Freshwater ecosystems, which provide critical services such as the purification of water by wetlands or forests are the most threatened and have been among the hardest hit, and;
- *Adverse climate change*—when combined with effects of climate change on dryland production systems, the International Food Policy Research Institute estimates that the aggregate effect of climate change is likely to be a significant reduction in total agricultural productivity. The greatest adverse impacts of climate change on people are expected in South Asia. In the next 40 years, child malnutrition is expected to increase by 20 per cent as a direct result of climate change.

3. Balancing Supply and Demand

In an attempt to understand the magnitude of this emerging water-scarcity challenge, the 2030 Water Resources Group has projected global demand for water and, under different scenarios, compared it with likely supply. They concluded that if there is no improvement in the efficiency of water use, in 2030 demand for water could outstrip supply by 40 per cent. Clearly, a gap of this magnitude cannot (and will not) be sustained.

Fig. 7.4 offers an alternative perspective on the magnitude of the emerging water-supply challenge. Under a business-as-usual scenario, improvements in water productivity can be expected to close around 20 per cent of the gap between global demand and supply. Increases in supply through the construction of dams and desalination plants, coupled with actions such as increased recycling, can be expected to close the gap by a similar amount. The remaining 60 per cent, however, must come from increased investment in infrastructure and water-policy reforms that improve the efficiency of water use. If the resources are not found to facilitate a significant increase in efficiency and if the water-policy reforms are not implemented, water crises must be expected to emerge. Figure 7.4 suggests that the average rate of improvement in water productivity and supply enhancement needs to increase at double the rate of improvement achieved in the past decade. Globally, the time for procrastination is past.

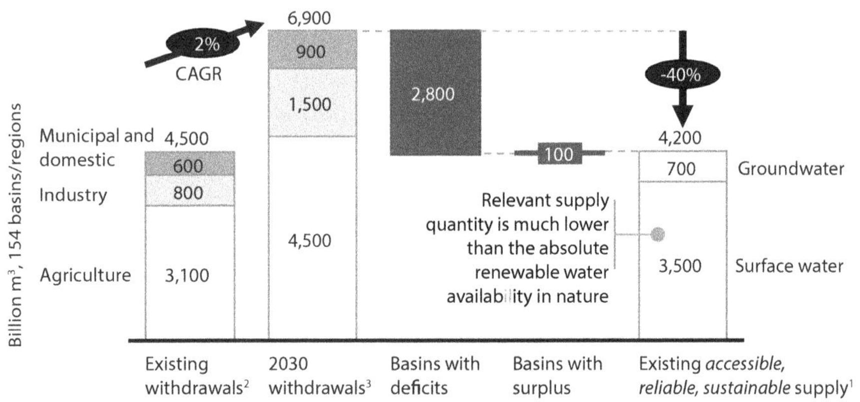

1 Existing supply which can be provided at 90% reliability, based on historical hydrology and infrastructure investments scheduled through 2010; net of environmental requirements
2 Based on 2010 agricultural production analyses from IFPRI
3 Based on GDP, population projections and agricultural production projections from IFPRI; considers no water productivity gains between 2005–2030

Fig. 7.3: Aggregated global gap between existing accessible, reliable supply and 2030 water withdrawals, assuming no efficiency gains.

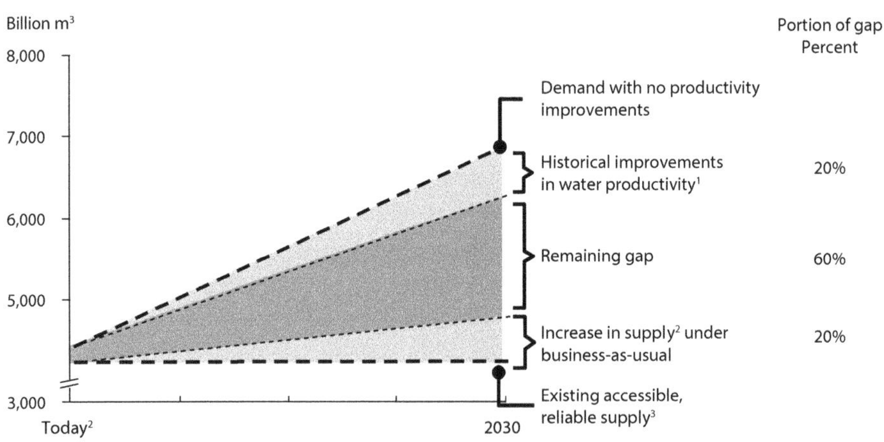

Fig. 7.4: Projection of the global demand for water and, under a business-as-usual scenario, the amount that can be expected to be met from supply augmentation and improvements in technical water use efficiency (productivity).

Figure 7.5 shows the nature of expected increase in demand for water throughout the world. As discussed, one of the more significant challenges is to find ways to supply more water to the industrial sector while increasing agricultural production. Significant transfers of water from rural areas to the industrial sector can be expected, especially in China and in North America. In anticipation of the pressure that these shortages will place on water-dependent business, a number of large companies are beginning to quantify and account for their water use and water related impacts and the nature of the water-related risks they face.

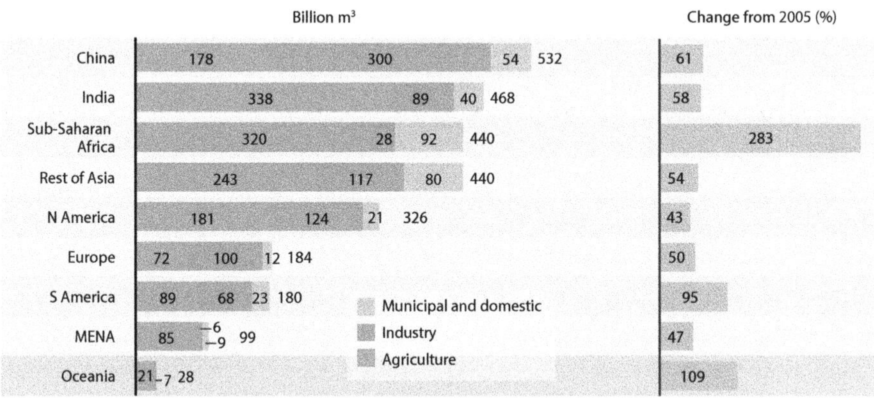

Fig. 7.5: Assessment of expected increase in the annual global demands for water by country and region (2005–2030).

7.4.2 Opportunities

1. Investing in Biodiversity and Ecosystem Services

In terms of ecosystem health and function, global assessments of the health of the world's water river systems and aquifers suggest that the aggregate trend is one of decline. Examples of this decline include:

- Barriers have been laid across China's Taihu Lake to stop regular algal blooms reaching the water treatment plant that supplies water to over 2 million people;
- From October 2002 until October 2010, the absence of flow has meant that dredges have been used to keep the mouth of the Australia's River Murray open to the sea;
- In Manila, the Philippines, groundwater extraction, primarily for industrial purposes, is lowering the water table at a rate of between 6 metres and 12 metres per year, and;
- In 1997, China's Yellow River flowed all the way to the sea only for 35 days. For much of the year the river's last 400-plus miles were dry.

There is a new recognition of the positive synergy that emerges between healthy environments and healthy communities. As documented by Le Quesne *et. al.*, some countries are now investing large amounts of money in the restoration of degraded river systems and the development of policies and administrative arrangements designed to prevent degradation of these systems.

2. Investment in Sanitation and Drinking Water Supply

In many developing countries, one of the biggest opportunities to expedite a transition to a green economy is to invest in the provision of water and sanitation services to the poor.

A recent estimate puts the cost of achieving the 2015 MDGs at US$ 142 billion per year for providing sanitation services and US$ 42 billion per year for drinking water supply to households. More investment is required for sanitation services than drinking water as the number of households without access to adequate sanitation services is much higher.

Although the amount of money needed to attain the MDGs for water is considerable, when spread over a number of years and divided by the number of people expected to benefit from such expenditure, the investment case is strong. In Ghana, for example, the OECD estimates that investment of US$ 7.40 per person per year over a decade would enable the country to meet its MDG target. Estimates of the required per capita expenditure in Bangladesh, Cambodia, Tanzania and Uganda range from US$ 4 to US$ 7 per capita per year.

Failure to invest adequately in the provision of affordable access to clean water and adequate sanitation acts as a barrier to development and that early investment in these areas is a necessary precondition to progress. Grey and Sadoff argue that a minimum amount of investment in water infrastructure is a necessary precondition to development; using a range of case studies, they identify a close association between adequate investment in infrastructure and environmental degradation.

3. Investing in Smaller, Local Water-supply Systems

As observed by Schreiner *et. al.*, the presence of economic water scarcity should not be interpreted as a recommendation for the construction of large dams. In many cases, greater returns can be achieved from the construction of smaller storages that are built by and serve local communities. At this scale, community engagement and management of infrastructure is easier and adverse environmental impacts tend to be fewer in both urban and rural settings.

4. Accessing New (Non-traditional) Sources of Water

One of the most common approaches to resolving water-supply problems is to build a large dam. Constructing them typically involves significant cost, the dislocation of many people and many adverse environmental problems. Schreiner *et. al.* observe that urban communities have historically relied on large dams for their water supplies. More recently, however, water-supply options have expanded to include the capture and storage of stormwater and desalination, fog interceptions in cloud forests (notably in the Andes mountains), transfers between islands, inter-basin water transfers, bulk transport such as by pipeline or Medusa bags (giant polyfibre bags holding up to 1.5 billion litres of potable water that are towed by ships). Other communities and countries are investing in sewage recycling. Singapore, for example, has invested in the development of systems that treat sewage to a standard allowing it to be used for drinking purposes. Most of these technologies, however, are reliant upon the use of increasing amounts of energy and, as a result, the costs of water provision are rising in most regions where there is physical water scarcity.

5. Producing More Food and Energy with Less Water

As the world's population increases, more water will be needed for household and industrial purposes with the consequence that in many areas, either more food will have to be imported, or more food produced with less water. When asked, "Is there enough land, water, and human capacity to produce food for a growing population over the next 50 years - or will we 'run out' of water?", analysis undertaken by the International Water Management Institute (IWMI) reveals that, "It is possible to produce the food - but it is probable that today's food production and environmental trends, if continued, will lead to crises in many parts of the world".

For example, in many developing countries, typical irrigated maize yields are in the vicinity of one to three tonnes per hectare, whilst they could be as high as eight tonnes per hectare. There is a significant opportunity to increase crop yields and avoid a global food security crisis. If this opportunity is realised, then not only will it also be possible to divert water to other uses, but it will also be possible for developing countries to produce a surplus for sale to others.

6. Institutional Reform

When coupled with more traditional hard approaches to investment in built infrastructure, the softer approach of developing more effective administrative arrangements and policies that encourage private investment can significantly reduce the amount of money that governments need to invest in the water sector to achieve the same outcome. Typically, soft approaches focus on incentives and the factors that motivate consumers to manage their water use.

7.5 THE ECONOMICS OF GREENING WATER USE

Research around the world suggests that there are no single-shot solutions to the world's mounting water access, sanitation and scarcity problems. Each circumstance has its own unique set of challenges and opportunities. At the most general level, it is becoming apparent that the best results come for the pursuit of mixed solutions. Simple single-shot solutions tend to be prohibitively expensive and, in many cases, are insufficient to solve known supply problems.

7.5.1 The Economics of Investing in Water and Ecosystems

Under the green investment scenario, global water use is kept within sustainable limits and all the MDGs for water are achieved in 2015. Water use is more efficient, resulting in increased agricultural, biofuel and industrial production. The number of people living in a water-stressed region is 4 per cent less under the green investment scenario by 2030 compared to business-as- usual, up to 7 per cent by 2050.

The results from this modelling are encouraging in terms of both economic terms and from the perspective of water management. For 2050, total employment and income is greater under the green investment scenario, whereas the number of people working in the water sector is lower. This counterintuitive finding occurs because the sector becomes much more efficient. Labour and other resources are freed for use in other sectors. In addition, as water is used more efficiently, more is available for manufacturing and other purposes with the result that more people are gainfully employed.

The overall conclusion from this assessment is that, where there is water scarcity or when large proportions of a population do not have access to adequate water supply and sanitation services, early investment in water is a necessary precondition to progress.

7.5.2 Selecting Projects and Initiatives for Investment

While it is useful and informative to examine the economics of investing in water at the global level, investments must be made primarily at the river basin, catchment and local level.

In areas where the costs of enhancing water supplies from traditional sources are rising, the 2030 Water Working Group is recommending the preparation of formal costs curves. These cost curves rank each potential solution to a problem in terms of the relative cost per unit of desired outcome achieved and can be used to assess the likely costs and benefits of each solution. One of the most striking features of this approach is that one often finds solutions that both make more water available and cost less money.

7.5.3 Flow of Benefits from Investment in the Water Supply and Sanitation Sector

Many returns to investment in the water sector are indirect. Build a toilet for girls in a school and they are more likely to go to school. This simple statement highlights the fact that investment in water opens up other opportunities for development. Assessing the case for more investment in water infrastructure in the Niger Basin, Ward et al. report that investment in providing access to potable water and in education are the only two variables that are consistently related to poverty reduction across the whole Niger River basin.

7.6 ENABLING CONDITIONS—OVERCOMING BARRIERS AND DRIVING CHANGE

Without significant water policy reform to enable the reallocation of water from one sector to another, financially reward those who make water use more efficient and so forth, a global analysis by the 2030 Water Working Group suggests that some nations will not be able to avoid the emergence of a water crisis in many regions. If wide ranging reforms are adopted, however, then the group's analysis suggests that most water crises can be averted. Investment in water policy reform and governance enables greater engagement and use of local knowledge and for investments to be made at a multitude of scales. When such approaches are taken, the 2030 Water Working Group estimates that the global amount of money that needs to be invested in the water sector can be reduced by a factor of four.

7.6.1 Improving General Institutional Arrangements

Arguably, the greatest impediment to investment in water infrastructure and management arrangements has been the difficulty in establishing high-level governance and political support for arrangements that support effective governance. Problems range from a simple lack of institutional capacity to the presence of widespread corruption and opportunities to gain political favour. Longterm solutions such as the establishment of reliable, stable governance arrangements for the supply of water are central to a green economy.

A parallel issue is the question of rights or entitlements to use land and water. When these rights are insecure, the incentive to take the long-term perspective necessary to encourage green approaches to investment is weak. When land tenure, water entitlements and other forms of property rights are well-defined, far more sustainable forms of resource use can be expected. Early investment in the development of land registers and other similar processes are simple ways to expedite the transition to a green economy.

Increases in the capacity of a nation to collect taxes will clearly make it easier to move to full-cost pricing arrangements and, where appropriate, provide rebates and other forms of assistance to the most needy without having to resort to inefficient cross-subsidies.

Another example of an enabling condition is the use of education and information programmes designed to increase awareness of opportunities to act in an environmentally responsible manner. If members of a community feel obligated to look after the environment then they are more likely to do so.

7.6.2 International Trade Arrangements

Whether or not freer trading arrangements will ultimately be to the benefit of water users depends upon the degree of trade liberalisation that occurs and what exceptions are made. As agriculture uses around 70 per cent of all water extracted for consumptive purposes, and large amounts of water are embodied in many of the agricultural products traded, this policy option deserves careful consideration. When trade is unrestricted and all inputs priced at full cost, communities have the opportunity to take advantage of the relatively abundant sources of water in other parts of the world. When trade in agricultural products is restricted, water use is likely to be less efficient. Fewer crops can be grown per drop of available water. As a whole, the world is generally worse off. However, some countries strive for food sovereignty for various reasons including security.

In an attempt to understand the likely impacts of freer trading arrangements on water use, a background paper prepared by UNEP uses a model to estimate the likely effectsof agricultural trade liberalisation on water use. It concludes that trade liberalisation:

- Increases the quantity of agricultural products traded and the capacity of nations to trade with one another with the consequence that global capacity to adjust to climate change is greater than it otherwise would be;
- Tends to reduce water use in water-scarce regions and increase water use in water-abundant regions, even though water markets do not exist in most countries; and
- Makes each nation more responsive to changing conditions and, as a result, reduces the negative impacts of climate change on global welfare by 2 per cent. Regional changes, however, are much larger than this.

7.6.3 Using Market-based Instruments

Market-based instruments that can be harnessed to foster a green economy include:

- Payments for Ecosystem Services (PES);
- Consumer-driven accreditation and certification schemes that create an opportunity for consumers to identify products that have been produced sustainably and pay a premium for access to them; and
- Arrangements that send a scarcity signal including the development of offset schemes, the trading of pollution permits and the trading of access rights to water.

Each of these approaches has direct application to the water sector and the degree to which communities are likely to become interested in maintaining and investing in the provision of ecosystem services.

7.6.4 Improving Entitlement and Allocation Systems

The last class of market-based instruments of particular relevance to water are those that use water entitlement and allocation systems to allow adjustment to changing economic and environmental conditions by allowing people to trade water entitlements and allocations.

In well-designed systems, water-resource plans are used to define rules for determining how much water is to be allocated to each part of a river or aquifer and a fully-specified entitlement system is then used to distribute this water among users. Under such an arrangement rapid changes in supply conditions can be managed efficiently.

7.6.5 Reducing Input Subsidies and Charging for Externalities

In some cases, subsidies can be justified but unless implemented with great care, they can have a perverse effect on progress towards the greening of an economy. In most cases, subsidies encourage the exploitation of water at unsustainable rates. In India's Punjab Province, for example, electricity for groundwater pumping is supplied to farmers either at a heavily subsidised price or for free. Experience is now showing that these subsidies encourage farmers to pump much more water than otherwise would be the case and, as a result, water levels in 18 of Punjab's 20 groundwater districts are falling rapidly. Officials are aware of the adverse effects of subsidising electricity to this extent but have been unable to find a politically acceptable way to phase them out.

Processes that attempt to reflect the full cost of electricity use include funding research on the adverse effects of providing these subsidies and stimulating public debate about the wisdom of continuing to do so. If this research is rigorous and the communication strategies well developed, it is hoped that ultimately there will be sufficient political pressure to enable these subsidies to be removed. As soon as this starts to happen, the money saved can be used to invest in other more sustainable activities. An alternative, much more expensive approach is to build a separate rural power supply system so that access to electricity can be rationed.

7.6.6 Improving Water Charging and Finance Arrangements

As noted by the OECD, water-supply pricing policies are used for a variety of economic, social and financial purposes. Ultimately, water policies need mechanisms that distribute water to where it is needed, generate revenue and channel additional sources of finance.

From a greening economy perspective, it is recognised, however, that there is little agreement about the best way to charge for access to water and sanitation services.

A. Sources of Revenue

Known as the "3 Ts," in essence, there are three ways to finance water infrastructure and the costs associated with operating that infrastructure:

1. Users can be charged a tariff for the water provided to them;
2. *Tax revenue* can be used to subsidise operating costs and cover capital costs; and
3. Grants and other forms of *transfer payment* can be sourced from other countries.

In OECD countries, it is now common for urban water-supply utilities to set a tariff that is sufficient to cover the full operating costs of supplying water.

B. Charging for Access to Water

Shifting to a green economy usually involves a commitment to begin charging for the full costs of resource use. With regard to water, however, there is a dilemma as access to clean water and adequate sanitation services is a human right. In a green economy, the efficient use of resources is encouraged, as is investment in built infrastructure. There is also an emphasis on equity.

C. Financing Access for the Poor

In an environment where a large number of children die as a result of lack of access to adequate water, what is the right tariff to set?

Seeking empirical evidence found that access to education and to clean water are the most consistent predictors of economic progress. Having analysed the data and, particularly, the high costs of delaying access because of revenue shortfall, one can observe that if countries cannot afford to make drinking water available at less than full cost of supplying it to all poor people, then an alternative approach is to focus on the efficient provision of water to all poor people at the cost of supply. From a green economy perspective, the strategy to pricing to adopt is the one that most speeds the transition.

D. Cross-subsidising (Selectively Taxing) Water Use

In many countries, the water tariff regimes are used to cross-subsidise the cost of supplying water to the poor. As a transitional strategy in countries with little other capacity to transfer wealth from the rich to the poor, a case can be made for the use of cross-subsidies, even though this approach distorts investment in water use. In developed countries, however, the use of a water charging regime to transfer income from one group of people or one region to another is extremely inefficient. For this reason alone, subsidies should be targeted as tightly as possible and accompanied by a transparent strategy for their removal. The result is the emergence of a regime that encourages investment and innovation. Infrastructure is located in places where its use can be sustained.

E. Increasing Private-sector Participation

As a transition to efficient supply of water at full cost occurs, opportunities for the involvement of private enterprise in the provision of water supply and sanitation services increase. The main reason for considering such arrangements is that research is showing that private-sector engagement can help to deliver benefits at less cost and thereby release revenue for green growth in other sectors. Once again, this opportunity is controversial. Several private-sector participation arrangements have failed. Nevertheless, there is little to suggest that the frequency with which these problems occur is less than that found among publicly-run systems.

■ ■ ■

8

Guidelines for Green Forest

8.1 INTRODUCTION

In 2006, the forest industry (defined as roundwood production, wood processing, and pulp and paper) contributed approximately US$ 468 billion or 1 per cent of global gross value added, of which pulp and paper represented about 40 per cent. Although this was an increase in absolute terms from 1990, the share of the forest sector declined due to the much faster growth of other sectors. Nevertheless, the forest industry is extremely important for some developing countries. Not captured in these figures on GDP share are the contributions made by forest ecosystem services to human wellbeing and the role of forests in sustaining livelihoods. With a broader concept of GDP, such as the GDP of the poor, which captures the reliance of rural populations on nature, the contribution of the forest sector is greatly increased.

Besides wood products and paper, the world's forests also produce a large amount of the energy used in developing countries, particularly among low-income households. About half of the total roundwood removed from forests worldwide is used for energy, including traditional heating and cooking and for heat and power production in industrial operations. More than 2 billion people depend on wood energy for cooking, heating and food preservation. Figures on biomass energy (wood plus crop residues and animal dung) from Openshaw give an indication of the economic and social importance of the energy derived from wood. According to the International Energy Agency (IEA), for the world as a whole, biomass energy accounted for an estimated 10 per cent of primary energy in 2005 (47.9 ExaJoule (EJ), of which 39.8 EJ were in Least Developed Countries (LDCs). But in many developing countries it dominates, with over 50 per cent of total energy use. Although much of it is used by the subsistence sector, in many countries biomass energy is the most important traded fuel, both in terms of employment and value. In sub-Saharan Africa, biomass fuels account for as much as 80 per cent of energy consumption.

Forests are also home to important non-wood forest products (NWFPs) that make a significant contribution to local economies and livelihoods; in some cases NWFPs are important exports. The main product categories are food from plant products, raw material for medicine and aromatic products and exudates such as tannin extract and raw lacquer. It has been estimated that in 2005 the value of NWFPs extracted from forests worldwide amounted to US$ 18.5 billion, but this was believed to cover only a fraction of the total value because of incomplete coverage of the statistics. Numerous studies have shown the importance of the subsistence use of NWFPs for people's livelihoods. In a review of 54 case studies, over half of which were from Eastern and Southern Africa, Vedeld *et. al.* estimated that the average annual forest environmental income amounted to 22 per cent of household income. While a large part of this was from fuelwood, wild foods and fodder for animals were also important.

Forests, which sustain more than 50 per cent of terrestrial species, play a vital role in protecting watersheds and regulating climate (ecosystem services) and they have great cultural and symbolic significance. Valuation studies of these services conducted in many different countries have shown a wide variation in results, reflecting the importance of location, the methodologies and assumptions about biophysical linkages, *e.g.* between forest cover and watershed services. Studies that concentrate on the value of the climate-regulation services of forests associated with reducing deforestation also produce substantial estimates.

Scaling up from such wide-ranging values is challenging, and estimations of values at a national or global scale have produced huge ranges. While there is still a high degree of uncertainty about the value of forest ecosystem services at a global level, even conservative estimates tend to be high, measured in trillions of US dollars. This indicates the importance of taking these services into account in decision-making on land and resource use.

Forests also provide significant employment, with the contribution of the formal sector greatly outweighed by that of the informal sector. About 10 million people are employed in forest establishment, management and use worldwide. Adding employment in primary processing, pulp and paper and the furniture industry brings the figure to about 18 million people. Despite growing informality and mechanisation, forestry is still a highly significant sector, with roughly 0.4 per cent of the global workforce. Outside of the formal sector there is greater uncertainty about the number of people dependent on forests for employment and livelihoods. As a result, the estimate for the total number of people dependent on forests ranges from 119 million to 1.42 billion. But even conservative estimates of people engaged in informal forest enterprises, indigenous people dependent on forests and people dependent on agroforestry, greatly exceed employment in the formal forest sector.

There are regional variations, however. The employment role of the sector has been declining, particularly in Europe, East Asia and North America, most probably because of gains in labour productivity. The only countries in Europe that have increasing employment in the forest industry sector are Poland, Romania and the Russian Federation. Latin America and the Caribbean and the developing Asia-Pacific region are the two regions where the forest industry sector has been expanding on all fronts over the last decade. This has been driven by various factors, including the abundance of low-cost, skilled labour, relatively abundant forest resources, a high rate of economic growth, specific polices to encourage development and investment in the sector and a general improvement of the investment climate.

The production and trade of fuelwood is also important for employment. Openshaw, while noting that there are no definite estimates, suggests that nearly 30 million people worldwide may be involved in the commercial production, transport and trade of biomass-energy products, generating around US$ 20 billion annually. More specifically, a survey in Malawi in 1996/97 found that 56,000 people were involved in tree growing, fuelwood and charcoal production, transport and roadside and urban trading in the country's four principal towns. This was many times greater than the number employed in kerosene, liquefied petroleum gas (LPG) and electrical production, transport or transmission and trading for the household sector, estimated at 350 to 500. A repeat survey carried out in 2008 found that employment in growing, production, transport and trade of biomass energy had increased significantly to 133,000.

8.2 WORLD'S LAST INTACT FORESTS ARE BECOMING INCREASINGLY FRAGMENTED

Intact forest landscapes (IFLs), or vast stretches of unbroken forest wilderness, are some of the most important ecosystems in the world. Their intactness makes them uniquely valuable to nature—they regulate temperature and rainfall across continents, provide homes to indigenous peoples, have some of the highest biodiversity rates and store huge amounts of carbon, among other benefits.

But they're also under threat. According to a new report in Science Advances, the amount of forest that meet IFL criteria decreased in size by 7.2 per cent between 2001 and 2013. This is an area more than twice the size of California that is no longer considered "intact forest." Furthermore, the rate of IFL area reduction increased during the study period.

Fragmentation by new road networks from industrial timber and oil and gas extraction, agricultural expansion and fires are the main causes of lost intactness. If the trend doesn't reverse, the world could lose much of its unbroken forest tracts this century, posing problems for nature, the climate and human well-being.

1. Where Are Intact Forests Being Fragmented?

Countries with the largest remaining areas of IFLs are found primarily in boreal and tropical regions. These areas also experienced the greatest amount of fragmentation over the past decade.

Russia had the greatest area of IFL reduction, thanks largely to oil and gas extraction linked to extensive human-caused fires. In Brazil, agricultural and pasture expansion drove IFL fragmentation. In Canada, fires were the leading driver, but unlike Russia, scientists believe they started due to natural causes. Paraguay lost nearly 80 percent of its IFL area.

Timber harvesting and agricultural expansion were the leading drivers of IFL area reduction in the tropics, while timber harvesting was the leading cause of loss in the southern boreal and temperate zones.

2. The Problem with Losing Forest Intactness

IFLs are unique and irreplaceable. Intactness cannot easily be regained—once altered, complex assemblages of biodiversity would be nearly impossible to restore. The extension of new roads into remote forest areas facilitates human access, and can lead to an irreversible cascade of changes—including wildlife habitat loss, unregulated hunting, increased human settlement and population density, forest clearing from logging, forest loss from agricultural expansion, fires, and other changes that degrade forest and may eventually convert vast forest areas to other land uses.

Intact forests also offer a unique opportunity to mitigate two of the greatest environmental challenges Earth faces: climate change and extinction. On average, tropical IFLs store triple the amount of above-ground carbon of non-intact tropical forests. Globally, scientists estimate that tropical rainforests contain half of the earth's species of plants and animals despite occupying only 2 per cent of global area.

Boreal forests have huge soil carbon stores. While they store less above-ground carbon than their tropical counterparts, boreal IFLs occupy a vast area and include the largest intact forest areas worldwide, making up 36 per cent of global IFL area. These IFLs play an important role in the protection of permafrost, critical to preventing the release of climate-warming carbon and methane locked in soils.

3. What to Do?

To date, international agreements and national forest commitments generally do not address preservation of intactness as a high priority. This lack of political will is a major factor facilitating the continued exploitation of IFLs.

While major international forest commitments like the Convention on Biodiversity Aichi Biodiversity Targets, UN New York Declaration on Forests and UN Sustainable Development Goals have put forth ambitious goals to halt forest loss in short time frames, they lack sufficient guidance on the means to achieve them. IFLs are arguably the most important forest areas for prioritization in achieving many of these targets, and offer a measurable benchmark to making progress towards goals. The latest research in Science Advances also shows that designating IFLs as protected areas has proven effective at limiting their fragmentation, so countries should consider IFLs as high-priority candidates when existing protected area networks are revised or expanded. Moving agricultural expansion to already-degraded lands, as well as increasing the yields of existing agricultural lands, can also help reduce pressure on remaining intact forests.

The Global Forest Watch (GFW) platform offers tools and data to monitor and share information on how forest change is impacting IFLs. IFL boundaries are available on GFW, and can be combined with near-real time tree cover loss and fire alerts to monitor threats in near-real time. Users can subscribe to receive email notifications of new alerts within an area of interest.

GFW researchers also working to identify recent forest loss within important forest areas, such as IFLs or protected areas. These "Places to Watch" will soon be available, and will be distributed regularly to local NGOs and journalists to help facilitate interventions.

8.3 VISION FOR THE FOREST SECTOR IN A GREEN ECONOMY

Greening the forestry sector implies managing it and investing in it as an asset class that produces a wide range of benefits to society. The wider economic roles of forests in a green economy include: as factories of production (producing private goods from timber to food), as ecological infrastructure (producing public goods from climatic regulation to water-resource protection) and as providers of innovation and insurance services (forest biodiversity being key to both).

The greening of the forest sector will be driven by societal demands for ecosystem services spread across several sectors, encompassing the traditional industries of wood processing and paper manufacture as well as tourism, energy, water management, carbon trading and new forest-based products. Forestry in a green economy will also meet critical livelihood needs of local communities by providing a stream of fuelwood, construction materials, food sources and medicinal plants. Effective local control and management of forests need to be improved but governments, through access and benefit-sharing (ABS), and new markets, such as ecosystem services, will ensure there are greater economic incentives to do so. These incentives would emerge from a robust and fair international system that ensures forest-related public goods, notably carbon storage and biodiversity conservation, are transferred between nations. Forests would also attract interest from financial institutions opening up forests as a new economic asset.

With greater understanding and recognition of the public goods generated by forests, and the increasing financial rewards for producing them, it becomes critical for forest managers and governments to account more effectively and transparently for forest stocks and flows. This entails being able to measure and value the forest sector's contribution to societal wellbeing in more sophisticated ways and capturing the full range of marketed and non-marketed goods and services, including the significant contribution they make to the livelihoods of the poor and marginalised.

8.3.1 Indicators

In order to assess how far the forest sector is shifting towards a green economy, it will be important to keep track of indicators that measure the following: (1) the changing proportion of consumption made up by forest goods and services, and particularly the rate of substitution of carbon-intensive products with forest products; (2) changing markets for forest ecosystem services; (3) investments in sustainable forest enterprise and production, especially those which aim at several ecosystem services and include sustainability conditions; (4) the changing ownership of forest land and forest enterprise, notably the inclusion of local forest stakeholder groups; (5) forest governance improvements; and (6) the sustainability of forest management, from stand to landscape to national levels, in environmental, social and economic terms.

8.4 CHALLENGES AND OPPORTUNITIES

The major challenges facing the forest sector include the loss of forest, competing land uses, market, policy and governance failures. These challenges are connected. Competing land uses, especially from agriculture, are immediate causes of forest loss. These competing land uses are, in turn, driven by market, policy and governance failures.

8.4.1 Trends in Forest Cover and Deforestation

There are clear signs that forests are not being sustainably managed. Table 8.1 shows that the world's forested area is declining both in absolute terms (deforestation) and in net terms (taking account of forest planting and natural expansion), although at a slower rate than in previous decades. Changes in total forest area at the global level, however, mask regional variations. Forest cover stabilised in North and Central America and expanded in Europe and Asia, in the latter case mainly owing to large-scale afforestation in China, which offset continued deforestation in Southeast Asia. Africa and South America underwent the largest net loss of forests in this period (2000–2010) and Oceania also experienced net loss.

In its latest Forest Resource Assessment, FAO revised upwards its deforestation estimate for the 1990s. In the Forest Resource Assessment 2005, deforestation in the 1990s was estimated at 13 million hectares per year.

Trends for different types of forests are also important. Of most concern is the decline in primary forests, 40 million hectares of which have been lost or modified since 2000. In contrast, planted forests are expanding more rapidly, with a 50 per cent increase in the growth rate over the previous decade, and now account for 7 per cent of the total forest area worldwide. This expansion—explained by the forest transition theory—is expected to continue. Carle and Holmgren predict that the area of planted forest in 2030 will reach between 302.7 million hectares and 345 million hectares, depending on assumptions about productivity increase. Three-quarters of all planted forests consist of native species, although introduced species are more common in a number of countries with large areas of planted forests across subSaharan Africa, Oceania and South America.

8.4.2 Competing Uses of Land

Agricultural expansion, often combined with timber extraction and the expansion of infrastructure, which facilitates access, has been found to be the main proximate cause of deforestation in tropical areas over the last two decades. Increasing population, increasing income and shifts in tastes to more meat-based diets are forecast to increase the demand for food by 70 per cent (in value terms) by 2050. To meet this demand, further clearing of forest will be required unless agricultural productivity can continue to rise significantly. Increasing demand for biofuels means they will compete with food crops for land, putting further pressure on forests. Climate change, where it has an adverse impact on agricultural yields, will add to the pressure for converting forests to agricultural land. It also affects forests directly through changes in their growth rate or in fire propensity.

8.4.3 Market, Policy and Governance Failures

Underlying the loss of forest and competing land- uses are governance and market factors that render deforestation a rational (and often legal) course of action, irrespective of the environmental and social costs. Governance drivers include the lack of forest rights for local stakeholders, which discourage local investment in intact forests and which enable appropriation of land and/or forest resources by more powerful outsiders. These are compounded by market failure, as not all of the important ecosystem services provided by forests are captured in markets. Those taking decisions on the practices used in timber extraction and conversion of forests to other land uses do not factor in the adverse effect on the provision of ecosystem services. Because maintenance of these other ecosystem services is not usually rewarded, there is very little incentive for forest managers to take them into account.

Table 8.1: Trends in forest cover and deforestation

	1990	2010
World forest area (hectares)	4.17 billion	4.03 billion
World planted forest area (hectares)	178 million	264 million
	1990–2000	**2000–2010**
Annual net forest loss (hectares/year)	8.3 million	5.2 million
Annual deforestation (hectares/year)	16 million*	13 million
Annual increase in planted forest (hectares/year)	3.6 million	4.9 million

Source: Compiled from data in FAO (2010)

** In its latest Forest Resource Assessment 2010 FAO revised upwards its deforestation estimate for the 1990s. In the Forest Resource Assessment 2005 (FAO 2005a), deforestation in the 1990s was estimated at 13 million hectares per year.*

8.4.4 Opportunities

Together with the challenges facing the forest sector, there are also opportunities for greening the sector. They include the establishment of sustainable forest management (SFM) criteria and indicators, the growth of protected areas, the concept of reducing emissions from deforestation and forest degradation (REDD+) and the growing acceptance of payments for ecosystem services (PES).

8.4.5 Sustainable Forest Management (SFM)

Although there is no consistent, routine and comprehensive assessment of forest management globally, considerable effort has gone into developing SFM criteria and indicators to describe comprehensively the elements of good practice. They cover the economic, social/cultural, environmental and institutional dimensions of SFM, based on scientific and technical knowledge of forest systems. Regional criteria include those of the International Tropical Timber Organization (ITTO), which apply to all its member countries. Recent initiatives led by civil society groups and some forest companies and industry associations have developed voluntary SFM codes of practice and management guidelines. Certification schemes provide an independent assessment of adherence to the standards and statistics on them provide an indication of the extent of best practice, although lack of certification does not necessarily imply bad practice.

Box 8.1: Forest transition theory

Globally, the area devoted to planted forests is growing. Planted forests are estimated to produce 1.2 billion m of industrial roundwood, which amounts to about two-thirds of all production. Further shifts in production to planted forests are expected. Improvements in technology mean that more and more can be produced per hectare of land. For example, eucalyptus plantings in Brazil have reached productivity levels exceeding 50 m per hectare. In view of such improvements, FAO predicts that growth in production from planted forests will keep pace with growth in demand for industrial roundwood. This can be expected to reduce the pressure on primary forest, although much of the latter could be lost by the time the switch to planted forest has taken place.

This growth of planted forests is explained by the forest transition theory and the stages of forest development (Hyde 2005, which draws on von Thunen's rent model; see also Angelsen 2007 who combines the von Thunen and forest transition theories). The theory suggests that countries start with high forest cover and as they develop, the forest is converted to other land uses, agriculture in particular. The process accelerates as infrastructure improvements open up frontier forest areas and makes timber extraction and agriculture economically viable. Over time, as timber becomes scarce, and as the economy develops, providing off-farm employment opportunities, a series of adjustments are made. It becomes profitable to manage forests and plant new ones. The area of forest cover starts to increase again.

This process has been followed by many developed countries and some developing nations, including Costa Rica, which is in the later stages of this transition. Similarly, Vietnam saw its forest cover decline from 43 per cent in 1943 to 20 per cent in 1993 as a result of agricultural expansion and migration into forested areas. Since then, considerable efforts have been made to increase forest cover, an ambitious programme of reforestation. By 2009 forest cover had increased to 39 per cent of the land area (FCPF 2010). In Vietnam, while forest cover has increased as a result of reforestation programmes, the quality of natural forests continues to be more fragmented and degraded (FCPF 2010). This is where valuation is important, as it would show the economic consequences of letting the standard forest transition takes its course.

There are other market adjustments in response to increasing scarcity of wood, in particular, increasing use of wood-processing residues and recovered paper and wood products. While global demand for wood and fibre is expected to almost double by 2030, global production of industrial roundwood is projected to increase by a more modest 40 per cent (FAO 2009).

Thus, taking this longer-term perspective, the concern about forests is not so much about the ability to provide the world's increasing demand for timber and fibre but about the ability to continue providing livelihoods for forest-dependent people outside of the formal economy and to continue providing non-marketed ecosystem services. The latter are currently unpriced and therefore largely ignored in management decisions to date. This raises the question of how to change the shape of this forest transition (Angelsen 2007). Is it an inevitable pattern of development or can a combination of policies ensure the retention of greater areas of primary forest cover? Neither the forest transition theory nor the land-rent model distinguish between forest cover of different types—*i.e.* primary forest and secondary forest, degraded forest and planted forest. The provisioning services of forest, such as timber and fibre, may be maintained through market adjustments, but other valuable ecosystem services could be lost.

Currently over 5 per cent of the world's production forests are certified under the Forest Stewardship Council (FSC) standard, at 133 millions hectares certified in 79 countries, including 77.6 millions hectares of natural forests, 12.5 millions of hectares of plantations and 43.3 millions of hectares of mixed natural/plantation landscapes (FSC 2010 data as of 15/04/10). Over 80 per cent of FSC-certified forests are boreal and temperate. Tropical and subtropical forests account for 13 per cent of the total FSC-certified area, with 16.8 million hectares.

The other major international forest certification scheme is the Programme for the Endorsement of Forest Certification (PEFC). Some 232 million hectares of forest are certified to PEFC's Sustainability Benchmark, nearly twice the area of FSC certification, although some forests are certified by the PEFC and FSC. Almost all the PEFC endorsed certified forests are in OECD countries, just under half in Canada with most of the rest in USA, Scandinavia and Brazil in the tropics. However, China is developing a national scheme and is expected to join the PEFC in 2011.

Table 8.2: Management status in tropical permanent forest estate (PFE) (2005, thousand hectares)*

	Africa	Asia and the Pacific	LA and the Caribbean	Total
Total closed natural forest (FAO 2001, thousand hectares)	208,581	226,984	788,008	1,223,573
Total area under permanent forest estate (PFE)	110,557	206,705	541,580	858,842
Percentage	53%	91%	69%	70%
Production PFE	71,286	135,726	190,331	397,343
	64%	66%	35%	46%
Natural production forests				
Total area	70,461	97,377	184,727	352,565
With management plans	10,016	55,060	31,174	96,250
Certified	1,480	4,914	4,150	10,544
Sustainably managed	4,303	14,397	6,468	25,168
Percentage sustainably managed	6%	15%	4%	7%
Planted production forests				
Total area	825	38,349	5,604	44,778
With management plans	488	11,456	2,371	14,315
Certified	–	184	1,589	1,773
Protection PFE	39,271	70,979	351,249	461,499
	36%	34%	65%	54%
With management plans	1,216	8,247	8,374	17,837
Sustainably managed	1,728	5,147	4,343	11,218
Percentage of PFE that is sustainably managed (excludes planted areas)	5%	12%	2%	4%

In 2005, ITTO found that only 7 per cent of its member countries' production forests (25 million hectares) were being sustainably managed. Whilst every ITTO producer-country's policies promoted sustainable management of forests in 2005, management plans existed for only 27 per cent of the 353 million hectares of production forests, and just 3 per cent were certified (Table 8.2). Despite the low level of sustainable management, however, this is a huge improvement on the mere 1 million hectares of all tropical forests that ITTO had assessed as sustainable in 1988. Furthermore, ITTO noted that some countries have made notable improvements, including Bolivia, Brazil, the Republic of Congo, Gabon, Ghana, Malaysia and Peru. There is still considerable room for improvement, in view of ITTO's conclusion that resources for enforcement and management are woefully and chronically inadequate, trained staff, vehicles and equipment are all in short supply, while systems for monitoring and reporting forest management are often limited or lacking.

In OECD countries, it is likely that there is a greater extent of sustainable management. The European Union estimates that 80 per cent of its forested area is under a management plan and 90 per cent of that area is managed sustainably: a large proportion of the area is managed by small private owners who have held the forest for generations. A majority of Canadian and many US production forests are certified. Although there are good examples of forest management in Russia, overlogging has occurred, especially in the Russian Far East, near the border with China.

It is also possible that a large proportion of small-scale informal forest enterprises (family forests, indigenous forests), which are beyond the scope of assessments like that of ITTO, are sustainably managed. This can be judged by the longevity of the forest resources, passed from generation to generation, and evident production of multiple goods and services. However, there is little information to go on, apart from the minority of forests that are certified.

1. Growth of Protected Areas

One apparently positive trend from the environmental perspective is that the area of protected forests is increasing. About 13.5 per cent of the world's forests are protected according to IUCN categories I-VI and 7.7 per cent (about 300 million hectares) for categories I-IV, involving more restrictions on land use. The area of protected forests has increased by 94 million hectares since 1990, of which two-thirds has been since 2000.

In Latin America designation of protected forests has been one of the most used strategies for the sustainable management of forests. It is estimated that there are 100 million hectares under IUCN categories I, II and III (which are the most restrictive) in Latin America and the Caribbean. Growth in protected areas has been particularly rapid since the 1980s. In sub-Saharan Africa, 32.5 million hectares of forests and woodland, corresponding to 5 per cent of the total forest area, are formally protected (IUCN categories I-VI) and as much as 8 per cent, if forestry reserves are included.

Box 8.2: The national PES scheme in Costa Rica

The Costa Rican Payments for Ecosystem Services programme (PSA, in Spanish) was created in 1996, through the Forestry Law 7575, which recognises the provision of ecosystem services from forests. Based on the beneficiary pays principle, it suggests that forest owners should be compensated for the following services:

- Mitigation of greenhouse gases (GHG) (reduction, sinking, fixing and storing carbon);
- Protection of water for rural, urban or hydroelectric use;
- Protection of biodiversity for conservation, scientific and pharmaceutical use; and
- Landscape beauty for tourism.

Forest owners are currently paid for several land-management practices, and all except agroforestry are paid per hectare over five years: forest conservation (US$ 320), offering higher payments in hydrologically-sensitive areas (US$ 400), areas identified as "conservation gaps" (US$ 375), reforestation (US$ 980), forest management (active before 2003 and again in 2010, receiving US$ 250); forest regeneration, which could be in areas that meet the additionality criteria (US$ 320), or not (US$ 205); and agroforestry (US$ 1.3 per tree, paid over three years).

In order to finance this program, FONAFIFO (Fondo Nacional de Financiamiento Forestal or National Forestry Financing Fund) receives funds from different funding sources: public funds in the national budget, donations, credits conceded by international organisms, private funds, own generated funds and timber and fuel taxes. Also, in 2001 FONAFIFO created the Environment Services Certificate (ESC), which is a financial instrument where FONAFIFO receives funds from companies and institutions interested in compensating forest owners for preserving forests.

Between 1997 and 2008 FONAFIFO distributed US$ 206 million, an average of US$ 17.2 million per year (Porras, 2010). The majority of funds were for forest protection (73 per cent), covering 460,000 hectares of forest, and almost 6,600 contracts were signed across the country.

It should be noted, however, that although there has been a marked expansion in protected areas, there is no guarantee that they will be well-enforced. This is evidenced by the continuing loss of forests and other natural ecosystems within protected areas. Effectively enforcing the land and resource-use restrictions in protected areas is challenging and many are being encroached on, particularly in densely populated countries. Unsustainable land uses within protected areas are another cause. Strassburg and Creed , in a study of 133 countries in Latin America, Africa, the Middle-East, Asia and Eastern Europe, estimate that only one-third of the protected forest area is effectively legally protected, corresponding to 6 per cent of the total forested area in these countries. Of the five regions examined, Latin America has both the highest proportion of legally protected forests (24 per cent) and effective legal protection (9 per cent).

2. Payments for Ecosystem Services (PES) and REDD+

New, incentive-based approaches to conserving forests have emerged over the last 10 to 15 years. The most high-profile of such initiatives are PES, which pay forest landowners for providing watershed protection, carbon storage, recreation, biodiversity, etc. These range from local-level schemes, such as the local government in the town of Pimampiro in Ecuador, which makes payments ranging from US$ 6-US$ 12 per hectare per year to a small group of farmers (19 in 2005), to conserve forest and natural grassland in the area surrounding the town's water source , to national schemes such as in Costa Rica, where farmers are paid US$ 64 per hectare per year in five year contracts (to protect biodiverse forests and global schemes *e.g.* a range of voluntary carbon offset schemes for planting or conserving trees to fix CO_2 and store it. Some environmental payments schemes also factor in social needs, attempting to persuade poor and marginalised groups to become engaged in providing the service, for example the schemes developed under the RUPES programme in Asia (Rewarding the Upland Poor in Asia for Environmental Services they Provide).

One of the most long-standing global payment schemes is the Noel Kempff Mercado Climate Action project in Bolivia, which was developed as a pilot project in 1997 under the Activities Implemented Jointly (AIJ) programme of the UNFCCC. A consortium of international and local NGOs, some US energy companies and the Bolivian Government bought out local timber concession holders and implemented a community development programme in order to extend the Noel Kempff Mercado Park. Through avoided deforestation the project was expected to avoid emissions of up to 3.6 million tonnes of carbon over 30 years.

While PES is primarily associated with developing countries, there are some well-known examples in industrialised countries. The New York City water utility - faced with the need to improve water quality—provides incentives to farmers and owners of forest land in the catchment areas to conserve the forest and adopt agricultural environmental management measures. This proved far less costly than building water-filtration systems (Landell-Mills and Porras 2002). In north-east France, the mineral- water producer, Vittel, paid local landowners to conserve the watershed.

Until recently, the main driver of investment in PES schemes involving forest conservation was the need to protect watersheds. The rules of the Clean Development Mechanism (CDM) limited eligible forest carbon activities to afforestation and reforestation. This meant that carbon projects based on forest conservation were confined to the voluntary carbon market. But as the contribution of deforestation and forest degradation to GHG emissions has become recognised, this approach to mitigation has moved up the agenda in international climate negotiations, first as REDD (reducing emissions from deforestation and degradation) and more recently as REDD+, which adds conservation, sustainable management of forests and enhancement of forest carbon stocks to the list of eligible activities. REDD+ has been likened to a multi-layer PES scheme, with transfers of finance between industrialised countries and developing countries in exchange for emission reductions associated with improvements in forest protection and management, and further transfers from the national level to forest landowners and communities (Angelsen and Wertz-Kanounnikoff 2008). Although PES will not be the only strategy used by governments to achieve forest-based emission reductions, it is likely to be important.

Unlike the project-based approach of international PES to date, REDD+ is likely to involve more national- level approaches, with finance being supplied by developed countries individually or as a bloc against the performance of national-level commitments to reduce deforestation and emissions. This is exemplified by Norway's contribution to the Amazon Fund in Brazil, which is conditional on the achievement of deforestation- reduction targets. In 2010 Norway announced a grant of US$ 1 billion to Indonesia in return for agreed measures to tackle deforestation and degradation. Indonesia, under the terms of the agreement, has accordingly announced a 2-year moratorium on new permits to clear natural forests and peatlands (Richardson 2010). The sums of money being estimated for full implementation of REDD+ amount to tens of billions of US$ worldwide. Already, the financial support committed for preparation activities and bilateral programmes greatly exceed what has been provided so far in PES, providing grounds for optimism that this new mechanism can capture and transfer important new resources for ecosystem services provided by forests.

8.5 THE CASE FOR INVESTING IN GREENING THE FOREST SECTOR

As indicated in the last section, there are promising developments such as certification of sustainable forest management, targets to increase protected areas and the growing momentum of PES and REDD+ schemes. But without a major change in the recognition given to the full suite of forest ecosystem services, in particular in climate negotiations, and in the absence of improvements in the agriculture sector, loss of primary forest is likely to continue. Protected areas will continue to expand but a large proportion will not be effectively enforced. The forest sector will meet the market demand for timber through planted forests and efficiency improvements in processing, but pressures on natural forests from other sectors, agriculture in particular, will continue, exacerbated by climate change. As a result, ecosystem services will continue to be lost.

Additional resources and policies are therefore needed to internalise the value of forest ecosystem services for forest landholders and ensure forests are worth more standing than cleared (Viana 2009). Investments targeted at increasing the profitability of sustainable harvesting techniques and making tree planting worthwhile can also make a contribution. This section reviews a range of investment options for greening the forest sector and identifies the economic, social, and environmental effects of these options.

8.5.1 Options for Green Investment in Forests

Some broad categories of green private and public investments can be distinguished for the main forest types, including agroforestry, as shown in Table 8.3. Green investment can be targeted at reversing the loss of forest area by conserving existing areas of primary forest or promoting expansion of forests through regeneration and reforestation. Green investment can also be directed to improving management in existing forests and agroforestry systems to ensure they continue to provide a wide range of ecosystem services. Such investment can only be considered green if it ensured that the forests conserved, established or restored meet principles of sustainable forest management, and

Table 8.3: Green investment options for various forest types

Forest type	Investment	
	Private*	Public**
	Ecotourism development	Create new protected areas
Primary forest	Private nature reserves	Improve enforcement of protected areas
	Pay landowners to protect watershed	Pay forest landholders to conserve forests
	Buy out logging concessions	
Natural modified forest	Reduced impact logging and other forest management improvements	Incentives for improved forest management
	Certification to sustainable forest management standards	Support establishment of certification systems
	Control illegal logging	
	Reforestation and afforestation for production	Incentives for reforestation/ afforestation
Planted forest	Improve management of planted forests	Incentives to improve management
	Reforestation to protect ecological functions	
	Extend the area with agroforestry systems	Incentives to landholders
Agroforestry	Improve management of agroforestry systems	Incentives to improve management Technical assistance

* Private could also include investments made by communities.
** Some of the public investments listed here may also be made by the private sector, often on a more limited scale.

balance the needs of different stakeholders. For example, creating a protected area that displaces forest-dependent communities would not meet the principle of supporting relevant socioeconomic functions. Moreover, creating a protected area does not guarantee enforcement. Similarly, extending the forest area through tree planting may be contentious if it uses a large amount of external inputs and directly or indirectly displaces local people from their land.

Some of the green investments listed in Table 8.3 are straightforward to quantify, although there will be considerable variation by location and species. Some of the public sector investments are not well-documented, in particular the amounts being spent on controlling illegal logging.

Because of the public-good nature of some forest ecosystem services, the private sector and holders of forested land are not always able to perceive a sufficient incentive to make green investments in forests, even if such investments often involve a positive rate of return for society as a whole. Investment by the public sector is therefore needed in some cases to provide forest ecosystem services directly, to provide financial incentives to the private sector to make green investment competitive and to prevent unsustainable forest management, *i.e.* by controlling illegal logging. The return on investment for the public sector is measured in terms of social and environmental benefits. Research carried out as part of TEEB on the costs and benefits of investing in ecological infrastructure indicates that the rate of return could be very high, with a benefit cost ratio of over 13 to 1 in the case of active restoration of eucalyptus woodlands and dry forest in Australia, and over 30 to 1 for restoration of Atlantic forest in Brazil.

8.5.2 Investing in Protected Areas

The creation of protected areas to restrict access and certain land-use practices has been the dominant approach used by governments to secure ecosystem services by controlling deforestation and forest degradation. In some cases the investment in protected areas may be made by NGOs. A well-known example is the conservation concessions whereby conservation organisations lease forest lands that would otherwise have ended up as logging concessions. Such concessions, mostly led by Conservation International but involving other major NGOs and donors, have been established in a number of countries, including Cambodia, China, Ecuador, Guyana and Madagascar. Private companies do sometimes operate protected forest areas, usually where there is a tourism interest or where the public sector is providing an incentive. In Brazil, for example, private landowners that set aside a protected area can receive a reduction in land tax.

The investment involved for the protected area authority, whether government, NGO or private sector, includes the administrative costs of demarcating and managing the area and keeping unauthorised users out. For the owners and users of the protected forest land it means forgoing timber royalties and giving up the net benefits from agriculture and other land uses that compete with forests. This latter cost has rarely been factored in, except where compensation schemes operate.

Balmford *et. al.* (2002) estimated current expenditure on protected areas at US$ 6.5 billion per year, of which half was spent in the USA. A more recent estimate suggests this could range from US$ 6.5 to US$ 10 billion per year. These estimates do not distinguish between forest ecosystems and other ecosystems in the protected areas. For example Mullan and Kontoleon (2008) cite an estimate by Bruner et al. (2003) of US$ 8 billion of total expenditure on protected areas, of which approximately 60 per cent covers forested land. This suggests a little under US$ 5 billion per year or US$ 16.7 per hectare (assuming IUCN categories I-IV) is being spent on protected forests.

Many protected areas do not receive adequate funds to ensure their effective management. Very little is spent on compensation to those local communities who lose access to land and resources when protected areas are created. Protected areas are a vital part of the management of forest ecosystem services, but they need to address concerns over ineffective enforcement and share benefits with local communities. Estimates made of the cost of effective enforcement of protected areas with compensation for local communities are two to three times the amount currently spent (Box 8.3). Increased investment is needed to ensure better integration of communities' interests and to improve effectiveness along with better buffer- zone management.

Box 8.3: Costs of effective enforcement of protected areas

The total annual cost of managing the existing network of protected areas effectively was estimated in 1999 to be around US$ 14 billion per year. This included increasing management costs (then estimated at US$ 6 billion) by over a third and introducing compensation payments to communities living in protected areas of some US$ 5 billion (James et al. 1999). A later estimate of US$ 20–28 billion (Balmford et al. 2002) added the cost of up-scaling protected areas to ensure protection of 15 per cent of land area in each region. Assuming that forests constitute 60 per cent of terrestrial protected areas, this would suggest a cost of US$ 12–17 billion per year for effective management of protected forests.

Investing in protected areas may bring economic benefits to the national economy in the long term. Some countries have been able to build up a lucrative nature-based tourism industry, which has brought in foreign exchange and generated employment. For example Costa Rica, where protected areas received more than 1 million visitors per year in the five years up to 2006, generated entrance-fee revenue of over US$ 5 million in 2005 and directly employed 500 people. Protected areas in Latin America receive large numbers of visitors and generate many associated jobs. For example, Mexican protected areas recorded 14 million visitors per year and 25,000 jobs.

Nature-based tourism is also a major economic activity in sub-Saharan Africa and the number of tourist arrivals is growing faster than the global average (in 2004 at 14 per cent compared with 10 per cent worldwide). In the Great Lakes region, revenue from tourism based on gorilla viewing and other activities brings in about US$ 20 million annually. But the tourism industry in Africa also has human and environmental costs, contributing to the displacement of communities, thus undermining rights and livelihoods.

Admittedly, setting aside forests as protected areas has often been controversial because it is seen as preventing more productive activities such as timber harvesting and agriculture and as being damaging to livelihoods and to human rights, particularly where indigenous people are involved. Adverse social impacts of protected areas identified by these authors include: displacement of local communities, changes in traditional land tenure, denied or restricted access to resources, loss of employment, crop damage and livestock predation.

Cost-benefit studies have been conducted for protected forests in different regions. These examine costs and benefits at local, national and global levels but are not able to monetise all of the social costs identified above. While there is some variation, a number of the studies conclude that global benefits and sometimes national scale benefits outweigh the overall costs including the tangible opportunity costs to local communities. For example, the protection of the Virunga and Bwindi afro-montane forests of Eastern and Central Africa—home of mountain gorillas - show positive benefits as opposed to costs, but most of them accrue to the international community. Overall, gorilla tourism generates US$ 20.6 million per year in benefits, with 53 per cent accruing to the national level; 41 per cent to the international level, and only 6 per cent locally.

Another study, one of six reviewed by Coad et al. (2008), examines the costs and benefits of the Ranomafana National Park in Madagascar, which was created in 1991. It finds that the opportunity costs to local communities amounted to US$ 3.37 million or US$ 39 per household per year, but were greatly exceeded by the global- and national-scale benefits. Earlier studies of the Mantadia National Park Madagascar and Mount Kenya National Park in Kenya reached similar conclusions.

These studies indicate that, in theory, those gaining from the protected areas should be able to compensate local communities and still be better off. Historically, this compensation to communities has rarely happened. This highlights a challenge and an opportunity in a green forest sector for capturing the global benefits and creating redistribution mechanisms that are able to compensate local communities and improve their livelihoods.

As far as environmental effects are concerned, although the creation of a protected area does not guarantee environmental effectiveness and many are being encroached on, there are positive examples suggesting that this investment option merits further attention. Protected areas are considered critical for conserving residual tropical-forest biodiversity. Studies in South-East Asia show that parks and reserves consistently recorded larger numbers of endemic bird species and higher population densities than surrounding human-modified areas.

Figueroa and Sanchez-Cordero evaluated the effectiveness of Mexican Natural Protected Areas (NPAs) for preventing deforestation. They constructed an effectiveness index, based on the protected areas' percentage of transformed areas, the rate and absolute extent of change in these areas, the comparison between rates of change observed inside the protected area and in an equivalent surrounding area, and between the NPA and the state(s) in which it is located. They found that over 54 per cent of NPAs were effective in preventing land-use or land-cover change.

8.5.3 Investing in PES

There are no precise statistics on the amount of money currently channelled into PES schemes, but Canby and Raditz estimate this as being hundreds of millions of US$. The majority of this money comes from governments directly or from international donor support. These funds cover two main types of cost: the payment to the landholder or forest concession holder, compensating for the opportunity cost of forgone land-use, along with the costs of any actions necessary for conservation such as fencing or employment of guards, and the transaction costs of designing, setting up and operating the payment scheme, including contract management, fund management, the transfer of funds and monitoring.

The evidence on the social and economic impacts of PES schemes is mixed, both in terms of the extent to which the poorest groups participate in the schemes and the extent of livelihood benefits for those who do. Evidence of impact on nonparticipants is particularly scanty, and largely confined to observations in Costa Rica where a high proportion of those receiving payments hire labour to carry out conservation-related work.

The two national PES schemes involving forest conservation in Costa Rica and Mexico provide contrasting experiences in terms of the nature of participants, reflecting to some extent differences in land and forest-tenure regimes. In Costa Rica, where most land is held privately, small farmers have very little participation in the PES scheme in spite of efforts made to prioritise the poorest regions. In Mexico, a high proportion of forest land is held as common property by local communities and even though criteria for selecting priority areas were primarily biophysical, the poorest groups were fairly well-represented. In 2003 and 2004, 72 per cent and 83 per cent respectively of the total paid out went to forests associated with marginalised population centres.

Local schemes such as at Pimampiro in Ecuador and Los Negros in Bolivia have achieved a fairly wide participation of local forest landowners, albeit over a small area, partly because they have been able to adapt to local circumstances (Porras *et. al.* 2008). In Los Negros, for example, the majority of landowners did not have clear land title, but the scheme went ahead on the basis of local recognition of farmers' landholding.

Analysis of the livelihood benefits of PES schemes in several Latin American countries has given varied results; in general they have been welcomed by participants. The cash payments, with some exceptions, appear to be relatively insignificant when compared with opportunity costs and household income (Porras *et. al.* 2008). This has led some researchers to conclude that the payments function more as support, providing recognition of existing good practice, rather than constituting a real incentive for land-use change.

Non-financial benefits, such as capacity building, strengthening of land and resource tenure are therefore often considered to be significant. For example, PES schemes have been found to strengthen resource management and social coordination capacities of the community institutions involved. Capacity building is commonly reported as a benefit from PES schemes (*i.e.* increasing agricultural productivity in Pimampiro, Ecuador; apicultural training in Bolivia measured at US$ 35 per participant. However, for Tacconi *et. al.*, there is little evidence available about the long-term impact of capacity-building activities, for instance whether new knowledge and skills were applied in practice.

The evidence on the effectiveness of PES in reducing deforestation is also mixed, reflecting difficulties in establishing a clear counterfactual of what would have happened in the absence of the scheme and in predicting the location of deforestation. The national scheme in Costa Rica reflects reductions in national deforestation rates after the scheme started, but much of the research on this scheme throws doubt on a causal link between the two (Box 8.4). The same can be said for the Mexico national scheme (PSAH). The only major study so far of this scheme, found that much of the land being put under payments was not at risk of being converted because of its low opportunity costs. In 2003, only 11 per cent of the participating hectares in the scheme were classified as having high or very high deforestation risk. This increased to 28 per cent in 2004 but fell again to 20 per cent in 2005.

A common thread in this research is the importance of targeting specific areas in improving the effectiveness of PES. Robalino *et. al.*, noting that in Costa Rica there was improvement in 2000–05 compared with the 1997–2000 period, argue that targeting areas affected by some deforestation pressure and including spatially-differentiated payments are two plausible next steps to improve the effectiveness of the scheme. This also points to the importance of developing monitoring and verification schemes and data collection (including the use of easily available GIS databases) that can help identify additional areas.

Box 8.4: Research on the Impact of PES on Deforestation in Costa Rica

In Costa Rica's Virilla watershed Miranda *et al.* asked PES participants about their motivations and found that many of them planned to retain their forests regardless of the scheme. But as forest clearance is prohibited by law, this may have influenced the responses of the landholders as they might not want to state openly that they would contemplate illegal activity. These responses only represent a snapshot in time. It is unclear how these motivations would change as macroeconomic and microeconomic conditions change. Another study examined the characteristics of land included in the PES scheme. In the isolated Peninsula of Osa, for example, it was found that land under protection contracts corresponds mainly to forest that may not be in direct danger of being converted because of its remoteness and difficult access.

Analysis by Sanchez-Azofeifa *et. al.*, at a national level found that although the average deforestation rate dropped from 0.06 per cent per year in 19861997, to 0.03 per cent per year in the first phase of the PES programme 1997-2000, there was no significant difference in the rate of deforestation between areas in the national PSA scheme and areas that were not. They suggest that this could reflect lack of targeting of areas under deforestation pressure and also the impact of previous forest conservation policies, including a 1997 legal restriction on forest clearing. Similar results were found in a more recent study by Robalino *et al.*, *i.e.*, the efficiency of PES in reducing deforestation between 2000 and 2005 was also low. Less than 1 per cent of the parcels of land enrolled in the programme each year would have been deforested without payments.

The PES experience also shows that while challenges have been faced in achieving environmental objectives and ensuring the participation of small-scale forest owners and marginalised groups, there has been considerable learning and adaptation to make improvements. In particular, ways have been found of including landowners without formal land title in PES schemes. The most important actions appear to introduce environmental and social criteria for targeting, actively promoting the PES option amongst groups that would not otherwise get involved and/ or to reduce transaction costs. The involvement of intermediaries or facilitating organisations that have a community development mission is also important.

The main constraint on the expansion of PES schemes has been the lack of funds to scale up from pilot projects. Even national-level schemes such that in Costa Rica have been constrained by lack of resources, with applications to enter the scheme greatly exceeding the funds available. If a REDD+ mechanism is negotiated, there will be a step change in the amount of funds available: the sums currently involved in the readiness phase are already significant.

However, if payment schemes are implemented at much larger scales and in locations where governance is weak, facilitators will have to guard against elite capture and more attention will have to be given to strengthening the land tenure of local communities. Attention to such safeguards will need to be a part of any investment in scaling up PES under REDD+.

8.5.4 Investing in Improved forest Management and Certification

This investment approach recognises the importance of the production of timber, fibre, and energy in natural forests; if managed well, they need not conflict with the provision of other ecosystem services. Moreover, the ability to generate returns from forests through timber harvesting that are high enough to compete with other land uses is an important factor preventing total conversion.

Since the early 1990s, various sets of timber-harvesting guidelines on Reduced Impact Logging (RIL) have been produced in different regions of the world, designed to reduce the adverse environmental impacts associated with tree felling, yarding and hauling. Some of the requirements of RIL imply higher costs for logging companies, in the form of new equipment, safety gear, technically qualified supervisors, reductions in the area harvested and/or the need to use helicopter or cable systems to log areas with steep slopes. Given the planning it entails, RIL should involve less wastage of saleable timber and there were high hopes when it was first promoted that it would be sufficiently financially attractive for logging companies to adopt it as part of their normal practice.

The evidence on its financial benefits is mixed though, reflecting the wide range of forest practices and conditions (see Box 8.5).

Box 8.5: Research on the Profitability of Reduced Impact Logging (RIL)

Studies of the costs and benefits of improved forest management produce conflicting results. Two studies in the Brazilian Amazon, in Tapajos National Forest and Paragominas have concluded that RIL can be highly profitable. But Putz *et al.* highlight other studies that have shown conventional logging to be more profitable or have given mixed results. They conclude that it is not possible to draw general conclusions about the financial viability of RIL because of the wide range of forest conditions and practices that influence profitability in the tropics.

An earlier review of cost information in over 250 RIL studies concluded that RIL does cost more, but not as much as expected. Activities where RIL involved higher costs included planning, where the median difference (10 observations) was US$ 0.28 per m, and felling, where RIL was US$ 0.56 per m^3 higher than conventional logging or 48 per cent higher. It is possible that the experience gained with RIL techniques since this review was carried out has led to a reduction in costs and a greater chance of profitability, as reflected in the more recent studies from Brazil cited above.

Reduced Impact Logging is just one aspect of SFM criteria and indicators used in national standards and in voluntary certification schemes which describe more comprehensively the elements of good practice. There are a number of cost-increasing requirements beyond RIL, which makes it unlikely that increased efficiency will be sufficient to offset these.

The experience from Africa and Gabon in particular has shown that meeting government SFM standards can be challenging (Box 8.6). SFM management plans are expensive and, as a result, there has been limited uptake.

Box 8.6: The High Cost of SFM plans in Gabon

Rough calculations show that to invest in a 15,000 hectare concession (for locals) a sum of US$ 4,505,000 is needed, of which US$ 2,850,000 (63 per cent) will go towards the development of a management plan and the rest into various associated studies and impact assessments, the most costly being those of fauna. These figures do not include management training and other costs such as licenses. Sustainable forest management has complex requirements. To formulate a Sustainable forest management (SFM) plan for a concession, an inventory of forest resources is needed and funds are required for associated mapping, in-forest measurement and assessment, and development of the plan and a process for implementation. These actions alone entail heavy investments. In addition, the Forestry Code for Gabon calls for low-impact logging practices; workers' compounds must be established for at least 25 years, and associated agricultural sites must be taken into account and studied in advance.

Many schemes have emerged to certify forest management against SFM standards, as well as wood tracking systems to ascertain sustainable and/or legal wood sources. Independent inspectors assess a mix of forest management documentation and actual field practice. There are two international approaches with widespread support: FSC and PEFC. Both also offer chain-of-custody certification, tracing products from SMFs and verifying they are not contaminated by other (potentially unsustainable) products. The logistics can be challenging, especially for pulp, where many wood sources are mixed. It usually operates through an electronic system of tagging logs with bar-codes and tracking subsequent products.

Companies opting for certification not only have to meet the costs of any improvements needed to meet the standards, but also the direct costs or transaction costs of the certification application. For small forest areas these can be relatively significant. The direct costs of FSC certification have been estimated to range between US$ 0.06 and US$ 36 per hectare certified, depending on the size of forest area, as unit costs decline with scale. In certification, links to markets and the possibility of premiums or improved access to high value markets provide the incentive for investment.

An analysis of the impact of forest certification by Cashore et al. used case studies from 16 countries in four regions (Asia-Pacific, Eastern Europe and Russia, Latin America and sub-Saharan Africa). Positive social effects were consistently reported, including improved pay and conditions for workers, the development of community infrastructure and the provision of training. There was less consistency in these case studies and other recent literature on the market benefits of certification for the companies concerned, raising concerns about its financial sustainability in some areas (Box 8.7).

Box 8.7: Costs and Benefits of Certification for Producers

In Uganda, there is no internal market for certified products and most exports are destined for other African countries that do not require certification. Paschalis-Jakubowicz (2006) reported that although FSC certification increased costs for private producers, this was not reflected in the price of lumber in Polish markets. In Guatemala and Mexico, economic benefits of certification have generally not lived up to expectations, despite major government initiatives encouraging its use in communities and industry. In Guatemala, the direct and indirect costs of certification in the Maya Biosphere reserve have been estimated to range between US$ 0.10 and US$ 1.90 per certified hectare per year, US$ 8–107 per hectare harvested per year, and US$ 4.2–52.9 per m^3 of harvested round timber. This indicates considerable variation but suggests that for some forest owners the costs are very high. While premiums have been obtained, they are not high (in the case of certified mahogany, US$ 0.05–0.10 per board feet, equivalent to less than 10 per cent of the sales price), and it was found that prices for non-certified wood soon caught up.

Malaysia has benefited from an average premium of 37 per cent on sawn timbers. Muhtaman and Prasetyo found that Perum Perhutani in Indonesia received a 15 per cent price premium, and Wairiu reported an increase in price per cubic metre for Solomon Islands Eco-forestry (SIEF) timber marketed through Village Eco-Timber Enterprises (VETE) in the Solomon Islands.

A survey of the furniture industry in South Africa found that although FSC certification does not lead to price premiums, there are other benefits in maintaining existing markets and contributing to quality control cited in Blackman and Rivera 2010).

In Finland, a survey of perceptions of certified and non-certified wood products companies found that certification was not considered to improve financial performance or to result in premiums but was important for signalling environmental responsibility and maintaining market share.

While a niche market may exist for some certified timber, many companies (especially in developing and transitional countries) produce for local and national markets. In these cases, tools such as FSC certification will not provide a significant impact on prices received. Studies of certification in Africa, Eastern Europe and Latin America provide support for this finding. Nevertheless, in three tropical-forest countries in Asia and the Pacific, there is some evidence of positive market benefits from certification. In other cases, in South Africa and Finland, certification is found to be beneficial in maintaining existing market share (Box 8.7).

Box 8.7 provides examples of both positive and negative cost-benefit ratios related to the uptake of certification.

Certification has so far been taken up by forest operations of all sizes in developed countries, as well as by larger companies (often plantation companies) in developing nations. None of the ten largest certified forests are in the tropics and few certified forests are community-run. This reflects challenges in interpreting and meeting social standards locally, addressing insecure rights and assets of tropical forest land-holders and managers, and poor access to capital, skills and markets.

However, there are some important exceptions that suggest these challenges could be overcome. Mexico contains more than 700,000 hectares of community-managed FSC-certified natural forest, spanning 33 communities with stands ranging from 56 hectares to 252,000 hectares. Most of these (26 out of 33) cover less than 20,000 hectares. The Mpingo Conservation Project in Tanzania was awarded an FSC group certification for its community forests in 2009 and Kikole village, one of the project's constituent rural communities, sold the world's first harvest of FSC-certified African blackwood in January, 2010.

In terms of the environmental impacts of certification, there is a general perception that certification has been taken up by forest enterprises that were already practising good forest- management. Some support to this perception is given by the geographic pattern of the uptake of certification, which is heavily concentrated (80 per cent in the case of FSC) in temperate and boreal areas. The evidence on the impact of forest certification on biodiversity has been reviewed by van Kuijk *et. al.* who concluded that while there is no conclusive quantitative evidence about the effects, the good forest-management practices associated with certification are beneficial for biodiversity. These include reduced impact logging, riparian buffer zones, green tree retention in clearcuts, protected areas within forest management units and biodiversity corridors. The review also showed that many species and ecosystems are negatively affected by any form of logging, highlighting the need for a mix of conservation areas and production areas of forest.

A more recent review and expert survey draws a heavily qualified conclusion that certification has helped reduce biodiversity loss in the tropics. The caveats to this conclusion relate to the limited area of certified natural forest in the tropics and the range of extra-sectoral threats to tropical forests which certification can do little to address.

In Uganda, there is no internal market for certified products and most exports are destined for other African countries that do not require certification. Paschalis-Jakubowicz reported that although FSC certification increased costs for private producers, this was not reflected in the price of lumber in Polish markets. In Guatemala and Mexico, economic benefits of certification have generally not lived up to expectations, despite major government initiatives encouraging its use in communities and industry. In Guatemala, the direct and indirect costs of certification in the Maya Biosphere reserve have been estimated to range between US$ 0.10 and US$ 1.90 per certified hectare per year, US$ 8–107 per hectare harvested per year, and US$ 4.2–52.9 per m^3 of harvested round timber. This indicates considerable variation but suggests that for some forest owners the costs are very high. While premiums have been obtained, they are not high (in the case of certified mahogany, US$ 0.05–0.10 per board feet, equivalent to less than 10 per cent of the sales price), and it was found that prices for non-certified wood soon caught up.

In short, while there are some positive examples of premiums being received by developing country producers, and good evidence of positive social impacts, the slow pace of expansion of forest certification in tropical and sub-tropical areas suggests that more proactive support is needed for scaling up. The evidence on environmental impact shows that there is potential, but that investment in certification needs to be accompanied by other measures aimed at protecting high conservation-value forest, controlling illegal logging and policies directed at other sectors.

8.5.5 Investing in Planted Forests

Investment in planted forest can take a number of forms. It can be for productive purposes and range from systems using native species to high-yield plantations. Alternatively, trees can be planted to promote ecological restoration and ecosystem services, as in the case of China (Box 8.8), although use of timber and fuelwood in such cases is often not precluded. A distinction is often made between reforestation and afforestation.

Box 8.8: Afforestation in China: The Sloping Land Conversion Programme

The Sloping Land Conversion programme (or Grain for Green programme) started in 1999 with a goal to convert around 14.7 million hectares of erosion-prone farmland to forest within critical areas of the watershed of the Yangtze River and Yellow River in China by 2010. This includes 4.4 million hectares of farmland on slopes greater than 25 degrees. There was also a goal to afforest a similar area of wasteland. Total investment has been US$ 4.3 million per year. By the end of 2003, 7.2 million hectares of cropland had been converted and 4.92 million hectares of barren or wasteland had been afforested. By the end of 2006, the area of cropland converted had reached 9 million ha. This was a considerable increase over previous trends for conversion of cropland to forests, estimated at just 1.2 million ha from the late 1980s to 2000.

Historically, governments have played a strong role in subsidising plantations, often providing as much as 75 per cent of total costs. This has been particularly significant in low- and middle- income countries, where governments have justified large subsidies in order to increase domestic timber supplies, supply industry with low-cost wood, and even to relieve pressure on natural forests. Global subsidies for plantations between 1994 and 1998 totalled US$ 35 billion, of which US$ 30 billion went to non-OECD countries.

In Brazil, for many years, industrial forest plantations were promoted for production purposes (fibre for pulp and charcoal) through national government financial incentives. But several programmes now promote reforestation for ecosystem services. For example, in Piragcaba in Sao Paulo state, the local authorities in charge of water supply provide assistance to farmers in the form of seedlings and technical assistance to restore riparian forests. A number of countries have invested in mangrove restoration in order to improve sea defences.

The cost of planting forests and the rate of return on investment varies according to the species, location, and whether planting is for productive or protective purposes. Differences in assumptions about the inclusion of opportunity costs of the land or the land price also lead to variations in reported costs. Table 8.6 gives an indication of the variation in costs. Taking the range of costs in Table 8.6 and an annual increase of 5 million hectares, the current level of investment in extending the forest area could range from US$ 1.25 billion to over US$ 40 billion per year.

Table 8.6: Costs of reforestation and afforestation

Activity	Location	Cost/ha	Reference
Restoring eucalyptus woodlands	S.E Australia	€ 285-(passive *i.e.* natural regeneration) -€ 970 (active *i.e.* replanting)	Dorrough and Moxham (2005) in Nel-Rhover *et. al.* (2009)
Restoration of degraded stands	Atlantic forest, Brazil	€ 2,600	Instituto Terra (2007)
Replanting of mangroves	Thailand	US$ 8,240 plus US$ 118/ha per year for maintenance	Sathirathai and Barbier (2001)
Reforestation for carbon sequestration and wood	Costa Rica	US$ 1,633	Based on payment in national PES scheme of US$ 980/ha (Robalino *et. al.* 2010) which covers 60% of costs (Miranda *et. al.* 2004)
Reforestation for carbon sequestration and wood	Ecuador	US$ 1,500	Wunder and Alban (2008)
Afforestation	India various regions	US$ 413 (2001 prices). Mean of 25 estimates from 21 studies ranging from US$ 12 to US$ 755	Balooni (2003)
Industrial forest plantation	Sabah, Malaysia (Acacia mangium)	US$ 921–1,052 (2001 prices)	Chan and Chiang (2004)
	Average for Southern hemisphere, USA and China - main species	US$ 957	
Industrial forest plantations	Uruguay (Eucalyptus globules)	US$ 500	Cubbage et al. (2009) excludes land costs, and uses 8% discount rate.
	US (Douglas fir)	US$ 1,300	
	Colombia (Pinus tecunumani and Eucalyptus)	US$ 1,800	

The rate of return on private investment in planted forest for productive purposes can be very high. Estimates made by Cubbage *et al.* of the financial viability of industrial plantations based on exotic species indicate that excluding land costs, returns for exotic plantations in almost all of South America—Brazil, Argentina, Uruguay, Chile, Colombia, Venezuela, and

Paraguay—could be substantial, with an internal rate of return (IRR) of 15 per cent or more. Yet the record of public incentives in plantations has been poor, with the wrong choice of sites, poor genetic material, poor maintenance and location too far from markets. Changes in local and global markets are also a major factor affecting rate of return. The depressed timber prices on world markets at the end of the 1990s and the early years of the last decade led to smallholder plantations in the Philippines becoming unprofitable.

The social impacts of reforestation can be very controversial, particularly where it involves large-scale plantations run by private companies because of concerns about land grabs, withdrawal of access to local communities to common-property forest resources and replacement of perceived degraded or low-value common property forest, or land important for food production, by forest plantations. Other reviews acknowledge these issues but point out that in some areas plantations can provide benefits to the local poor. Garforth, Landell-Mills and Mayers highlighted the employment generated by the plantation sector in South Africa, directly and indirectly in small-scale processing and retailing and supporting industries, estimating that about 7 per cent of the population depend on the sector. Bull *et al.*, pointed to extensive outgrower schemes and social programmes of HIV AIDs, education and job training as benefits from plantations in the Southern Hemisphere. But Garforth *et al.*, stressed that significant investment in local bargaining power is needed for outgrower schemes to offer routes out of poverty.

Small-scale reforestation on the part of communities or small farmers has been less controversial because it is often an important livelihood option introduced with a poverty-reduction aim. Farmers in India have become important suppliers of wood as a result of such programmes. A number of reforestation schemes have been targeted at the provision of ecosystem services, notably carbon sequestration. While some case studies have been generally positive, *e.g.* Miranda *et al.*, on Costa Rica and Wunder and Alban on PROFAFOR in Ecuador, concerns have been raised about the long time scales involved for benefits to accrue to farmers and the need for capacity building. The Sloping Land Conversion Programme in China was welcomed by farmers in its early years because the compensation offered outweighed the loss of agricultural return. However, surveys in five provinces found that there were shortfalls for a significant proportion of farmers from 7 per cent to 77 per cent.

The environmental impacts of reforestation and afforestation vary considerably. Plantations can be contentious owing to their more intensive use of water and chemicals, as well as introduction of exotic and genetically modified tree species. There has been much criticism of monoculture plantations of exotic species. Recognising plantations' high potential to produce wood, potentially taking pressure off natural forests, their sustainability is often conferred at the landscape level rather than within the plantation—siting plantations on less biologically and culturally important land within a land-use mosaic, so that the landscape as a whole provides the range of goods and services required.

Even where tree planting is for protective purposes rather than production, much depends on the way programmes are carried out. The mangrove-planting programme in Vietnam has been widely hailed for its environmental benefits. It involved an investment of US$ 1.1 million in planting (carried out by volunteers) and protecting 12,000 hectares of mangroves but saved US$ 7.3 million per year on dyke maintenance. In contrast, mangrove restoration in the Philippines produced poor results because trees were planted in the wrong places leading to low survival rates.

Similarly, the Sloping Land Conversion Programme in China, although effective in bringing about tree planting on large areas of land, has problems of low survival rates and lack of technical support. The suitability of this approach for drier regions of China has also been questioned, for example by Zhang *et al.*, who estimated that in the sub-alpine region of south-western China, afforestation would reduce water yield by 9.6–24.3 per cent, depending on the type of species and the climatic conditions. Another study which applied a simplified hydrological model across the diverse regions of China, estimated higher annual water yield reductions from afforestation from 50 per cent in the semi-arid Loess Plateau region in the north to 30 per cent in the tropical south.

To conclude, private investment in reforestation has a place in a green forest sector to ensure sufficient supplies of wood. But it needs to take place within management of the landscape and should not replace natural forests, nor land that is important for subsistence food production. The economies of scale of planted forests, particularly high-yield, fastgrowing, single-species plantations are such that market forces will drive expansion. But incentives are often given in forms that lead to their replacing natural forests. The CDM also was restricted to reforestation and afforestation, putting natural forest management at a further disadvantage in developing countries. As stressed by Bull *et al.*, incentives to plantations should be directed instead at promoting forest ecosystem services and social development. Governance conditions are also required that will tilt the balance away from those planted forests that do not support many ecosystem services towards those that do. It is important that certification schemes continue to provide criteria for planted forests, including high- yield plantations, to encourage best practice while not putting sustainable timber harvesting from natural forest at a disadvantage.

8.5.6 Investing in Agroforestry

Agroforestry encompasses a wide range of practices as demonstrated by a definition given in a recent assessment: "Agroforestry systems range from subsistence livestock silvo-pastoral systems to home gardens, on-farm timber production, tree crops of all types integrated with other crops, and biomass plantations within a wide diversity of biophysical conditions and socioecological characteristics. The term has come to include the role of trees in landscape level interactions, such as nutrient flows from forest to farm, or community reliance on fuel, timber, or biomass available within the agricultural landscape."

Zomer *et al.*, estimate that as much as 1 billion hectares of agricultural land could currently be considered as agroforestry if a threshold of 10 per cent tree cover is taken. With a higher threshold of 30 per cent tree cover, the area of agroforestry would be considerably lower at 375 million hectares, but still significant. They conclude that trees are an integral part of the agricultural landscape in all regions except North Africa and West Asia. Agroforestry is relatively important in Central America, South America and South-east Asia, where there are many long-standing management traditions as well as new scientific forms of agroforestry, but agroforestry is also practiced on large proportion of Africa's land area.

As with reforestation, the costs and rates of return of agroforestry systems vary considerably depending on location, species and management type. FAO cites a review by Current and Scherr of agroforestry practices in Central America and the Caribbean which found that in 2/3 of the cases, Net Present Value (NPV) and returns to labour were higher than for the main alternative practices. Some more recent studies in different locations that have compared the profitability of agroforestry systems with conventional farming systems are shown in Table 8.7. They are generally consistent with the conclusions in Current and Scherr but show the importance for the results of time horizons, discount rates and the range of benefits included. A common conclusion of the studies that find in favour of the profitability of agroforestry is that it requires considerably higher investment in the early years. This constitutes a major obstacle to its adoption.

The Food and Agriculture Organization of the United Nations' review of the benefits of agroforestry cited a number of positive impacts for farmers, an additional source of cash income, provision of products such as fodder for livestock, fuelwood and fertiliser in the form of nitrogen-fixing trees, that the farmer would otherwise have to buy, decreased risk because of the wider range of products on the farm, and the ability to earn income throughout the year and accrue benefits at different times, over the short, medium and long term.

Research on the payments for agroforestry scheme introduced in Costa Rica in 2004 as an additional eligible activity in the national PES scheme, provides some evidence on the social impact of providing incentives for agroforestry. A high proportion (78 per cent) of the farmers interviewed reported an increase in income. This was not from sale of harvested timber but from money left over after planting and maintenance costs were covered. This was particularly important in indigenous communities because of their strong dependence on subsistence farming and little other opportunity for outside

Table 8.5: Rate of return of agroforestry compared with conventional farming

Type of agroforestry system	Location	Rate of return/comparison with conventional farming	Reference
Silvo-pastoral	Central and South America	4–14%	Pagiola *et. al.* (2007)
	Peruvian Amazon	Lower return than shifting agriculture with short time horizon but higher return over a longer period	Mourato and Smith (2002)
Three strata: 1) fruit trees, 2) banana, papaya, lemon 3) spices	Northern Bangladesh	Agroforestry is more profitable than conventional farming with or without the inclusion of family labour costs and less risky	Rahman *et. al.* (2007)
Mixed agroforestry, timber, horticulture, agriculture—timber harvested after 15 years	Chittagong Hill Tracts, Southern Bangladesh	Agroforestry gives lower annual return per land unit than shifting cultivation in year 1, 5, 9 and 13 and higher in other years. Agroforestry has a higher NPV over 15 years at 10% discount rate	Hossaiin *et. al.* (2006)
Contour hedgerows	Eastern Visayas, Philippines	Through soil conservation and improved yields increases agricultural profits by average US$ 53/household or 6% of total income but outweighed by opportunity costs of land and labour. Excludes on-farm benefits such as fuelwood and fodder as well as long run and external benefits	Pattanayak and Mercer (1998)
Fertiliser tree fallows	Zambia	Over 5 years at 30% discount rate, agroforestry is more profitable than continuous maize with no mineral fertilisers	Ajayi *et. al.* (2006)
Rotational woodlots	Tanzania	Agroforestry has an NPV of US$ 388/ha, six times that of conventional maize	Franzel 2004 cited in Ajayi *et. al.* (2006)

income. However, farmers commonly viewed the plantings as a savings account for future generations and saw little short-term benefit. While the payments were concluded to be effective in overcoming initial economic and technical obstacles, the need for ongoing capacity building and support from strong local organisations was highlighted.

A number of projects and programmes have promoted the wider adoption of agroforestry on the basis of its significant on-site and off-site environmental benefits. The Alternatives to Slash and Burn programme showed that tree-based farming systems, whether mixed or monocultural, had significant carbon storage benefits, in part due to its limited soil cultivation and consequent oxidation of soils, in part due to making use of many vertical layers of vegetation. It has been estimated that in Sumatra, Indonesia, rubber agroforestry systems store about 116 tonnes of carbon per hectare, 45 per cent of the amount stored by undisturbed natural forests (254 t/C per ha), whereas continuous cultivation of cassava stores only 39 tonnes of carbon per hectare. The Food and Agriculture Organization of the United Nations (2005b) cites evidence of various types of environmental benefits from agroforestry. In Sumatra showed that households with diversified agroforestry systems depend less on gathering forest products from protected areas than farmers cultivating wetland rice. In the USA, trees planted as wind breaks have been estimated to increase crop yield significantly, for example, by 23 per cent for winter wheat. More recently, the GEF-funded Silvopastoral project in Colombia,

Costa Rica and Nicaragua, which targeted areas of degraded pasture, provides some rigorous evidence of the environmental benefits of creating incentives for agroforestry (Box 8.9).

Box 8.9: Evidence on the Impact of Incentives for Silvo-pastoral Practices

Around US$ 4.5 million was invested in payments to farmers in Central America and Colombia to fund a transition to greater use of silvo-pastoral practices in cattle ranching. The payments to farmers were based on a scoring system for environmental services.

Research on the implementation of this scheme in Quindio, Colombia shows a significant difference between participants and the control group after four years of payments. Only 13 per cent of the land area in the control group experienced any change in land use and the effect of this change was to increase the environmental service score by 7 per cent. In contrast, changes in land-use practices extended to 44 per cent of the area occupied by participants in the payment scheme and the environmental service score increased by 49 per cent. Similar conclusions based on casual observation of neighbouring areas are drawn for the silvopastoral scheme in Matiguas-Rio Blanco, Nicaragua.

Although water-related services were not a focus of the payment scheme, some positive impacts were also found. The silvo-pastoral scheme in Quindio, Colombia monitored water quality upstream and found a rapid drop in turbidity, biological oxygen demand (BOD) and coliforms after measures had been taken to reforest riverbanks and protect them from livestock entry.

In general, agroforestry has potential to be both beneficial to farmers and to provide offsite-benefits in the form of carbon sequestration, reduced sedimentation in surface water, and maintenance of a wider basis of biodiversity than agriculture. But the economic evidence shows that farmers need both financial assistance and technical assistance in making the transition to modern forms of agroforestry. Investment in incentive schemes combined with longer-term technical support can be effective in promoting its expansion.

8.6 MODELLING GREEN INVESTMENT IN FORESTS

In this section we examine the impacts at a global level of increasing investment in two of the options discussed in the previous section: private investment in reforestation and public investment in payments to avoid deforestation. This is because both are highly likely to play a role in climate-change mitigation and will form part of a post-2012 international climate agreement.

8.6.1 The Green Investment Scenario

Under the global model developed for the Green Economy Report by the Millennium Institute, the green investment scenario (G2) allocates 0.034 per cent of global GDP to reforestation and incentives for avoiding deforestation/forest protection between 2011 and 2050. This equates to US$ 40 billion (in constant 2010 US dollar prices) per year on average, with 54 per cent or US$ 22 billion directed to reforestation and 46 per cent or US$ 18 billion per year to avoided deforestation.

This is similar in order of magnitude to estimates made in the 1990s of the amount of investment needed for sustainable forest management in production forests of US$ 33 billion per year (Tomaselli 2006) and estimates made in recent years for the cost of avoiding deforestation, which range from US$ 5 billion to US$ 15 billion per year (Stern 2007; Grieg-Gran 2006) to US$ 1728 billion.

8.6.2 The Baseline Scenario: Business-as-usual

In the model, the baseline scenario or business-as-usual (BAU) for the forest sector replicates the historical trend from 1970 and assumes no fundamental changes in policy or external conditions going forward to 2050. Under business-as-usual, the projection is for a steady decrease in forest cover from 3.9 billion hectares in 2010 to 3.7 billion hectares by 2050. As a result, carbon storage in forests will decline from 523 Gt in 2009 to 431 Gt in 2050. The contribution of the forest sector to global GDP and employment is projected to grow at 0.3 per cent per year between 2010 and 2050 to reach US$ 0.9 trillion and 25 million jobs by 2050. This is in line with growth rates in the sector between 1990 and 2006.

8.6.3 Investing to Reduce Deforestation

The cost of avoiding deforestation is assumed to start at US$ 1,800 per hectare, increasing to US$ 2,240 per hectare by 2050. This is based on the global average value added per hectare of crop production plus the value added of forest products per hectare (measured in constant 2010 US$ prices), which is taken to represent the opportunity cost if forests are conserved with no extraction of forest products or clearing. This approach to estimating opportunity cost is somewhat different from that taken in a number of studies on this topic, which add together the present value of agricultural revenues net of cost discounted over several years and the stumpage fees for timber, but the result is within the range of most such estimates. It can be considered a generous estimate of the opportunity cost as in many locations the returns to converting forests to smallholder agriculture, subsistence and cash crops and to cattle ranching are considerably lower than US$ 1,800 per hectare. This figure is more representative of higher-value land uses such as oil palm.

Nevertheless, the cost of designing and administering a payment scheme, the so-called transaction costs, can be considerable, particularly in developing countries and in remote forest areas. While existing national-level PES schemes in Costa Rica and Mexico have administration costs of well below 10 per cent of the overall amount spent, analysis of the Bolsa Floresta scheme in Amazonas state in Brazil indicates a much higher proportion, around 40 per cent. The cost figure used in this model is high enough to incorporate some provision for transaction costs.

The investment would enable payments to be made to forest landholders over a steadily expanding area, with the yearly increase reaching 6.76 million hectares by 2030 and then decreasing to 6.66 million hectares by 2050, in effect reducing the annual rate of deforestation by just over 50 per cent, as shown in Fig. 8.2. This is consistent with other studies, which have predominantly estimated the cost of reducing deforestation by 50 per cent.

8.6.4 Investing in Planted Forest

The cost of planting forests is assumed to be US$ 1,630 per hectare based on the costs of reforestation in Costa Rica's national PES scheme, which pays farmers US$ 980 per hectare to cover 60 per cent of the costs of establishment. As shown in Table 8.4, this is within the range of costs estimated for production planted forests, which is the type of reforestation under consideration here.

The modelling examines the full cost to a landowner of establishing a planted forest rather than the incentive payment that might make such a land use competitive. On average, the investment allocated will cover the cost of reforesting an additional 9.6 million hectares per year or 386 million hectares over the 40-year period.

8.7 IMPACTS OF INVESTMENT IN REDUCING DEFORESTATION AND IN PLANTED FOREST

In the short term the reduction in deforestation leads to a decrease in the value added of the forest sector (wood, wood processing and pulp and paper) so that it is 1.7 per cent below the baseline in 2013. Similarly, employment is 2 per cent below the baseline level in 2013. However, this does not take account of the economic impacts on other sectors such as tourism, which may benefit from the reduction in deforestation and also the economic value of the reductions in carbon emissions. In the longer term, as the area of planted forest increases, value added in the conventional forest-based industries rises to US$ 10.4 trillion, some 19 per cent above BAU. The increase is accompanied by growth in employment from 25 million to 30 million worldwide, or 20 per cent above business- as-usual.

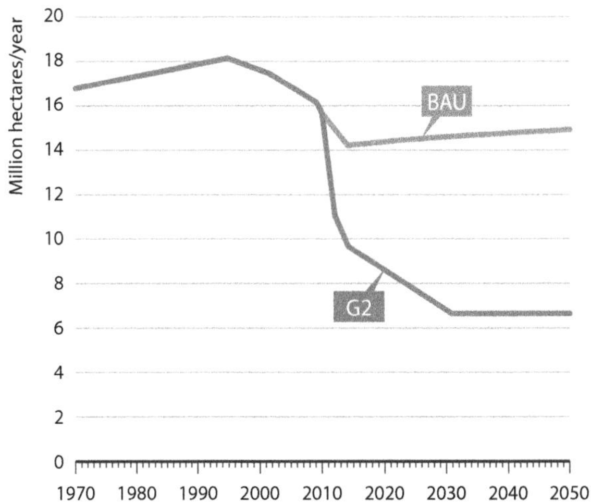

Fig. 8.1: Deforestation reduction under the green investment scenario (G2).

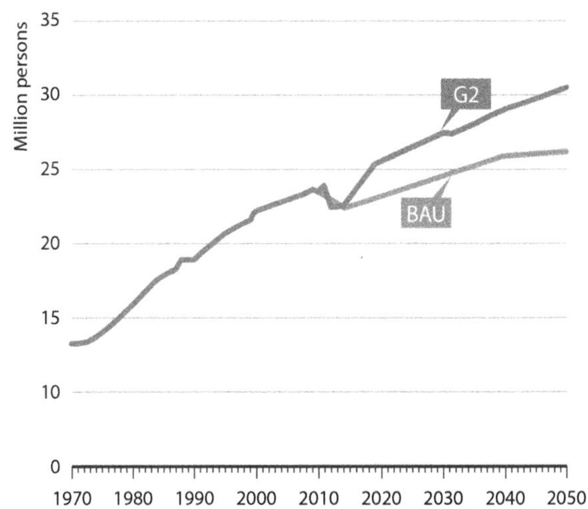

Fig. 8.2: Employment under the green investment scenario (G2) and business-as-usual (BAU).

The main environmental impact is on the area of natural forest, which in 2050 is 8 per cent more extensive in the green investment scenario than under BAU, and on the total area of forest (natural and planted) which in the green investment scenario is 21 per cent more extensive in 2050 than under BAU and 14 per cent higher than the current forest area. This has positive implications for biodiversity and carbon storage and results in reduced greenhouse gas emissions. The increase in the forest area is made possible by the investments in improved agricultural productivity. This means that demand for agricultural production can be met from a smaller area of land, freeing up land for reforestation or afforestation. It also means that there is less pressure on natural forest.

These projections indicate the potential of increasing green investment in the forest sector. But much depends on how the investment is made and in what policy and institutional context. As discussed above, reforestation programmes do not always work financially, socially or environmentally, and the small amount of investment in avoiding deforestation so far, mainly in the national PES schemes in Costa Rica and Mexico, has struggled to demonstrate cost-effectiveness. Large investment programmes on the scale modelled here will be more challenging although they can draw lessons from the existing experience. Global aggregate projections of this nature cannot, owing to limitations of their design, capture the differences in response between tropical countries and non-tropical countries, or between countries with high forest cover and low forest cover, or between high income and low income countries. They do, however, indicate what can be achieved at a global level in the appropriate policy and institutional conditions.

Box 8.10 15% of India's Power by 2030 to be Green, says NITI Aayog

The National Institution for Transforming India (NITI) Aayog has told the environment ministry that renewable sources can make up at least 15% of India's energy mix by 2030 if present policies are given a push. This analysis was done as part of the government's strategy to enumerate India's climate action plan to be submitted to United Nations.

The action plan, called Intended Nationally Determined Contributions (INDCs), is awaiting Cabinet approval and is expected to be made in the first week of September. Environment minister Prakash Javadekar had said that India's submission to the UN will be the most exhaustive, covering all crucial climate change areas—mitigation, adaption, finance, technology.

As energy is the most important component of the INDCs, the NITI Aayog has conducted a detailed analysis of the futuristic energy needs depending on different scenarios. "We have presented various scenarios and told the environment ministry that by 2030, India's energy mix can have 15% renewable share easily."

The analysis shows it would be a big leap for India as renewable energy currently contributes to less than 6% of the country's energy mix. Prime Minister Narendra Modi has announced setting up a green power capacity of 175 gigawatts (GW) by 2022 to ensure every household gets adequate power.

India's power demand by 2030 is expected to be about 10 lakh MW. According to the panel, the biggest contributors to enhanced green power will be solar and wind energy, contributing about 70% to total power generation by renewable, followed by nuclear energy. India plans to generate 20,000 MW of power from nuclear energy by 2030, when renewable is expected to contribute 2,00,000 MW of energy.

Box 8.11 Growth and Emissions and their Role in Environment Policy

On 2 October, 2016 India promised to reduce the emission intensity of its gross domestic product (GDP) by about a third from 2005 levels by the year 2030. That doesn't mean overall emissions will come down, but that the rise will be at a slower pace, as the charts below show.

The rise in emissions is not only a function of how much the emission intensity (as a ratio of GDP) is cut, but also how fast the economy grows. If India sticks to the current pledge and the economy grows at a constant 7.5% from now to 2030, total emissions can rise to 8.3 billion tonnes of carbon dioxide equivalent (tCO_2e).

How likely is India to achieve its 35% cut target?

India has already reduced emission intensity by 13.5% between 2005 and 2012, Mint calculations show. According to the 2014 Emissions Gap report released by the United Nations Environment Programme, India is on track to cut emission intensity by 20-25% by 2020.

Apart from reducing the emission intensity of GDP, India has set out two other goals in its pledge document. One is to ensure that 40% of electricity-generation capacity is fuelled by non-fossil fuel based energy resources by 2030. Funding will be crucial for this.

The other pledge is to increase tree and forest cover and create an additional carbon sink of 2.5 to 3 billion tCO2e by 2030. A carbon sink is a natural or artificially created source that absorbs more carbon than it releases. For example, a forest.

India's forest cover has increased from 19.44% of total geographical area in 1991 to 21.23% in 2013, development and industrial activities such as mining and construction, and population pressure have seen this increasingly come under threat.

■ ■ ■

9

Guidelines for Green Renewable Energy

9.1 INTRODUCTION

World primary energy demand is expected to continue growing. The International Energy Agency's (IEA) Current Policies scenario, which assumes no major change in policies as of mid-2010, projects a growth rate of 1.4 per cent per year up to 2035 (Table 9.1). The fastest growth is expected in non-OECD countries with a projected rate of 2.2 per cent per year, particularly in China and India and other emerging economies in Asia and the Middle East. Many non-OECD countries are also expected to see large increases in imports of oil or gas or both.

Table 9.1: Primary energy demand by region in the IEA Current Policies scenario

	Total energy demand [Mtoe][a]		Growth rate [%]	Share in total energy demand [%]	
	2008	**2035**	**2008–2035[b]**	**2008**	**2035**
OECD	5,421	5,877	0.3	44.2	32.6
Non-OECD	6,516	11,696	2.2	53.1	64.8
Europe/Eurasia	1,151	1,470	0.9	9.4	8.1
Asia	3,545	7,240	2.7	28.9	40.1
China	2,131	4,215	2.6	17.4	23.4
India	620	1,535	3.4	5.1	8.5
Middle East	596	1,124	2.4	4.9	6.2
Africa	655	948	1.4	5.3	5.3
Latin America	569	914	1.8	4.6	5.1
World[c]	12,271	18,048	1.4	100.0	100.0

[a]. Million tons of oil equivalent. b. Compound average annual growth rate. c. World includes international marine and aviation bunkers (not included in regional totals), and some countries/regions excluded here.

Box 9.1: The Rise of Renewable Energy

Renewable energy reached an important turning point last year with record new installations of emissions-free power surpassing sources that burn fossil fuel, according the International Energy Agency. New installations of renewable energy overtook conventional power for the first time in 2015, the agency said in its Medium-Term Renewable Energy Market Report. Employment in the renewable energy sector increased by 5% in 2015, to 8.1 million jobs (direct and indirect), as estimated by IRENA. Here is a look at the growth of the one of the most dynamic sectors:

Jobs in Renewable Energy

Estimated direct and indirect jobs in renewable energy worldwide, by industry.

	World	China	Brazil	US	India
Solar	2,772	1,652	k	19«»	103
Liquid biofuels	1,678	71	821	277	35
Wind power	1,081	507	41	88	k8
Solar heating	939	743	41	10	75
Solid biomass	822	241	–	152	58
Biogas	382	209	–	–	85
Hydropower	204	100	12	8	12
Geothermal energy	160	–	–	35	–
CSP	14	–	–	4	–
Total	**8,052**	**3,523**	**918**	**769**	**416**

Energy demand is growing against the backdrop of fluctuating, but generally increasing fossil-fuel prices. Expenditure on oil alone increased from 1 per cent of global GDP in 1998 to around 4 per cent at the peak in 2007, and is projected to remain high in the period to 2030.

Findings indicate that the share of renewables in total energy supply is expanding and that the greening of the energy sector can contribute to the growth of income, jobs and access by the poor to affordable energy, which are other objectives of sustainable development. Worldwide investment in renewable energy assets—without large hydropower—grew by a factor of seven from US$ 19 billion in 2004 to US$ 143 billion in 2010. For OECD countries the share of renewable energy sources in total primary energy demand has risen from 4.6 per cent in 1973 to 7.7 per cent in 2009.

Renewable energy is derived from natural processes that are replenished constantly. In its various forms, it derives directly or indirectly from the sun, or from heat generated deep within the earth. Included in the definition is energy generated from solar, wind, biomass, geothermal, hydropower and ocean resources, and biofuels and hydrogen derived from renewable resources.

9.2 CHALLENGES AND OPPORTUNITIES

The global community and national governments are faced with four major challenges with respect to the energy sector: (1) concerns about energy security; (2) combating climate change; (3) reducing pollution and public-health hazards;

and, (4) addressing energy poverty. Greening the energy sector, including by substantially increasing investment in renewable energy, provides an opportunity to make a significant contribution to addressing these challenges.

9.2.1 Energy Security

Increasing energy demand together with rising energy prices raise concerns about energy security, a topic which covers a range of issues but primarily is associated with the reliability and affordability of national energy supply. Such concerns are particularly relevant for low-income countries, but also for emerging and developed economies, where a relatively high dependence on a limited range of suppliers can mean higher risks to the security of national energy supply due to geo-political and other developments. Risks to national energy security can also carry downwards to impinge on energy security at local levels.

Oil importing countries (especially developing countries and emerging economies) are expected to become increasingly dependent on OPEC countries for oil. While total non-OPEC output is expected to remain about constant until 2030, production in OPEC countries is projected to increase, especially in the Middle East. OPEC's share in the world oil market consequently rises from 44 per cent in 2008 to 52 per cent in 2030, above its historical peak in 1973. For natural gas, increases in exports are mainly projected to come from Russia, Iran and Qatar, which would increase the world economy's energy dependency on these countries.

The increase in oil prices since 2002 has increased pressure on the balance of payments of developing countries. To protect consumers from increased fossil-fuel prices, some countries have increased their fuel subsidies putting additional strain on government budgets, and underpinning the demand for fossil fuel imports. Investing in renewable sources that are available locally—in many cases abundantly—could enhance energy security for such countries. Energy security would then be influenced more by access to renewable technologies, including both their affordability as well as the capacity to adapt and deploy those technologies. Diversifying the energy matrix thus presents both a considerable challenge and opportunity for oil importing countries.

9.2.2 Climate Change

The Intergovernmental Panel on Climate Change's (IPCC) fourth assessment report (IPCC 2007) underscored the importance of mitigating future human-induced climate change—mostly driven by the combustion of fossil fuels—and adapting to the changes that occur. Estimates of the damages of climate change and costs of mitigation and adaptation vary widely. Substantial damages will occur even with a rapid greening of the energy system, but will be much higher if no action is taken. The annual global costs of adapting to climate change have been estimated by the United Nations Framework on Climate Change Convention to be at least US$ 49—US$ 171 billion by 2030. About half of these costs will be borne by developing countries. Moreover, climate change is likely to worsen inequality because its impacts are unevenly distributed over space and time and disproportionately affect the poor.

The Intergovernmental Panel on Climate Change and International Energy Agency (IEA) estimate that in order to limit the rise of average global temperature to 2°C, the concentration of GHGs should not exceed 450 parts per million (ppm) CO_2-eq. This translates to a peak of global emissions in 2015 and at least a 50 per cent cut in global emissions by 2050, compared with 2005.

A shift from fossil fuels to renewable energy in the energy supply can contribute to achieving ambitious emissions-reduction targets, together with significant improvements in energy efficiency. To reduce emissions to a level that would keep the concentration of GHGs at 450 ppm in 2050, the IEA projects that renewable energy would need to account for 27 per cent of the required CO_2 reductions, while the remaining part would result primarily from energy efficiency and alternative mitigation options such as carbon capture and sequestration (CCS). A major part of the CO_2 reductions resulting from the promotion of renewables would take place in developing countries.

Box 9.2: Solar Could Beat Coal to Become the Cheapest Power on Earth

Solar power is now cheaper than coal in some parts of the world. In less than a decade, it's likely to be the lowest-cost option almost everywhere.

In 2016, countries from Chile to the United Arab Emirates broke records with deals to generate electricity from sunshine for less than 3 cents a kilowatt-hour, half the average global cost of coal power. Now, Saudi Arabia, Jordan and Mexico are planning auctions and tenders for this year, aiming to drop prices even further. Taking advantage: Companies such as Italy's Enel SpA and Dublin's Mainstream Renewable Power, who gained experienced in Europe and now seek new markets abroad as subsidies dry up at home.

Since 2009, solar prices are down 62 per cent, with every part of the supply chain trimming costs. That's help cut risk premiums on bank loans, and pushed manufacturing capacity to record levels. By 2025, solar may be cheaper than using coal on average globally, according to Bloomberg New Energy Finance.

"These are game-changing numbers, and it's becoming normal in more and more markets," said Adnan Amin, International Renewable Energy Agency 's director general, an Abu Dhabi-based intergovernmental group. "Every time you double capacity, you reduce the price by 20 per cent."

Better technology has been key in boosting the industry, from the use of diamond-wire saws that more efficiently cut wafers to better cells that provide more spark from the same amount of sun. It's also driven by economies of scale and manufacturing experience since the solar boom started more than a decade ago, giving the industry an increasing edge in the competition with fossil fuels. The average 1 megawatt-plus ground mounted solar system will cost 73 cents a watt by 2025 compared with $1.14 now, a 36 percent drop, said Jenny Chase, head of solar analysis for New Energy Finance.

That's in step with other forecasts. GTM Research expects some parts of the U.S. Southwest approaching $1 a watt today, and may drop as low as 75 cents in 2021, according to its analyst MJ Shiao. The U.S. Energy Department's National Renewable Energy Lab expects costs of about $1.20 a watt now declining to $1 by 2020. By 2030, current technology will squeeze out most potential savings, said Donald Chung, a senior project leader.

The International Energy Agency expects utility-scale generation costs to fall by another 25 per cent on average in the next five years. The International Renewable Energy Agency anticipates a further drop of 43 percent to 65 per cent for solar costs by 2025. That would bring to 84 per cent the cumulative decline since 2009.

The solar supply chain is experiencing "a Wal-Mart effect" from higher volumes and lower margins, according to Sami Khoreibi, founder and chief executive officer of Enviromena Power Systems, an Abu Dhabi-based developer.

The speed at which the price of solar will drop below coal varies in each country. Places that import coal or tax polluters with a carbon price, such as Europe and Brazil, will see a crossover in the 2020s, if not before. Countries with large domestic coal reserves such as India and China will probably take longer.

In China, the biggest solar market, will see costs falling below coal by 2030, according to New Energy Finance. The country has surpassed Germany as the nation with the most installed solar capacity as the government seeks to increase use to cut carbon emissions and boost home consumption of clean energy. Yet curtailment remains a problem, particularly in sunnier parts of the country as congestion on the grid forces some solar plants to switch off.

Sunbelt countries are leading the way in cutting costs, though there's more to it than just the weather. The use of auctions to award power-purchase contracts is forcing energy companies to compete with each other to lower costs.

An August auction in Chile yielded a contract for 2.91 cents a kilowatt-hour. In September, a United Arab Emirates auction grabbed headlines with a bid of 2.42 cents a kilowatt-hour. Developers have been emboldened to submit lower bids by expectations that the cost of the technology will continue to fall.

9.2.3 Impacts of Energy Technologies on Health and Ecosystems

There are high indirect costs associated with the pollution arising from combustion of fossil and traditional fuels. The release of both black carbon particles (from incomplete combustion of fossil fuels) and other forms of air pollution (sulphur and nitrogen oxides, photochemical smog precursors, and heavy metals, for example) have a detrimental effect on public health. Indoor air pollution from burning solid fuel accounted for 2.7 per cent of the global burden of disease in 2000 and is ranked as the largest environmental contributor to health problems after unsafe drinking water and lack of sanitation. Burning fossil fuels costs the United States about US$ 120 billion a year in health costs, mostly because of thousands of premature deaths from air pollution. This figure reflects primarily health damage from air pollution associated with electricity generation and motor vehicle use. According to the IEA, the costs of air pollution controls worldwide amounted to about € 155 billion in 2005 and are estimated to triple by 2030. Renewable energy can mitigate or avoid many of these public health risks caused by the mining, production and combustion of fossil fuels.

The use of fossil and traditional energy sources in both developed and developing countries also impacts global biodiversity and ecosystems through deforestation, decreased water quality and availability, acidification of water bodies, and increased introduction of hazardous substances into the biosphere. These impacts also reduce the natural capabilities of the planet to respond to climate change.

Renewable energy technologies are not without impacts and careful planning to address possible environmental and social impacts are essential. Production of biofuels, for example, can have negative impacts on biodiversity and ecosystems, while the environmental and social impacts of large-scale hydropower can be significant. The World Commission on Dams has provided guidelines for reducing possible negative impacts of hydropower development. First-generation biofuels have also received substantial attention for their impacts due to land-use change and agricultural production practices, leading to the development of biofuel sustainability standards. Increased mining activity and deforestation could result from increased use of renewable energy sources requiring rare earth elements, and this is an area getting increased attention to reduce possible negative impacts as much as possible.

9.2.4 Energy Poverty

Expanding access to energy is a central challenge for developing countries. Reliable and modern energy services are needed to facilitate poverty reduction, education and health improvements, as reflected in a number of studies identifying access to energy services as crucial for the achievement of most of the Millennium Development Goals (MDGs).

The scale of the challenge is massive with 1.4 billion people currently lacking access to electricity, and 2.7 billion depending on traditional biomass for cooking in developing countries as calculated by IEA, UNDP and UNIDO. In Sub-Saharan Africa 80 per cent of people rely on traditional use of biomass for their cooking, making it the region with the highest dependence on this energy source. While 53 per cent of urban populations in sub-Saharan Africa have access to electricity, the figure for the rural population is only 8 per cent. This rural-urban electrification imbalance contributes to a highly uneven spatial distribution of economic activity, encouraging larger and more rapid rural-urban migration. On average, 26 per cent of people have access to electricity in sub-Saharan Africa, ranging from 3 per cent in Burundi, Chad and Liberia, to 75 per cent in South Africa and to 92 per cent in Togo at the top. Unless dedicated new efforts are implemented, the IEA estimates that by 2030 1.2 billion people will still lack access to electricity and the number relying on biomass will even rise slightly to 2.8 billion. In some African countries, the share of the population without access to electricity might even increase. Renewable energy sources offer some cost-effective solutions to solving energy poverty; one of the opportunities is explored in the next section.

9.2.5 Solutions for Energy Access

There are various technological options to addressing the energy-poverty challenge described above. Implementing most of these options requires additional, publicly-financed investment, including development assistance, as the commercial market potential is likely to remain limited in some cases. Public-private partnerships may be one option along with promising alternative financing mechanisms, including cost- recovery from users.

In terms of technologies for electricity delivery, there are potentially three broad options for expanding access. First, existing centralised grids can be expanded to non-served areas, potentially based on new renewable sources of energy. Second, decentralised mini-grids can be installed to link a community to a small generating plant. Third, off-grid access can be facilitated by producing electricity for a single point of demand. The optimal mix of these options for any given country is determined by the availability of energy resources, the regulatory and policy environment, the institutional and technical capacity, geographic considerations, and relative costs. Intelligent planning should allow for the flexibility to integrate these systems as countries develop.

Grid expansion is generally the lowest-cost option in urban areas and in more densely populated rural areas. Successful expansion has been achieved recently on a large scale in China, South Africa and Vietnam. Grid expansion at a regional level in Africa could facilitate hydropower trading among countries, thereby supplying low-cost power while reducing the continent's vulnerability to varying oil prices and its carbon emissions.

9.3 INVESTING IN RENEWABLE ENERGY

Both the challenges and opportunities facing the energy sector call for scaling up investment in renewable energy. We summarize recent investment trends in renewable energy and the associated evolution of the competitiveness of renewable energy technologies.

9.3.1 Recent Trends in Renewable Energy Investment

During the past 10 years the growth of investment in renewable energy has been rapid, albeit from a low base. From 2004 to 2010, total investments into renewable energy exhibited a compound annual growth rate of 36 per cent. There were a number of reasons for this performance:

- The relatively easy access to capital for project developers and technology manufacturers in the developed world and major emerging economies and low interest rates supported the growth of renewable energy technologies;

- For some renewable energy technologies, technological developments have led to a significant decline in costs and increased reliability of the technology, which have made investments more attractive;
- High oil prices contributed to the interest in renewable energy investments; and
- Regulatory support for renewable energy technologies increased over the past 10 years. Between 2004 and early 2011, for example, the number of countries that have supportive renewable energy policies in place rose from about 40 to almost 120.

The global financial crisis that began in 2008 appears to have temporarily reduced investment in renewable energy, with growth in new investments slowing in 2008 and 2009. Despite more difficult access to capital, especially the availability of debt finance, the sector as a whole has so far proven to be fairly resilient.

9.3.2 Technical Advances and Cost Competitiveness

As renewable energy technologies have matured their costs have come down, making many of them increasingly competitive with other energy technologies. This section briefly reviews such developments, drawing on recent reviews of relative maturity and costs of different energy technologies.

Overall, the review of renewable energy technologies by UNEP concluded that technical potential, at a global level, does not present a constraint to continued growth in the use of these technologies. In its assessment, the review also found that a growing number of these are technically mature and are being deployed at significant scale.

Box 9.3: Coal, Not Solar, is Energizing Power Capacity Addition

In a century belonging to renewable energy, traditional, non-renewable sources are continuing to shape India's power landscape in a big way. In the past three years, India added 77,636 megawatts (MW) of installed capacity in power generation, an increase of 34%. As much as 68% of this new capacity was powered by coal, whose prices have halved globally since January 2011. As a result, coal's share in installed capacity has crossed 60%. Renewables, led by wind and solar, accounted for 21% of the increase. India added 16,051MW of new power capacity via renewables, where solar is the centre's focus. The National Solar Mission is targeting 100,000MW of solar capacity by 2022.

At present, India has 7,805MW of solar capacity. Bridge to India, a clean-tech consultancy, is projecting 37,000MW by 2020. Unless the sector can address the challenges of securing financing and grid stability, and adding rooftop capacity, it will struggle to come close to that 100,000MW target.

Solar energy technologies for heating purposes (low temperature solar thermal), are commercially mature and commonly used in many parts of the world. Solar PV for electricity in small-scale applications is approaching commercial maturity, such as solar rooftop home systems or solar lanterns in off-grid areas, but is generally still dependent on subsidies or price support mechanisms. Concentrating solar thermal power has been in the demonstration and deployment phase for some time and diffusion has recently begun in a few locations. Geothermal energy can be harnessed for heat in almost any temperate climate, and in some locations also for power generation.

Diffusion and commercial maturity of many renewable energy technologies reflects ongoing, and in some notable cases rapid, improvements in their cost competitiveness.

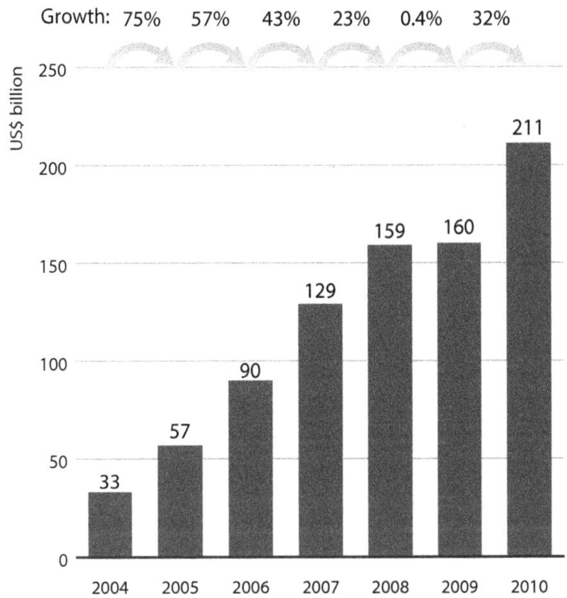

Fig. 9.1: Global new investment in renewable energy in US$ billion.

9.3.3 Externalities, Subsidies and Cost Competition

The considerable externalities generated by fossil fuel energy sources include both the current and future health impacts of various air and other pollutants, as well as the costs necessary to adapt to climate change and ocean acidification resulting from CO_2 emissions. In many cases, there is a lack of political willingness to apply mechanisms to price these externalities. Failure to do so distorts the relative costs and returns of investing in renewable energy compared to fossil fuel alternatives.

The health externalities from fossil fuel energy usage are widespread and difficult to translate into monetary terms. In a recent study on global health, the World Health Organisation found external environmental risks accounted for up to 10 per cent of the global death and disease burden; over half of which is a direct result of fossil fuel use.

The size of the externalities calculations indicates that various renewable technologies would already be competitive if important external costs were internalised to producers and consumers, but are primarily illustrative as there are acknowledged uncertainties in climate-change modelling and the calculation of the resulting damage costs. Because these external costs are not adequately reflected in energy prices, consumers, producers and decision-makers do not receive accurate price signals that are necessary to reach decisions about how best to use resources.

Governments should, though, consider these externalities in formulating policy and strategy for the energy sector.

Box 9.4: Carbon Markets

Carbon markets are an instrument for reducing carbon emissions and targeting greenhouse-gas externalities from fossil-fuel use. They are essentially a group obligation to limit the total emissions of specified sources. A limited amount of tradable emission allowances are sold or given gratis, thus creating an artificial market from which a carbon price can emerge. This price imposes extra costs on the use of fossil fuels, making non-fossil based alternatives more competitive. These alternatives can include not only renewables, but also energy-efficiency measures, nuclear power generation, carbon capture and storage (CCS) and the reduction of non-CO2 greenhouse gases. As of 2010, the two most prominent schemes for developing markets for carbon emissions are the EU Emissions Trading Scheme (EU-ETS) and the Clean Development Mechanism (CDM). These are actually interlinked as the ETS is the principal market in which CDM credits are traded. Owing to the low current carbon prices and uncertainty about their future levels, however, carbon pricing mechanisms have not yet led to large-scale deployment of renewables.

The return on investment for renewable energy projects, relative to fossil fuel alternatives, is sensitive to both the carbon price and market power prices, in addition to the specific support measures for renewables. The carbon price is in turn sensitive to policy decisions. For example, that wind energy, assuming set capital and operating costs, can go from being an expensive carbon mitigation option at low natural gas prices, to a costeffective technology in its own right at higher natural gas prices.

The competitive position of renewable energy would be strengthened if subsidies for fossil fuels were also phased out. In many developing countries, government support to the energy sector is used to decrease the price of energy consumption to below market levels in the belief that this will reduce poverty and spur economic growth. Economically, the most efficient approach to making renewable energy attractive for large-scale market penetration is to remove all subsidies on fossil fuel and impose a price on carbon (for example through fossil-fuel taxes), and then to use the proceeds to subsidise renewable energy for a set duration and to provide targeted subsidies to poor households. Phasing out fossil-fuel subsidies is difficult because doing so has impacts throughout the economy and affects those with vested interests. Any politically-viable reform would thus have to be well planned and probably phased in gradually.

9.3.4 Employment Potential in Renewable Energy

Employment in the renewable energy sector has become substantial—in 2010 more than 3.5 million people worldwide were estimated to be working either directly or indirectly in the sector. A small group of countries currently account for the majority of jobs, especially Brazil, China, Germany, Japan and the United States. China accounts for the largest number, with total employment in renewable energy in 2010 estimated at more than 1.1 million workers. In Germany, the industry employed 278,000 people in 2008, with 117,500 new jobs having been created since 2004. These five countries are also those with the largest investments in renewable energy assets, R&D, and production.

Further growth in employment in renewable energy generation will depend on such factors as the size of investment, the choice of available technologies to invest in, further maturing of technologies, overall progress in economic development, market size, national regulation, and the quality and cost of the labour force. The Green Jobs Report estimated that, with strong policy support, up to 2.1 million people could be employed in wind energy and 6.3 million in solar PV by 2030.

9.3.5 Investment Required for Renewable Energy

Forecasts for future investment needs are based on estimated costs of meeting climate change mitigation targets, while still satisfying the growing demand for energy.

There are a number of analyses with varying estimates of the investments required in renewable energy. The World Economic Forum (2010) suggests that to limit the global average temperature increase to 2°C, global investment in clean energy needs to reach US$ 500 billion per annum by 2020, while current policies imply that this figure would likely only reach US$ 350 billion per annum by 2020. Greenpeace and the European Renewable Energy Council estimate that a total additional investment in renewable energy over 2007–2030 of US$ 9.0 trillion (averaging US$ 390 billion per year) is required for the "Advanced Energy [Revolution scenario". The target of this scenario is the reduction of CO_2 emissions down to a level of around 10 Gt per year by 2050, and a second objective of phasing out of nuclear energy.

9.4 OVERCOMING BARRIERS: ENABLING CONDITIONS

Current levels of investment in renewable energy are still below what is needed to address the challenges facing the energy sector outlined earlier in the chapter.

Box 9.5: India's Solar Capacity May Double to 18 GW This Year

Solar energy capacity in India could nearly double to 18 gigawatt (GW) this calendar year as large projects get commissioned.

During 2017, the solar sector is likely to add close to 9 GW of capacity—taking its overall capacity to 18 GW and the country into the league of nations such as China, the US and Japan in terms of solar capacity, according to consultancy Mercom Capital Group Llc.

However, the expected capacity addition is much lower than the government's own target of 12GW for fiscal 2017 and 15GW for fiscal 2018 and also lower than what the tendering pipeline suggests.

India has a target of setting up 100 GW of solar and 60 GW of wind energy capacity by 2022. The growing energy sector requires $250 billion in investments to reach this target.

A cause of worry for the sector has been several instances of state power distribution utilities (discoms) unplugging their generating capacity from the grid and delaying both payments and the signing of power purchase agreements (PPAs) with renewable power producers.

Incomplete infrastructure at solar parks in India has been hurting the development of some large-scale solar projects. A sharp fall in solar module prices, however, has helped renewable energy producers, who have won solar projects at aggressive tariffs. Despite these hurdles, a large number of overseas investors including pension funds and infrastructure-focused funds have committed sizeable investments in the sector.

The major barriers and respective policy responses may be grouped under the following headings: (1) energy policy framework; (2) risks and returns associated with renewable energy investments, including fiscal policy instruments; (3) financing constraints for renewable energy projects; (4) electricity infrastructure and regulations; (5) market failure related to investments in innovation and R&D; (6) technology transfer and skills; and (7) sustainability standards.

9.4.1 Policy Commitment to Renewable Energy

In general, the growth in investment and deployment of renewable energy technologies, documented above, has been driven by an increasing number and variety of policies. Individual policies to overcome various barriers to renewable energy development and deployment are most effective when they are part of a broad enabling policy framework, which builds on complementarity between a range of measures operating at multiple stages of the chain from research and development, through to deployment and diffusion. An enabling policy framework for renewable energy includes clear commitments to long-term development of the sector. Such commitment can be manifested by targets for investment in additional capacity and penetration rates within the energy mix. When supported by other enabling policies, setting targets to achieve these goals can send a strong signal to potential investors.

9.4.2 Risks and Returns

As is the case in other sectors, the nature of risks, relative to expected returns, influences the incentive to invest in renewable energy. If a project or company has an expected risk-adjusted rate of return on investment that is sufficiently high, it is considered an interesting opportunity for financing. Taking first the risks in renewable energy projects, these can be categorised as follows:

(*i*) **Technical and project-specific risks,** including risks associated with lead times, construction costs, novelty of the technology, fuel and resources, and operations and management. Newer technologies have higher risks than traditional ones. Risks will therefore differ at regional or national levels.

(*ii*) **Country-specific institutional risks,** such as stability of the government, reliability of the legal system, transparency of business dealings, currency risks, and general instability due to wars, famine and strikes. For large-scale investments in a specific country, a long-term stable policy regime with a sound legal basis is needed;

(*iii*) **Political risk and regulatory risks,** such as unexpected changes in policy or uncertainty about the future direction of policy. Given the long pay-back periods, the contribution of policies to predictability, clarity and long-term stability in the investment climate, are viewed as critical in being able to stimulate more investments; and,

(*iv*) **Business and market risks,** including: (1) financial risks relating to the capital structure of the project such as high upfront capital intensity and the project's ability to generate enough cash flow; (2) economic risks relating to interest rates, exchange rates, inflation, commodity prices, counterparty credit risk; and (3) market risks associated with, forexample, futureelectricityand carbon prices (which may also be influenced by political and regulatory risks). Most renewable energy technologies are less vulnerable to the price and availability of fuel during the operation of a project. Those technologies that are dependent on biomass, however, do face potential market-price risks if the opportunity cost of biomass production is related to agricultural commodity prices and also because a reduction in fossil-fuel prices can make renewable energy less competitive in fuel and power markets. Such risks may be reduced with second generation, relative to first generation biofuels.

Various government initiatives, including regulatory policies, fiscal incentives and public financing mechanisms, can reduce many of these risks and thus increase expected returns. Such measures include offering long-term policy commitment to increased deployment of renewable energy investment, helping to mitigate political and regulatory risks. Shorter-term political commitment is similarly important. Owing to the long-lead times for project development, clarity over the development of regulation in support of renewable energy over a five-year horizon is desirable.

9.4.3 Financing Mechanisms

Over the past decade, a variety of formal and informal financial institutions and financing arrangements have emerged that offer facilitate small-scale products for the energy-poor in rural areas. The smallest projects are found in consumer-driven renewable energy solutions in developing countries, such as solar home systems. The high transaction costs involved call for innovative consumer finance mechanisms that address the particular needs of rural developing country customers. These mechanisms can make renewables attractive and cost effective for addressing energy poverty in off-grid situations.

9.4.4 Electricity Infrastructure and Regulations

The increased use of renewable energy in power generation faces specific barriers due to the demands it makes on existing electricity infrastructure.

In some situations, vested interests and control of access to the grid by incumbent power companies can pose barriers for independent providers of power from renewable sources. Similarly, oil companies may impede the distribution of biofuels through networks, such as pipelines, that they control. The construction sector may be reluctant to integrate renewable heating and cooling technology in their practices and building codes. Authorities have to be alert to signals from renewable energy companies and move quickly to address such market entry barriers.

Regulations may thus be needed to promote the types of investments in infrastructure necessary for further development of electricity from renewable sources.

9.4.5 Innovation and R&D

The technological development of renewable energy faces barriers due to the market failures inherent in innovation. Knowledge spillovers from research and development activities to create better products at lower costs benefit both consumers and other enterprises, but the potential innovator may not receive sufficient share of these to justify the investments. Furthermore, new technologies can be intuitive and easily learned, which contributes to cost reductions, which others are also able to apply. Both situations result in a general under-investment along all stages of the innovation chain.

9.4.6 Technology Transfer and Skills

Technology transfer is the flow of knowledge, experience and equipment from one area to another. Often, technology transfer is exclusively seen as being from an industrialised country to a developing country, but it can also be between developing countries or even from urban areas to rural areas.

Like other new technologies, renewable energy faces barriers that relate to technology transfer. Before a technology can be transferred successfully, enabling conditions need to be fulfilled, such as institutional and adaptive capacity, access to finance, and both codified and tacit knowledge of the technology. In developing countries, especially in remote rural areas, however, such conditions are often not present. Even when the economic feasibility of renewable energy options in those areas is favourable, these barriers can prevent their application.

9.4.7 Sustainability Standards

Renewable energy is not synonymous with sustainability. The term renewable refers largely to the naturally regenerative nature of the energy source, whereas sustainability has a broader scope, including economic, social and environmental considerations. Although renewable energy technologies are generally perceived as being more sustainable compared to non-renewable sources, due to smaller environmental impacts, there is still a need to develop agreed standards to reduce and manage these impacts. The environmental and social impacts of large hydropower reservoirs are one prominent example, including their potential to release carbon dioxide and methane from decaying biomass in tropical locations.

Increasing investments in renewable energy, as part of a green economy strategy spanning all major sectors, can contribute to reducing health and environmental impacts from energy production and use, while ensuring the basis for long-term economic growth. Such a strategy is based on the substitution of fossil fuel energy with renewable energy, savings from energy efficiency in manufacturing, buildings and construction, and transport, and behavioural change. Such an integrated strategy can increase national energy security and reduce carbon emissions while providing new employment opportunities that may, in global terms, more than compensate for jobs that disappear. This, however, should not prevent policymakers from recognising that in specific countries, depending on the extent to which fossil-fuel subsidies are phased out and negative externalities addressed, there could be net declines in employment, at least in the short term. The focus should be on specific countries and on practical ways of building capacity and skills to facilitate a transition to a green economy.

In order to play a role as part of an integrated strategy to reduce energy poverty, specific aspects of renewable energy development needs to be tailored to the circumstances in rural areas where the majority of the poor in developing countries live. Mini-grids and off-grids may provide a cost-effective means of delivering electricity to the poor, while also reducing growth in GHG emissions. This requires additional financing flows, as well as continued development of new financing models.

■ ■ ■

10

Guidelines for Green Manufacturing

10.1 INTRODUCTION

Manufactured products are a key component of human consumption, whether as finished or semi-finished goods. Manufacturing processes are a key stage in the life-cycle of material use, which begins with natural resource extraction and ends with final disposal. Basic industries such as cement, aluminium, chemical and steel supply the semi-finished, or intermediate goods, used to build houses, cars, and other appliances used in daily life. Other industrial sectors produce finished goods such as clothing, leather, fine chemicals, electrical and electronic products.

In *Our Common Future*, the Brundtland Commission foresaw industrial operations that are more efficient in resource use, generate less pollution and waste, are based on the use of renewable resources, and that minimise irreversible impacts on human health and the environment. This vision became the drive for concepts such as Cleaner Production promoted by UNEP and others since the 1980s. It remains a challenge for manufacturing industries world-wide, highlighting a need for more fundamental change in which the purpose of products and side-effects of manufacturing become a source of inspiration for re-design and beneficial output.

In order to implement a strategy of sustainable use of natural resources based on integrated resource management and resource efficiency, policy interventions supplemented by voluntary initiatives are needed at each stage of the life-cycle of production and use. The balance between upstream and downstream interventions is up for policy debate. Upstream policy interventions, for example, at the stage of mineral extraction or forest harvesting, to minimise adverse environmental impacts or to charge users appropriately for depletion or appropriation of resource rents, would have the effect of raising input prices to manufacturing companies.

Policy interventions targeted at manufacturing companies with the aim of reducing pollution to air and water, safeguarding health from exposure to toxic chemicals, and emitting greenhouse gases can also have the effect of increasing the cost of using resource inputs. These, together with other measures, can be powerful drivers in encouraging manufacturing industries to become more efficient in their use of natural resources and energy. Measures intended to improve the performance of markets for secondary raw materials and to encourage recycling can help to improve the performance of manufacturing companies in reducing their use of virgin raw materials. These are all building blocks for moving us closer to the vision described in *Our Common Future.*

10.2 GREEN MANUFACTURING

Historically, GDP has grown more rapidly than material, energy and labour inputs required to produce it. This has been owing to a combination of structural change, as service consumption sectors have grown faster than material consumption, technical change, which, has reduced material and labour inputs (*e.g.* automation) per unit of production, and more stringent environmental policies, which have driven up the cost of using some pollution-intensive inputs. This resulted, among others, in *relative* decoupling of resource input from output and *absolute* decoupling of some of the associated environmental pressures. Yet resource-efficiency gains have been offset by economic and population growth: overall emissions, energy use and material use continued to grow despite lower emissions, energy and material use per unit output. Without *absolute* decoupling, continuous economic growth implies continuously higher energy and resource demands, to levels that put the health of our natural resource base at risk.

The greening of manufacturing is essential to any effort to decouple environmental pressure from economic growth. Green manufacturing differs from conventional manufacturing in that it aims to reduce the amount of natural resources needed to produce finished goods through more energy- and materials-efficient manufacturing processes that also reduce the negative externalities associated with waste and pollution. This includes more efficient transport and logistics, which can also account for a significant percentage of the total environmental impact of manufactured products.

10.3 CHALLENGES—THE RISKS AND COSTS OF INACTION

The new economic reality for manufacturing industries today include key structural changes such as the globalisation of production with transnational supply and demand, strong economic growth in Asia (notably China) and an increase of raw material prices.

1. Natural Resource Scarcity

Resource scarcity is an increasing threat to future economic growth and a real challenge to the manufacturing industries, especially scarcity of fresh water, oil and gas, and some metals. Secure resource provision needs to be supported by healthy ecosystems, the vitality of which depends on biodiversity. The TEEB for business report highlights what is called the "impacts and dependencies" of the manufacturing industry on biodiversity and ecosystem services, reflecting the footprint of facilities and the pollution arising from production processes, as well as the role of suppliers of raw materials or semi-finished goods. These linkages are often complex and sector-specific. In the case of direct impact and dependency on biodiversity, the industries most implied include the pulp and paper industry as well as the textile and leather industry. If one considers high dependence on specific ecosystem services, this points to a wider range of industries. What they face is dependencies that pose risks associated with operations, markets, finance, regulations and reputation. A clear operational risk is that of increased scarcity and cost of natural resources.

Land use is mainly a problem related to agriculture and food production, rather than industrial production. The exception may be the future production of biomass for energy and feedstock purposes in industry. But industry is likely to face a significant challenge with regard to water in some countries or regions although it is responsible for less than 10 per cent of water use globally. Agriculture dominates with 70 per cent, followed by the energy sector and domestic uses, each with 10 per cent. Owing to expected high growth of industrial production, **water use** by industry is expected to grow to over 20 per cent of global total demand by 2030. At the same time, by 2030 a potential water shortage of 40 per cent of expected demand, compared to maximum sustainable supply, is projected at the global level. The extent to which industry drives water demand is highly differentiated by region and river basin (Fig. 10.1). The implications of this are that industries operating in regions of high water stress, and regions where industrial water demand is relatively important compared with other water demand, must improve their water productivity greatly or relocate to more water-abundant locations. This is particularly true for industries with high water use, such as the paper and pulp, textiles and leather and the steel industries.

Demand for water by industry (and for the electric power sector) increasingly competes with water demand by agriculture and urban consumers. In addition, all of this needs to be balanced with water demand by ecosystems and biodiversity. Water treatment is a necessary precondition for industrial (or consumer) water use. About half of industrial water use is for cooling purposes, and about a fifth of this water is lost as vapor, but much of the other four-fifths can be used downstream for other purposes (although the discharge of heated water can be harmful to aquatic ecosystems). The best way to reduce water loss for cooling large central power facilities is to find productive uses for the heat. This strategy, called co-generation or combined heat and power (CHP), is applicable in urban areas, industrial parks and in buildings generally, but its widespread application requires a major change in the structure of the electric power grid. Other industrial water uses include quenching of hot coke or red hot steel ingots, wood pulping, washing, rinsing and dyeing of textiles, tanning of leather and surface finishing of metals (including electroplating). These uses leave polluted and sometimes toxic waste streams that need treatment (which uses even more water), and whose costs in many instances are not reflected in the cost of production.

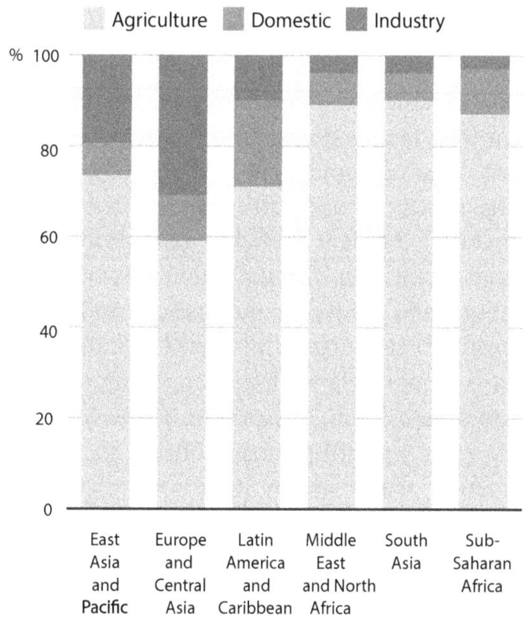

Fig. 10.1: Water demand in end-use by region.

Reserves of easily recoverable oil are diminishing, stimulating technological innovation to extract oil from deep ocean underwater reservoirs and non-conventional sources, such as oil and tar sands, and natural gas from shale, as a close substitute for many uses of petroleum. Since the early 1980s the amount of new oil discovered each year has been less than the amount extracted and used (Fig. 10.2). The overall peak is only a question of time. However, market forces including high prices may reduce demand and increase the use of substitutes, causing demand to peak before supply. Some think peak oil may still be 20 years in the future. Others think it has happened already.

The energy and other costs of replacing oil exploration and development are rising. The energy return on investments in energy (EROIE) of oil discovered in the 1930s and 1940s was about 110, but for the oil produced in the 1970s it has been estimated at 23, while for new oil discovered in that decade it was only 8. Decades ago, only 1 per cent of the energy in oil discovered was needed to drill, refine and distribute it, but since then the EROIE has declined drastically. In the case of deep-water oil, the EROIE is not above 10.

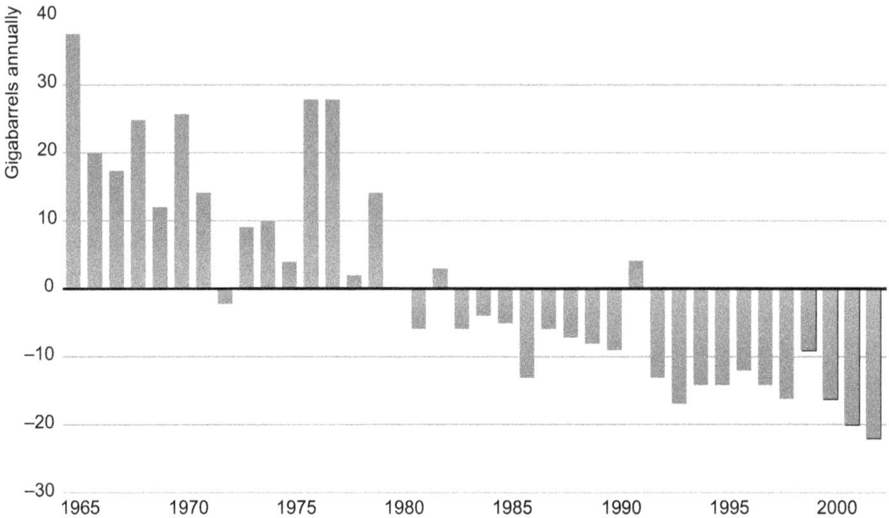

Fig. 10.2: Discovery rate of oil trend, 1965–2002.

High quality metal ores are also gradually being depleted. While absolute scarcity is not yet perceived as an immediate problem for most metals, the indicators on the life expectancy of reserves show that lower grade ores must be used. However, in order to do so more energy is needed to extract the useful metal content, adding marginally to GHG emissions.

Against this background, resource-intensive sectors face a multitude of challenges. First, rapidly industrialising economies are building their infrastructure rapidly and requiring large amounts of resources. Competition over access to resources is likely to grow. Second, high quality metal ores are gradually being depleted. This leads to use of lower grade ores, which require much more energy to extract its useful metal component. Third, at local level resource extraction can have significant impacts on ecosystems and landscape. Mitigating these impacts through environmental policy or industry initiatives can also increase the cost of extraction. Fourth, there are risks of security of supply and price volatility.

Not all industrial production sectors are equally affected by these challenges, and not all materials are equally important in terms of economic or environmental impacts. Many minerals that dominate consumption by mass are of marginal importance for global warming, human toxicity, land use, or an integrated "Environmentally Weighted Material Consumption" index. Indeed, environmental impacts are dominated by fossil fuels, their derivatives (such as plastics) and biotic materials.

Resource scarcities—absolute or relative, actual or perceived—impact the prices of commodities and manufacturing inputs. Since the mid-2000s, commodity prices have shown an increasing volatility, which is mainly owing to a series of energy, financial and food crises. Economic recession, in turn, reduces demand for oil and can be followed by an equally drastic price decline that is further exaggerated by speculation. Thus, price volatility can seriously inhibit long-term green investment.

Since the early 2000s, other commodity prices, especially non-ferrous metals, have also been sensitive to shortterm factors such as the boom in China coupled with recession in the US, depreciation of the US dollar (all commodities are priced in US$) and speculative activity. In 2008, commodity prices exceeded previous records from the 1970s. Higher prices induce investment in alternatives, but excessive volatility tends to have the opposite effect, because it prevents rational planning.

It is important to differentiate between short and long-term impacts and trends. When prices for natural resources rise because long-term trends in demand begin to exceed long-term trends in supply, or when governments internalise some of the environmental costs of natural resource extraction or use to business, the response of market participants can facilitate the adjustment process. Manufacturers are more likely to adopt innovative technologies that can improve resource efficiency. To the extent that this is not fully sufficient to absorb the increase in costs, the selling price of their products will increase, providing an incentive for consumers to search for less costly substitutes in the market place. Meanwhile, exploration and development of additional resources will occur, and markets will reach a new equilibrium at a higher price that stimulates innovation.

2. Hazardous Substances and Waste

Other significant environmental externalities at a global scale include impacts associated with hazardous substances and waste. The waste sector produces pressure on the environment through releases from landfills, domestic and commercial waste-water treatment and industrial waste-water.

Releases of toxic substances cause health and safety problems and ecosystem degradation. Some countries have made significant progress by applying cleaner production, product substitution and end-of-pipe measures. In developed countries, toxic emissions have been one of the few success stories, with releases and exposure diminishing while production and GDP grew. This is related to the fact that most toxic substances are emitted as small mass flows, and for which substitution or emission reducing measures are relatively easy to achieve. Production patterns have changed radically, with industries based in developed countries focusing on high-value chemicals and pharmaceuticals. The manufacture of high production volume (HPV) chemicals, on the other hand, has been progressively migrating to developing countries, where regulatory frameworks are often lacking and where costs for the sound management of industrial (hazardous) waste are rarely internalised.

In the absence of good waste management, the following industries in particular may face toxicity challenges:

- Textile industry and leather industry in relation to dying and tanning products;
- Paper and pulp industry in relation to bleaching processes and related water emissions;
- Chemical and plastics industry, depending on the type of chemicals produced; and
- High-temperature processes such as in the cement and steel industry, where the formation of by-products or emissions of metals can be a problem.

Of the hundreds of thousands of chemicals on the market, only a small fraction has been thoroughly evaluated to determine their effects on human health and the environment. Some chemicals that have been used in large quantities for many years are now suspected of carcinogenicity or teratogenicity. Some of the most toxic and dangerous chemical products (such as DDT) have been phased out, at least in the OECD countries. Adverse human health effects of chemicals include acute and chronic poisonings, neurodevelopmental disorders, reproductive/developmental disorders and cancer. Preventing chemical pollution at the source avoids generating harmful wastes and emissions while reducing and eliminating costs of cleanup.

Gaps in applying standards for industrial safety and accidents give historical examples of the risks and societal costs that can be associated with industrial production, in particular where hazardous substances are involved.

The cost of industrial accidents represents a great source of public and private expenditure and social distress. Over the past three decades, a rough cost assessment of only a few of the major industrial accidents worldwide shows that a minimum of US$ 40 billion have been spent on addressing the damages. If smaller incidents are taken into account, the real economic cost is likely to double, while deaths and injuries would be on the scale of several hundreds of thousands. Some major incidents are listed in Table 10.1. Clearly, there are global benefits in human and environmental health associated with cleaner and safer industrial production, which has to be part of a transition to green manufacturing.

10.4 OPPORTUNITIES—STRATEGIC OPTIONS FOR THE MANUFACTURING SECTOR

In its *Vision 2050* report, the World Business Council on Sustainable Development describes a world in which the manufacturing industries follow life-cycle approaches that enable dematerialisation and expanded service systems.

Table 10.1: Examples of major industrial accidents and related economic and social costs

Location	Date	Cost (US$)	Number of fatalities and injured
Chemical industry			
Bhopal, India	03/12/1984	US$ 320 million in claims and compensation; US$ 10 million in economic, medical, social, environmental rehabilitation. However, the Indian government estimated the cost of the Bhopal disaster at US$ 3.3 billion.	2,800 fatalities and estimated 170,000 long-term adverse health effects
Toulouse, France	21/09/2001	€2 billion (environmental and social cost)	31 fatalities and 4,500 injured
Oil and Gas industry			
North Sea	06/07/1988	US$ 3.4 billion (mostly clean-up cost)	167 fatalities
Gulf of Mexico	20/04/2010	US$ 6.1 billion (as of 09/08/2010), (containment, relief, grants to the US Gulf states, claims paid, and federal costs); creation of a US$ 20 billion escrow account for clean-up and other obligations.	11 fatalities (oil platform workers)

Source: Adapted from Mannan (2009), Grande Paroisse–AZF (2010), Kuriechan (2005) and BP (2010).

In a sustainable world of about 9 billion people by 2050, a complete range of new products and services is offered, based on high longevity, low embodied water, as well as low-energy and material content. This transition will not happen overnight, and it will require substantial investment. A major challenge is one of transition in industrial production, to become less carbon and material intensive while at the same time preserving jobs or reinvesting in completely new employment opportunities. This is particularly relevant for developing and emerging economies that currently invest heavily in conventional production infrastructure. Both at the country and industry sector level, improved resource-efficiency and decoupling offers the opportunity of competitive advantage and a sustainable future.

To what extent will green investments in efficiency have a more favorable payoff than conventional investments? Big companies normally set their hurdle Rate of Return on Investment (ROI) at around 25 per cent, pre-tax. There is overwhelming evidence of significant opportunities for efficiency investments that yield much higher rates of return, even under current economic conditions. The economic opportunities increase dramatically at higher carbon prices.

10.4.1 Decoupling and Competitive Advantage

As indicated earlier, historical evidence shows that declining energy intensity in industry and relative decoupling have typically been offset by increases in energy demand associated with higher levels of GDP. In addition, there may have been additional demand for energy as an input owing to a decline in its relative price and to the increase in economic growth owing to the gain in resource efficiency itself; the two effects together are sometimes called the rebound effect. Overall emissions, energy use and material use have kept on growing despite lower emission, energy and material use per unit output. Resource extraction per capita has been stable or increasing only slightly. What economies world-wide need is absolute decoupling of the environmental pressure associated with resource consumption from economic growth. This will be easier to achieve to the extent that resource use itself becomes more efficient.

In recent decades, OECD countries have decreased their extraction intensity per US$ of GDP, reflecting some decoupling of primary resource extraction from economic growth. This trend is expected to continue. The main drivers are increased applications of more material-efficient technologies (technology effect), shifts from the primary and secondary sectors towards the service sector (structural effect), and associated increases in material-intensive imports (trade effect) owing to outsourcing of material-intensive production stages to other world regions. For the world as a whole, of course, there is no trade effect because one country's imports are another country's exports.

The decoupling ofmaterial use from GDP growth has been less pronounced in fast-growing transition economies that need to build infrastructure, which requires more resources (in mass terms) than in economies with low growth rates. Similarly, the energyintensive industrial sectors are not equally affected. The cement industry drives large material flows, but of relatively non-scarce resources such as limestone and clay. Iron ore and bauxite are not particularly scarce, and near substitutes are available. The paper and pulp and the natural fibre-based textile industry use renewable resources where the challenge is to avoid using them beyond the maximum sustainable yield. The challenges for the electrical and electronic industry may be more fundamental. High grade (> 1 per cent) and easy-to-refine copper ores are becoming scarcer and low-grade ores need more energy in the extraction and refining stages. Rarer metals such as silver, indium and tellurium are mostly extracted from other metallurgical wastes.

One of the major effects of the globalised nature of the world economy is the increasing shift of the manufacturing base from developed to developing and transition economies. This means that associated environmental damages from local pollution are also shifting. Accordingly, decoupling energy use and CO_2 emissions from GDP growth needs to be considered in the international context, rather than in terms of individual countries.

Improving the environmental efficiency of production at the global level can occur through technology and knowledge transfer from developed economies or through technology spillovers that occur as a result of international investment and globalised supply chains. With demand increasingly being driven from outside the advanced economies, these transfers and spillovers have dual benefits—not just reducing the extent of environmental damage exported from developed countries, but also helping developing economies shift to a more resource-efficient growth path.

10.3.2 Innovation in Supply and Demand

Making society more efficient with regard to the use of energy, water, land and other resources is a challenge that requires changes along the full chain of production and consumption. Scholars have suggested that one way to realise "Factor X" improvements in resource productivity would be a radical change in end-use products, new ways of (*e.g.* shared) using products (*e.g.* sharing), and changes in consumption habits. This includes consideration of concepts such as "sufficiency" and asking critical questions about the function and service of proposed products.

It also requires a life cycle approach. This concept focuses on those resource efficiency measures that also generate a positive rate of return to business on the required investments. Eco-efficiency provides a graphic tool for combining different measures, yet still has shortcomings in allowing quantification and comparison based on empirical indicators. The guidelines behind eco-efficiency include reducing the material and energy intensity of products, enhancing material recyclability, extending product durability and increasing the service intensity of products.

■ ■ ■

11

Guidelines for Green Waste Management

11.1 INTRODUCTION

The environmental and social (including health-related) benefits from greening the waste sector have been stressed already for a long time. The impact of this has, however, been limited, as environmental and social concerns are often seen as competing with economic imperatives. Environmental and social aspects of greening the waste sector are discussed, but the emphasis is on making an economic case based on the available data.

11.2 GREENING THE WASTE SECTOR

Greening the waste sector refers to a shift from less-preferred waste treatment and disposal methods such as incineration (without energy recovery) and different forms of landfilling towards the three Rs: Reduce, Reuse and Recycle. The strategy is to move upstream in the waste management hierarchy, based on the internationally recognised approach of Integrated Solid Waste Management (ISWM) (Fig. 11.1).

The ISWM is a strategic approach to managing all sources of waste; prioritising waste avoidance and minimisation, practicing segregation, promoting the 3Rs, implementing safe waste transportation, treatment, and disposal in an integrated manner, with an emphasis on maximising resource-use efficiency. This marks a departure from the usual approach where wastes are managed mainly from a compliance point of view characterised by end-of-pipe treatment such as incineration (without energy recovery) and landfilling.

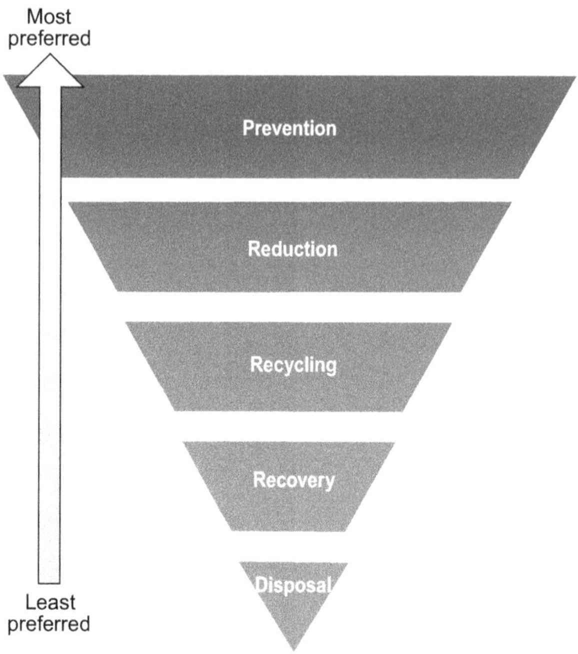

Fig. 11.1. The waste management hierarchy.

Under ISWM, activities of greening the sector can include:

- Resource conservation, which avoids excessive resource consumption;
- Waste reduction through resource use optimisation that minimises resource wastage;
- Waste collection and segregation, ensuring appropriate waste treatment;
- Waste reuse, which circulates waste and avoids the use of virgin resources;
- Waste recycling, which converts waste into useful products;
- Energy recovery, which harnesses residual energy from waste;
- Landfill avoidance, which conserves land and avoids risks of contamination; and
- Construction and maintenance of infrastructure for waste collection, recovery of materials from waste streams (collection and segregation) and application of 3R technologies and associated activities.

Indicators to measure the progress of greening the sector can include:

- Resource consumption rate (material use in kg per capita);
- Waste generation rates (kg per capita/year, overall and by economic sector);
- Proportion of waste being collected and segregated;

- Proportion of materials in waste streams being reused or recycled;
- Proportion of virgin material displacement in production;
- Proportion of waste used for energy recovery;
- Proportion of materials in waste streams diverted from landfill;
- Reduction in greenhouse gas (GHG) emissions due to avoided landfilling;
- Proportion of total waste disposed in landfill; and
- Extent of capture, recovery and/or treatment of polluting emissions such as leachate and landfill gas.

In relation to an overall green economy, indicators of greening the waste sector can include the value of—and jobs related to—the goods generated through the greening of the waste sector such as remanufactured products, recovered energy, and the services in terms of waste collection, segregation, and processing. Economic and social benefits in terms of health, property values, tourism as well as direct and indirect job creation should also be included. Not all of these indicators may, however, be readily available. Proxies are used where possible in this chapter to gauge and estimate the economic significance of greening the sector.

11.2.1 A Vision for the Waste Sector

The long-term vision for the waste sector is to establish a circular global economy in which the use of materials and generation of waste are minimised, any unavoidable waste recycled or remanufactured, and any remaining waste treated in a way that causes least damage to the environment and human health or even creating additional value by recovering energy from waste. To achieve this vision, radical changes to supply-chain management, especially to the product and industrial design part of the supply chain, are needed. Specifically, the 3Rs need to guide industrial design—with implications for materials at all stages—and be overlaid on the entire supply chain. This requirement is, in turn, expected to motivate innovation. The chapter on manufacturing further elaborates on life-cycle approaches, including closed loop and circular systems in manufacturing.

11.3 CHALLENGES IN THE WASTE SECTOR

The waste sector is facing three sets of challenges: (1) increasing growth in the quantity and complexity of waste streams associated with rising incomes and economic growth; (2) increasing risk of damage to human health and ecosystems; and (3) the sector's contribution to climate change.

11.3.1 The Growing Volume and Complexity of Waste

The exploitation of the earth's resources continues apace; material use increased eight-fold in the last century. According to the Wuppertal Institute, an average European consumes about 50 tonnes of resources a year, around three times the amount consumed per capita by emerging economies. Furthermore, on average, Europeans dispose twice as much as citizens from emerging economies. Per-capita resource use in emerging economies is also increasing considerably while the world's Least Developed Countries (LDC) are beginning the transition towards an industrial type of societal metabolism, as incomes rise and purchasing power is deployed in consumer spending.

Government Tightens e-waste Rules, Offers Sops to Consumers

Waking up to the challenge of managing hazardous e-waste, the Centre has notified revised rules to collect, dismantle and recycle all electronic and electrical waste by expanding its ambit, extending incentives to consumers who hand over waste to dealers or retailers and introducing a financial penalty clause for violators.

The government, under E-waste (Management) Rules, 2016, has, for the first time, also brought under its purview the management of the waste generated from disposal of Compact Fluorescent Lamp (CFL) and other mercury containing lamps. These items were not covered under the earlier rules, framed five years ago.

Though the responsibility to implement these rules will lie with the state governments, new economic instruments of `e-waste exchange' and `deposit refund scheme' will act as incentives for consumers to voluntarily adopt the system of waste management.

Under the `deposit refund scheme', the manufacturer or producer will charge an additional amount as deposit at the time of sale of electronic or electrical equipment. It will be returned to the consumers along with interest when the end-of-life equipment is returned.

It will, however, be an optional financial mechanism for effective implementation of `extended producer responsibility' that makes the producers or manufacturers responsible for collection of e-waste as well as its disposal in an environmentally sound manner. If the manufacturers or producers fail to adhere to their roles, they will be financially penalised as per the amount specified by state governments.

On the other hand, the concept of `e-waste exchange' has been introduced as an independent market instrument offering services for sale and purchase of e-waste. It means, if a consumer deposits his/her end-of-life equipment to such service providers, he/she will get specified amount for the scrap.

Under the new rules, the dealer, if given the responsibility of collection on behalf of manufacturers or producers, will need to collect e-waste by providing the consumer a box and channelise it to the producer for sending it to authorised dismantling centres.

New rules say the department of industry in state will ensure earmarking of industrial space for e-waste dismantling and recycling in the existing and upcoming industrial parks.

Currently, 3.4–4 billion tonnes of municipal and industrial waste are produced every year, of which non-hazardous industrial waste accounts for 1.2 billion tonnes. A major share of the waste generated is MSW originating from urban settlements (1.7–1.9 billion tonnes, or 46 per cent of the total waste generated) with 0.77 billion tonnes of this being produced by 25 Organisation for Economic Cooperation and Development (OECD) countries alone.

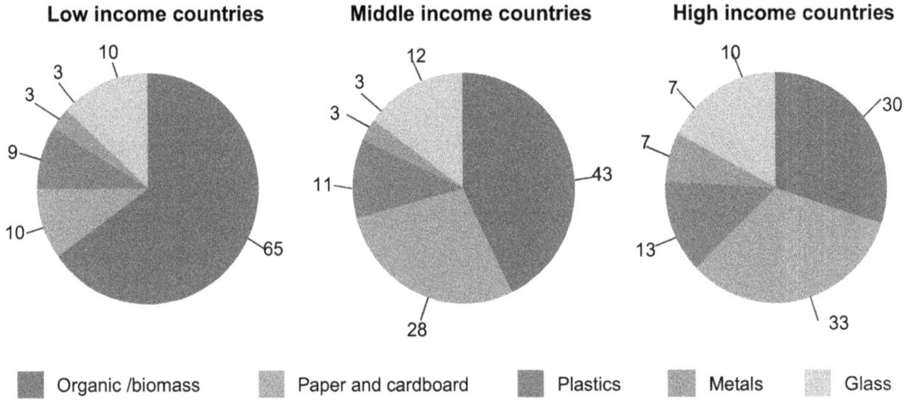

Fig. 11.2: Composition of MSW by national income.

As a country develops and becomes wealthier, the composition of its waste stream typically becomes more varied and complex. Figure 11.2 illustrates the high proportion of organic-rich MSW in middle and lower income countries with a gross national income per capita of less than US$ 12,196, while the high-income countries' MSW streams contain a large proportion of paper and plastics.

Apart from MSW, other major types of waste streams are listed below:

- Construction and Demolition (C&D) waste represents 1015 per cent of total waste generated in developed countries and some countries have reported a much higher proportions. For example, OECD estimated that Germany generates 178.5 million tonnes of C&D waste, which is about 55 per cent of the total waste reported. C&D waste can be classified as high-volume waste with relatively low impact compared with other types of waste.
- End-of-life Vehicles account for 8–9 million tonnes of waste in the European Union (EU) with Germany, UK, France, Spain and Italy responsible for approximately 75 per cent of EU-25 vehicles de-registrations. Japan generates about 0.7 million tonnes of Automobile Shredder Residues (ASR) every year - materials such as plastic, rubber, foam, paper, fabric, glass, etc. that remain to be recycled after the reusable parts of the automobile are removed from shredded EoLV. In the United States of America, ASR amount to 5 million tonnes annually.
- Biomass waste includes agricultural and forestry waste. It is estimated that globally 140 billion tonnes of agricultural residue is generated every year. Like C&D, biomass waste is a high-volume waste with a relatively low impact.
- Health-care waste is sometimes classified as a subcategory of hazardous waste. No global estimates are available. On average, however, low-income countries generate between 0.5 kg and 3 kg of health-care

waste per capita per year, which includes both hazardous and non-hazardous components. High-income countries have been reported to generate up to 6 kg of hazardous waste per person per year from healthcare activities.

- Electronic waste (e-waste) continues to increase dramatically amid growing global demand for electronic and electrical goods. Like personal computers and mobile phones Jinglei Yu *et. al.* predict that obsolete PCs in developing regions will exceed those of developed regions by 2016–2018 and that by 2030 they could amount to 400–700 million units (compared with 200–300 million units in developed countries).
- Hazardous waste requires special handling and treatment even in low quantities. They may also mix up with the stream of waste generated in the municipal or agricultural sector, for *e.g.* used batteries, spent paints and residual chemical pesticides as well as Ozone Depleting Substances (ODS) such as refrigerators, air- conditioners, fire extinguishers, cleaning products, electronic equipments and agricultural fumigants. Reports submitted to the Basel Convention suggest that at least 8.5 million tonnes of hazardous waste have been crossing international boundaries every year.
- Packaging waste and its management has become a major issue in high-income countries. According to the European Environmental Agency, an increase in packaging waste has been observed in both older and newer European Union (EU) member states.
- Marine litter consists of material discarded directly or indirectly from recreational/shoreline, ocean/waterway, smoking-related, dumping and medical and personal-hygiene-related activities and sources. Marine litter has been reported to have significant impacts on wildlife and sensitive ecosystems, human health and safety and the economies of coastal areas.

Soon, 'Green' Tyres Made of Biomass

'Green' car tires—made from trees and grasses—may soon be a reality , thanks to scientists who have developed a new technology to produce a key molecule in automobile tires using renewable resources found right in our backyards.

Conventional car tires are viewed as environmentally unfriendly as they are predominately made from fossil fuels.

The new car tires produced from biomass that include trees and grasses would be identical to existing car tires with the same chemical makeup, colour, shape and performance. "Our team created a new chemical process to make isoprene, the key molecule in car tires, from natural products like trees, grasses, or corn," said Paul Dauenhauer, an associate professor at University of Minnesota in the US.

"This research could have a major impact on the multi-billion dollar automobile tires industry," Dauenhauer said.

Currently, isoprene is produced by thermally breaking apart molecules in petroleum that are similar to gasoline in a process called "cracking."

Waste generation is linked to both population and income growth. Of the two, income level is the more powerful driver.

Figure 11.3 shows estimates of MSW generation in different parts of the world. It rose in the US and the EU by 21 per cent and 14 per cent respectively from 1995 to 2007. However, due to increased awareness and policy interventions addressing waste management (for example, EU regulations stimulating recycling of obsolete vehicles since 2000 and electrical and electronic waste since 2002), the rate of MSW generation slowed in the EU and (to a lesser extent) in the US in the period from 2003 to 2007. The linkage between affluence and waste generation remains quite strong, in spite of improvements in efficiency, and represents a significant challenge for developing countries as they become wealthier, particularly in Asia. At best, relative decoupling has begun in OECD countries, with a stabilisation of per-capita waste generation in the last decade, as shown in Fig. 11.4.

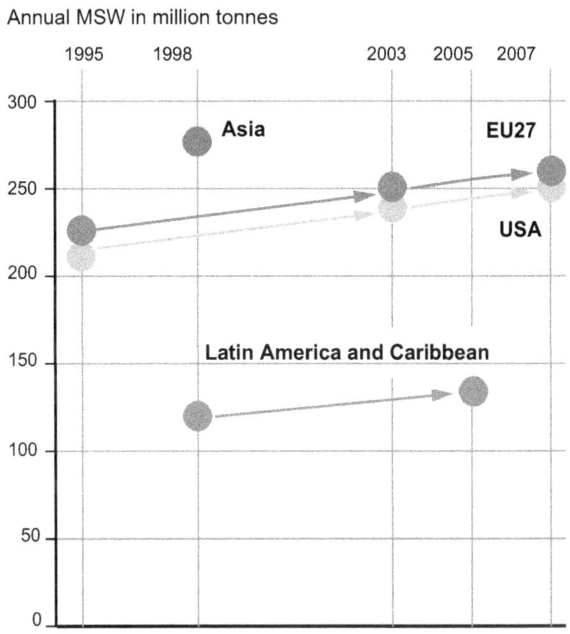

Fig. 11.3: Estimated generation of MSW across regions of the world.

The recent awareness on benefits of waste minimisation, but also the shifting of waste-intensive production to developing and emerging countries may have contributed to this development. Landfill remains the predominant method of disposal in these countries.

Waste volumes are not necessarily the most important challenge ahead. Mixed MSW, hazardous health-care waste, and industrial waste streams can impose serious health and ecological risks if these wastes remain uncollected or dumped in uncontrolled and unsecured landfill sites. In low- income countries, for example, collection rates are lower than 70 per cent, with more than

50 per cent of the collected waste disposed through uncontrolled landfilling and about 15 per cent processed through unsafe and informal recycling. Given the amount of valuable components in MSW, the mixing of wastes also means a lost opportunity to recover components that could be recycled and used as new resources.

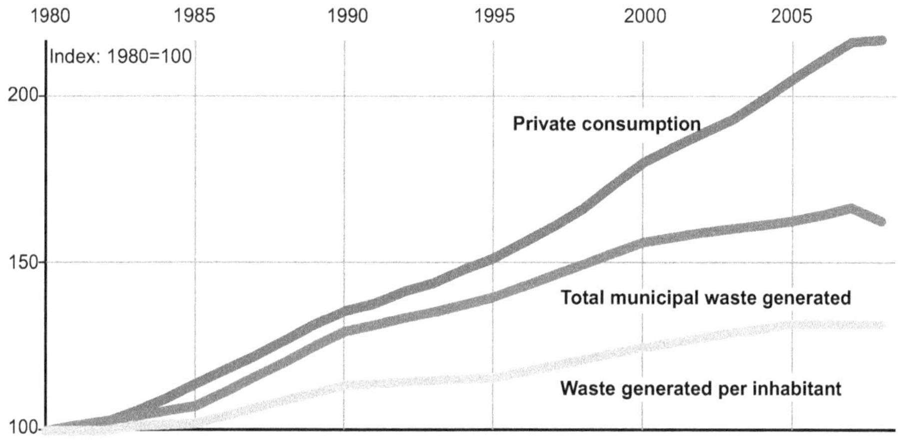

Fig. 11.4: Relationship between private consumption and municipal waste in OECD countries.

Globally, UNEP and the United Nations University (UNU) estimate that 20 to 50 million tonnes of e-waste are disposed of each year, which accounts for 5 per cent of all MSW. E-waste also has a significant role to play in the recycling sector in developing countries even though it is not necessarily generated in those countries. With sales of electronic products in China and India and across Africa and Latin America predicted to rise sharply in the next ten years, the challenge is only set to grow.

Adding to the complexity of waste streams is the impact of increasing trade on waste. Lack of information on the constituents of waste products, such as valuable raw materials and toxic pollutants, makes trading of such waste challenging and risky. There have been increasing packaging requirements to minimise damage to goods in transit. Packaging requirements have also increased to meet tightened food health and safety standards.

The waste problem has been accentuated by the issue of waste trafficking. Several developed countries have been illegally dumping hazardous waste and exporting significant quantities of used electrical and electronic products to developing countries that do not have adequate infrastructure to manage them. Such illegal shipments are a matter of global concern. The Basel Convention requires its members to report the aggregated numbers, but there is ambiguity in the available data on hazardous shipments and difficulty in dealing with illegal activities. Another issue is the difficulty in classifying used electronic or electrical products as second-hand products and hazardous waste. These shortcomings heighten the threat that the hazardous waste poses to the environment and human health.

Solid Waste Policy in India

The "policy", specifies the duties and responsibilities of cities and citizens, for hygienic waste management. The starting point was a report prepared in 1999, and submitted by a committee to the Supreme Court, which asked that its recommendations be followed. What resulted was the framing of the policy in 2000.

As with all "time-bound" programmes, a strict time-table was laid down. Existing landfill sites were to be "improved" by end-2001, new landfill sites were to be identified by end-2002, and waste-processing facilities were to be set up by end-2003. Municipalities were to ensure "community participation", there was to be no manual handling of waste, and dry waste collection was to be left to the "informal sector." There was more, all based on high principle and all of it politically correct. Municipalities were to charge for waste collection (except in slums), on the "polluter pays" principle, with the charges indexed to the cost of living. And finally, the Supreme Court would ask the various High Courts to monitor compliance.

11.4 OPPORTUNITIES

The opportunities for greening the waste sector come from three inter-related sources: (1) growth of the waste market, driven by demand for waste services and recycled products; (2) increased scarcity of natural resources and the consequent rise in commodity prices, which influence the demand for recycled products and WtE; and (3) emergence of new waste-management technologies. These developments have opened up significant opportunities for greening the waste sector.

11.4.1 Growth of the Waste Market

Despite data limitations, there is a clear indication that the market for waste management is growing. The world market for municipal waste, from collection to recycling, is worth an estimated US$ 410 billion a year. This estimate can only be indicative since assessing the exact market size is difficult given the paucity of reliable data, particularly in developing countries, and existing data being limited to the formal component of the waste-management sector.

Four factors are driving this growth: (1) the overall increase in the volume and variety of the waste generated; (2) rising political awareness of the need to better manage waste in the context of avoiding ecological and health risks and climate change; (3) urbanisation in emerging economies, which is typically accompanied with a growing interest in a better living environment including better waste management; and (4) development of formal and informal trade in secondary raw materials recovered from waste.

Change in the consumer demand is a major determinant underlining the potential greening of the waste sector. With increased environmental awareness, more and more consumers have started demanding recycled products and waste-derived compost.

Throwaway Living

In August of 1955, Life magazine published an image of a man, woman and child cheerfully tossing in the air a bunch of sundry household items—a frozen food container, an all-purpose bucket, a sectional plate, paper napkins, a disposable diaper, a dog-feeding bowl, and an ash tray, among others. The accompanying text explained that all the items in the air would take about 40 hours to clean up but no housewife needed to bother about that any more because all the items were disposable. The piece was titled "Throwaway Living".

Published at a time when disposable items were gaining currency in the consumer market, partly because of the proliferation of plastics, it reflected on the ethos of a newly booming industrial economy. Over the next couple of decades, this ethos would trickle down in an increasingly inter-connected world from the Global North to the Global South.

So much so that today, even in a developing nation like India, a mobile phone that is more than two years old is considered obsolete and, increasingly, consumers prefer to just buy a new device rather than repair a broken screen or a malfunctioning motherboard. Yet, hardly a generation ago, durability was the norm, household goods were used for decades, and hand-me-downs were commonplace. But even as India, particularly its aspirational middle and neo-middle class, takes to the values and trends of wealthy consumerist societies, in the West, a reversal of attitudes seem to be underway.

There is also a mounting effort to move from a linear economy, wherein resource-intensive goods end up in landfills after a one-time use, to a circular economy, wherein products and processes are designed for reuse, refurbishment or even remanufacturing. For example, popular brands like H&M, American Eagle Outfitters, Levi Strauss, The North Face and Zara now collect old garments at their stores, which are repurposed. Interestingly, India imports large quantities of used clothing from the US, some of which is resold while the rest is pulped and recycled to make blankets, doormats, prayer rugs and pillow stuffing.

The growth of the waste market, however, does provide an opportunity for greening the sector. As the market evolves and becomes mature, consumers are likely to demand stringent standards in order to avoid any health and environmental risks. In the waste sector, existing standards focus mostly on the protection of environmental and human health, but working conditions and standards for recycled products are increasingly receiving attention. Market development in this direction thus provides a platform on which to systematically introduce green standards into waste management systems.

11.4.2 Scarcity of Resources

Rapid population growth and economic expansion have led to escalating demand for energy, basic industrial commodities and consumer goods.

Amid this rapid consumption of the earth's resources there appears to be great potential to create new markets by recycling and reusing existing metals, minerals, plastic, wood and other materials. Currently, however, only a quarter of the 4 billion tonnes of municipal waste produced each year is recovered or recycled. For example, scrap metals, paper and cardboard, compost, plastics are all valuable, relatively easy to recover from waste streams and can displace raw materials that are likely to become less readily available. One tonne of electronic scrap from PCs, for example, contains more gold than that recovered from 17 tonnes of gold ore and 40 times more concentrated copper than that found in copper ore.

The increasing scarcity of resources and the rising cost of extracting raw materials, which feeds into higher commodity prices, are turning waste into a new source of resources to be mined. Examples include the reprocessing of metal waste, composting, waste to energy, recycling of e-waste and C&D waste.

11.4.3 New Technologies

The greening ofthe waste sector is also facilitated by significant breakthroughs in technologies required for collection, reprocessing and recycling waste, extracting energy from organic waste and efficient gas capture from landfills. Compactor trucks, fore-and-aft tippers, container hoists and open-or-closed top tailers are now available for the collection and transportation of waste. Recovering energy and other useful products from waste has been enabled by considerable technological breakthroughs.

11.5 MAKING AN ECONOMIC CASE FOR INVESTING IN GREENING THE WASTE SECTOR

A case for investing in greening the waste sector may be made on various grounds. In the past, cases have been made largely on environmental and health-related grounds, based on costs that can be avoided by proper collection and disposal. These arguments, particularly health-related ones, continue to be important for motivating policy actions.

In order to scale up the greening of the sector, however, environmental and health-related arguments alone are inadequate or may be seen as competing with economic imperatives. For policy-makers to channel significant resources towards the greening of the sector, they need to appreciate how such actions are likely to contribute to economic performance and job creation relative to BAU scenarios. Adequate economic arguments are, therefore, needed to motivate fundamental changes in the management of the sector.

To make a primarily economic case for investing in greening the waste sector, three steps are needed, which are elaborated on in this section. First, we need to have an idea of the extent to which the sector could be greened. Second, we need to have some ideas about the financing gaps for priority areas. Third, given the priorities of greening the sector, we need to show the potential gains if green investment is made in those areas.

There are no established international targets for greening the waste sector, apart from the control of specific hazardous substances as governed by international conventions. Most goals are national or even local. For example, in northern Europe, the Republic of Korea and Singapore, over 50 per cent of waste is subjected to material-recovery processes. Japan has set material-flow indicators that fall under three categories: input, cycle and output, to compare developments in recycling rates with those of previous years. China has adopted the Circular Economy (CE) approach in a move towards achieving a more balanced growth. The Republic of Korea planned to increase its waste-recycling rate of MSW. Under the directive on packaging and packaging waste, the EU increased the target for overall recycling from 25 per cent (min.) and 45 per cent (max.) in 1994 to 55 per cent (min.) and 80 per cent (max.) in 2004. South Africa has set out a minimum set of target parameters for use by municipalities in the provision of waste services.

It is, difficult to have one-size-fits-all goals for the greening of the waste sector. Generally speaking, however, in greening the waste sector, all countries should seek to: (1) avoid waste in the first place through sustainable community practices, (2) minimise the generation of waste; (3) recover materials and energy from waste and re-manufacture and recycle waste into usable products, where waste is inevitable; and (4) treat any remaining unusable waste in an environmentally friendly or in the least damaging way. For developing countries, one of the goals should be the formalisation of the waste sector, following environmental guidance and labour- protection measures.

■ ■ ■

12

Guidelines for Greening Building

12.1 INTRODUCTION

According to the International Energy Agency, green buildings are characterised by increased energy efficiency, reduced water and material consumption, and improved health and environment. The International Organisation for Standardisation's definition of sustainable buildings combines a minimum adverse environmental impact with economic and social aspects across various geographic scales.

The last 40 years have seen much experimentation and significant progress with low-energy building design strategies and technologies. However, in most countries, green buildings are still at a nascent phase of development. Yet they are expected to become the norm in future. Experimentation with net-zero-carbon buildings, passive houses and energy-plus buildings are emerging worldwide.

12.2 SIZING THE BUILDING SECTOR

Driven by population growth and urbanisation, the building sector itself is a significant contributor to economic growth, both globally and at the national level. Globally, it is estimated to be worth US$ 7.5 trillion per year or approximately 10 per cent of global GDP and the construction sector employs more than 111 million people. At the national level, the sector generates 5–10 per cent of employment.

There are important differences between developed and developing countries in both the current building stock and projected building-sector growth. *Developed country* populations are broadly more urbanised and more economically reliant on the service sector than on industry or agriculture. They also have higher household incomes than developing country populations. Developed countries currently account for the majority of the world's existing building-related energy demand and CO_2 emissions.

In contrast, developing countries are fast-growing, rapidly urbanising and are projected to add 2.3 billion to global population over the coming four decades. Of the 9 billion people predicted to live on Earth in 2050, 70 per cent are expected to live in urban areas.

> The commercial construction industry is booming globally. Office towers and residential high-risen are popping up in Manhattan, Singapore, Dubai and everywhere. Sky-lines are changing fast in all metros in India. The United Nations says cities around the world are growing faster than ever, with 54 percent of the world's population now living in urban areas, and 66 percent expected to do so by 2050. According to the UN World Urbanisation Prospects report, urbanization combined with overall world population growth could add another 2.5 billion city dwellers by 2050. The report says that there were 10 megacities with 10 million people or more in 1990 and 28 today; it predicts 41 by 2030.
>
> But as the world's metropolises get bigger and reach higher into the sky to accommodate more people, the earth's natural resources supply chain is being pushed to the limit. That strain is not from increasing demand for gold, diamonds, or copper, but for sand, the primary material for construction and hence for economic development.
>
> Concrete and glass are made mostly of sand, a certain type of sand found deep below the earth's surface, underwater, and on beaches. Sand mining to meet increasing demand over the years has become a thriving multibillion-dollar industry, but research by the United Nations Environment Programme (UNEP) shows that rate to be unsustainable.
>
> "Sand and gravel represent the highest volume of raw material used on earth after water. Their use greatly exceeds their natural renewal rates" (UNEP, 2014).
>
> Each grain of sand originates from rock on a mountaintop. The grains, formed by erosion over thousands of years, make a long journey through springs, streams, and rivers to the ocean, where the tides and waves distribute them across the ocean floor and eventually carry them onto beaches.
>
> The damming of rivers during the past century has dramatically impeded this natural process, and so roughly half of the estimated 40 billion metric tons of sand and gravel extracted every year for the construction industry, glass manufacturing, and other uses—such as land reclamation and oil exploration—will never be replenished.
>
> The seemingly endless supply of sand in the deserts just won't cut it. Desert sand granules have been rounded by wind over time and no longer bind together, an essential characteristic of sand used in construction.

Historical trends demonstrate that increasing wealth leads to higher space standards and an increase in household appliances, with implications for energy consumption. Another critical factor in developed countries is demographic and societal change, with a significant increase in one-person households.

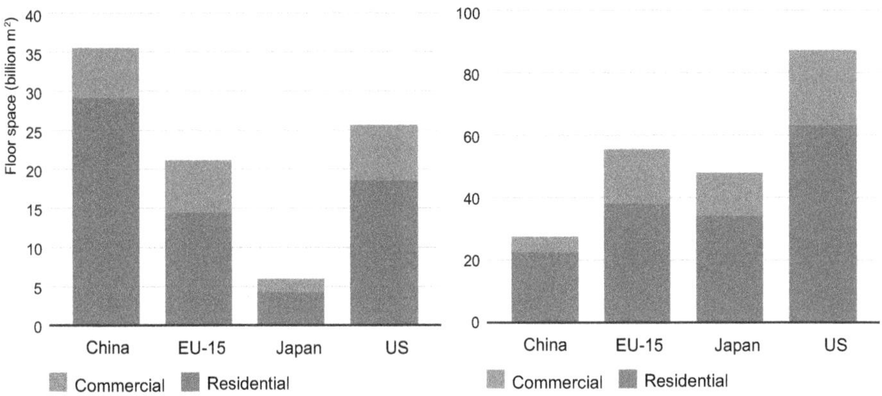

Fig. 12.1: Commercial and residential floor space in China, the EU, Japan and the USA.

12.2.1 Developmental Challenges

Developing countries are urbanising at a rate two to three times faster than developed countries, resulting in massive informal settlements and slums. In the majority of the developing world, the scale of informal and low-cost housing is vast. In some cities, the informal city is bigger than the formal city. In Indonesia, an estimated 70–80 per cent of housing construction is informal. In Brazil, more than half of all low-cost homes are built by the informal sector.

In this context, providing affordable green housing for the poor is a considerable challenge when so many already face major economic barriers to afford conventional housing. Analysis of social housing, however, does not lead to clear results as to whether green social housing is more expensive at the point of construction; environmental design features may be but do not have to be, more expensive than the conventional features. Further, if the environmental features are more expensive at the point of construction, they may yield benefits in terms of savings on water and energy during the occupation of the building.

Poverty and housing raises other unique challenges for sustainable building and construction in developing societies. Slums, be they informal settlements or rundown and overcrowded housing estates, are associated with social and environmental challenges including lack of access to electricity, fresh water, health-care and effective waste management. Marginal locations poorly connected to public transport services are an additional obstacle in that they constrain access to employment opportunities.

Greening of buildings can be one of a series of strategies that improve access to basic services and reduce vulnerability and, more broadly, contribute to better living conditions of the poor. Facing this challenge, India, for example, is experimenting with three approaches, namely vernacular building (which focuses on local solutions and traditional knowledge), green building (supported by the internationally recognised Indian GRIHA rating systems, developed by TERI) and; energy- efficient building (focused on energy-use in commercial buildings). New approaches can contribute to providing electricity to the 1.5 billion people in the developing world currently living without it, and to lifting 100 million people from slum conditions and providing them with safe water and sanitation—a distinct Millennium Development Goal.

12.2.2 Energy and Environmental Challenges

Whether existing building stock or projected growth of building stock, this sector is already the single largest contributor to global greenhouse gas emissions. Approximately one-third of global energy end-use takes place within buildings. Nearly 60 per cent of the world's electricity is consumed in residential and commercial buildings, although this usage varies widely according to geographical location, climate and consumption patterns. For developed countries located in cooler regions of the world, space heating, on average, represents 60 per cent of residential energy consumption, followed by water heating at 18 per cent.

Projections for 2030 based on IPCC scenarios suggest CO_2 emissions from buildings will continue to account for around one-third of total CO_2 emissions. Apart from energy use and emissions, the building sector is responsible for more than a third of global resource consumption annually, including 12 per cent of all fresh water use. The manufacture of building materials consumes about 10 per cent of the global energy supply. Building construction and demolition waste contributes about 40 per cent of solid waste streams in developed countries, with most waste associated with the demolition phase.

12.3 OPPORTUNITIES

The major opportunities for greening the building sector are the relatively low cost of the process, be it retrofitting or new construction, the availability of technologies, and the green evolution of energy supply and demand. These trends are encouraging the effort to transform the building sector.

12.3.1 Low Net Cost

Although the building sector is the largest contributor to human-related GHG emissions, it also holds the greatest potential to reduce these emissions. Based on 80 studies spanning 36 countries, the IPCC report suggests that a 29 per cent reduction in projected baseline emissions by 2020 is achievable at zero cost (below 0 US$/t$CO_2$-eq), while further improvements could be made with relatively low levels of investment.

12.3.2 Adapting Behaviour Patterns

People spend most of their lives in their homes, places of work and other buildings. For this reason, understanding the economic and psychological rationale of decisions made by individuals and institutions is increasingly recognised as fundamental to achieving energy-efficiency improvements in buildings.

The core concept of "thermal comfort" is more of a state of mind (reflecting different cultural, class and geographical conditions) than a technical certainty. Assessing the right level of thermal comfort is critical to setting performance standards for buildings but requires not just an understanding of what a human body can bear, but also to what extent people are ready to make behavioural changes in the way they experience comfort in their environment. This affects the way building occupiers interact with their environment in very precise ways—from choosing to pull down external blinds to limit sun penetration at certain times of day (rather than switching on the air conditioning) to putting on a sweater when the external temperature drops (rather than turning up the thermostat). On balance, green buildings require a more proactive engagement between occupier and the environment, which reflects the degree of "active" or "passive" environmental design techniques available in individual buildings, to which the report now turns.

12.3.3 Design and Technology

The greatest opportunities to achieve a higher environmental performance for buildings can be found in the early stages of their design. There are two basic paradigms of green building. The first is based on the concept of "passive" design where buildings respond to their local site context by using natural elements (such as air-flow and sunlight) to limit the effect of external conditions on the internal environment. Many traditional buildings with thick walls and small windows in hot climates, or with natural through-ventilation with courtyards and terraces in humid areas, belong to this category. Passive design aims to provide a comfortable environment while eliminating or reducing the need for space heating, cooling, ventilation or artificial lighting. The second paradigm is based on a more "active" approach that uses newer technology and state-of-the-art building management systems to reduce the energy load of buildings. Solar screens, lighting scoops, environmental flues, photovoltaic cells (PV), wind turbines and other devices are found in most state-of-the-art high-tech buildings. Both paradigms can be applied to new buildings as well as retrofitting existing building stock.

Many passive design techniques are finding their way into a new generation of building designs across the developed world, while new forms of green energy generation are being integrated in building projects in the developing world.

Greater attention is now given to the impact of sustainable environmental design solutions on the running costs of buildings and how much energy is embodied in construction materials and processes. Increasingly, life-cycle assessments (LCA) are being applied, which include not only operation and maintenance, but also the manufacture of construction materials. In addition, a new generation of building guidance is focussing on the total energy costs of buildings, from the design stage through to completion, including considerations about their recyclability.

12.3.4 Managing Energy Supply and Demand

Energy use and emission patterns are affected by a building's environmental performance and its energy load (on the demand side) or by the extent of its use of green sources of energy (on the supply side). Recent developments in design and technology offer significant potential to change the way energy demand and supply is managed in buildings.

On the demand side, there is growing evidence that energy consumption can be reduced by modifying the specification of technologies, appliances and fittings within buildings—in addition to designing the built form in a more sustainable way. Leading Information and Communication Technology (ICT) Infrastructure firms produce software for command centres, which can actively help to reduce a building's carbon footprint by monitoring and controlling all components of a building's energy use, from heating/cooling demand, to lighting and printing.

On the supply side, there has been a significant shift in some countries in favour of renewable energies with bio-fuel and solar heating technologies becoming competitive with conventional sources. Photovoltaic (PV) technology is still relatively expensive but with the increasing volume of installed capacity and improvements in production, prices are lowering steadily. District heating and cooling systems that link buildings are also proving effective at reducing energy costs, notably in Iceland, where 94 per cent of heat demand is now provided by these technologies.

12.3.5 Retrofitting and New Construction

In developed countries, opportunities for greening the building sector are found mainly in retrofitting existing buildings to render them more environmentally efficient by reducing energy demand and using renewable energy sources.

There are significant opportunities for retrofitting buildings in some of the bigger cities of the developing world by adopting energy efficiency design measures such as solar technology, clean water supplies and reducing dependency on air-conditioning through technical improvements. In India, for example, potential energy savings of 25 per cent have been estimated through cost-effective retrofitting of existing commercial buildings.

The pros and cons of constructing a new building or retrofitting an existing structure have to be individually examined and compared. In some cases, retrofitting allows a further reduction of the energy load by preserving building materials, which can contain high levels of embodied energy, expended in the extraction of resources, the manufacture of materials and their transportation. Both new building construction and retrofitting are fundamental for catalysing a green building transformation. Retrofitting in developed countries can yield significant energy savings as the design, construction and technology of older buildings is often significantly less efficient than current best practices. In addition, retrofits that address daylight or on-demand ventilation to improve air quality can bring benefits through lower health care costs and higher productivity levels.

The Social dimension of Green Buildings: Implications for Decent Work and Poverty Reduction

The building sector has high potential for propoor economic growth through its high labour absorption capacity in developing countries. The sector employs a wide range of workers with different levels of education and has the ability to absorb the excluded (de Souza 2000). This has strong implications for income generation and poverty reduction. Take the example of the Johannesburg Housing Company project in South Africa. This project involves the introduction of energy-efficient light bulbs and day-night sensors, solar energy systems for heating water and the insulation of boilers. It provides jobs for over 1,000 contractors in maintenance, cleaning and security services and even more in specialised functions such as plumbing and electrical services. The Watergy Soweto project for the rehabilitation of plumbing fixtures has provided 1,500 temporary jobs.

Despite this potential, workers of the construction industry are often subject to poor working conditions. High informality, low wages, instability, gender discrimination, frequent accidents and occupational diseases characterise the working conditions of a large share of workers in the building sector around the world, especially in developing economies where construction work is more precarious and less formalised.

Where the employment relationship of contractors, subcontractors and workers is casual or informal, workers' rights are often unclear and they enjoy less protection from the law than those directly employed. In recent years it has become the norm for workers to be employed on a short-term basis, and instability of work is one of the major problems facing the building industry.

Construction is also one of the most dangerous occupations. Workers in this sector are 3–4 times more likely than other workers to die from accidents at work. Many others suffer and die from diseases arising from exposure to dangerous substances at the workplace, such as asbestos. In regard to social protection, there is evidence that many employers do not pay into social security funds for workers who are on temporary contracts, depriving them from health care, holiday pay, and compensation owing to unemployment, ill health, accidents or old age.

For a long time the ongoing dialogue with employers as well as the government has been a successful approach for workers to collectively negotiate better wages and working conditions. However, nowadays a large workforce of temporary, casual, informal and unemployed workers find it very difficult to organise themselves to engage in such dialogue. The greening of buildings may provide a new opportunity for social dialogue. Many employers and government authorities have shown enthusiasm for green construction. This may open a new door to dialogue with workers on labour issues in the context of greening of the industry, involving workers in green management, resource efficiency and safety improvements.

In the area of working conditions, greening the building sector will have an impact on health and safety. Green construction is however not safer per se, as is shown in research by the American Society of Civil Engineers. With data collected through a structured questionnaire survey, the study tested the presence of a difference in Occupational Safety and Health Administration (OSHA) recordable incident rates (RIRs) and lost time case rates (LTCRs) between green and non-green projects. There was suggestive, but inconclusive evidence of a statistically significant difference in the RIRs of the green and non-green building projects that were examined. Also, no statistically significant difference was found between the respective LTCRs.

These considerations provide further cause to turn the role of labour inspectors to one of education and prevention, as opposed to mere inspection and prosecution. The greening of the industry brings the opportunity to create synergies between inspection about the environmental and the health and safety components of construction.

13

Guidelines for Green Transport

13.1 INTRODUCTION

Transport is central to the lives of citizens across the world, yet the current patterns of transport, dictated mainly by fossil-fuel driven motor vehicles, generate a range of environmental, social and economic costs. It is estimated, for example, that transport is responsible for nearly a quarter of global energy-related carbon dioxide (CO_2).

There is a growing consensus on the need for more sustainable patterns of transport activity but investment patterns are still heavily skewed towards supporting the "motorisation" model of development.

Given the pivotal role of transport in the global economy, much of the analysis of the potential for greening the sector is interwoven with other chapters, notably cities, energy, manufacturing and tourism. The chapter was compiled through extensive collaboration with experts from around the world, whose background papers are available in the accompanying Full Technical Report.

13.2 CHALLENGES IN THE TRANSPORT SECTOR

A. Unsustainable Trends

The challenges for the transport sector in becoming green are made obvious by observing current trends, whereby:

- Overall demand for transport activity (for both passenger and freight) is growing rapidly, and it is predicted to roughly double between 2005 and 2050;

- Transport activity is increasingly motorised (private cars for passenger transport and lorries for freight, almost all of which are propelled by internal combustion engines);
- The global vehicle fleet is set to multiply three or fourfold in the next few decades, with most of this growth set to occur in developing countries. In 2050, two-thirds of the global vehicle fleet is expected to be in non-OECD countries; and
- Technological improvements such as fuel-efficient vehicles and alternative power sources have not been rapid enough to offset the impacts of this growth.

These trends translate directly into various costs for the environment, society and economy including:

- Energy consumption and greenhouse gas emissions (GHG);
- Congestion (and associated losses in productivity of urban areas);
- Resource depletion and land grab;
- Degradation of human health (through air pollution, noise, vibration, etc.);
- Reduction in human security (through traffic accidents);
- Reduction of accessibility and severance of communities; and
- Loss of biodiversity.

It should be acknowledged that such costs vary significantly between regions, and that priorities may differ between regions and by urban and/or nonurban area.

B. Fuel and Natural Resources

The transport sector's impact on natural resources is wide-ranging, including through manufacturing of vehicles and/or rolling stock (*e.g.* metals and plastic) and the construction of infrastructure (*e.g.* concrete and steel). Fossil fuels, engine oil, rubber and other consumable material (including biofuels, which in certain circumstances may deplete farmland for food production) are consumed through the operation and maintenance of vehicles.

Transport consumes more than half of global liquid fossil fuels and it is expected to account for 97 per cent of the increase in the world's primary oil use between 2007 and 2030.

C. Greenhouse Gases

The transport sector's consumption of fossil fuels translates into around a quarter of global energy-related CO_2 emissions, which is projected to increase by 1.7 per cent a year from 2004 to 2030. Land transport accounts for roughly 73 per cent of the sector's total CO_2 emissions, followed by aviation (11 per cent) and shipping (9 per cent). Passenger transport accounts for the lion's share of overall emissions, with freight transport—predominantly road-based trucks—comprising 27 per cent of all transport energy use (and therefore emissions). More than 80 per cent of the predicted growth in transport emissions is expected to come from road transport in developing countries.

Moreover, it is estimated that around 15 per cent of the total CO_2 emissions generated from the car are a result of manufacturing and disposal, when a full life-cycle analysis is conducted.

D. Pollution and Health

Transport-related pollution, noise and vibration can pose serious threats to human health and well-being. Local air pollution is caused by exhaust emissions produced by traffic.

Such air pollutants are a cause of cardiovascular/ pulmonary and respiratory disease. For example, exposure to lead can cause increased blood pressure, liver and kidney damage, impaired fertility, comas, convulsions, and even death. Children are particularly vulnerable; they can suffer from reductions in IQ and attention span, learning disabilities, hyperactivity, impaired growth and hearing loss. Hatfield *et. al.* estimate that the removal of lead from vehicle fuels has resulted in more than 1 million avoided premature deaths per year with annual financial benefits over US$ 2.4 trillion.

E. Human Security and Accidents

The latest report from the World Health Organization confirms that road accidents remain a serious public health issue. Every year more than 1.27 million people die in road accidents, of which 91 per cent occur in low and middle income countries. About half of those who die in road accidents worldwide are pedestrians, cyclists and motorcyclists, for whom infrastructure provision is often neglected. In Europe, traffic accidents are a major cause of fatalities for young people, particularly men aged between 15 and 25.

It is estimated that the cost of traffic accidents amounts to US$ 518 billion, and represents between 1 per cent and 1.5 per cent of GDP in low-and middle-income countries and 2 per cent of GDP in high-income countries.

F. Accessibility and Severance

Traffic-filled roads can become physical and psychological barriers that can sever communities and divide entire cities. There are various ways in which accessibility and severance can be quantified and monetised. Transport systems dominated by motor vehicles have been shown to hinder access to jobs, markets and essential facilities, particularly for the poorest and most vulnerable members of society.

G. Land Use and Loss of Biodiversity

Roads, railways, airports, harbours and other transport infrastructure can have a severe impact on the natural environment, from the removal of vegetation during construction or the subsequent fragmentation of habitats. Fragmentation, without proper ecological infrastructure planning can severely disturb wildlife and reduce biodiversity.

Traffic Chaos Costs ₹60,000 cr Annually

Traffic congestion on Delhi roads costs around $10 billion or about ₹60,000 crore annually. This is on account of fuel waste due to idling of vehicles, productivity loss, air pollution and road crashes, according to a study done by IIT, Madras.

As the vehicular population rapidly grows in the capital, the study projects the congestion cost would increase to the tune of around ₹98,000 crore by 2030 unless steps are taken to tackle the daily mayhem. It has recommended dedicated lane for buses since "the productivity loss due to congestion delays of commuters who use buses accounts for about 75% of total cost of congestion."

Congestion has become a routine on Delhi's roads despite a large number of flyovers and elevated stretches coming up in the past several years. The definition of peak hours has changed as almost all major roads remain clogged from 9 am to 9 pm and it's worst in the case of roads linking Gurgaon, Faridabad, Noida and Ghaziabad.

"One study by IIM-Calcutta in 2012 had shown how ₹60,000 crore was being lost because of congestion on the entire NH network. Our study showed how the cost of congestion is similar only in the national capital," said Harry Raymond Joseph, one of the authors. The other two authors Gaurav Raina and Krishna Jagannathan are faculty members at IIT, Madras.

Harry said about 50% of these costs can be offset by making a single infrastructural change—dedicated bus lanes—and using smart traffic management systems. The study mentions buses are the most popular means of road transport catering to about 60% of Delhi's total demand and the average occupancy in buses is 20 whereas the average occupancy in cars and two-wheelers is 2.2 and 1.2 persons respectively.

A study conducted by Maharashtra's transport department last year had shown that one bus can replace 35 private cars and thereby reduce congestion. Another study by a global automobile major in 2015 had shown that congestion was the main reason for anxiety for at least six out of every 10 Indian drivers on the roads.

Pitching for the government to put greater emphasis on buses, the study has recommended, "In order to make dedicated bus lanes effective, it would be important to have more frequent and more comfortable buses. This could also help in shifting a fraction of the motorists to buses. Additionally, it would be prudent to employ state-of-the-art scheduling policies for buses."

It has also suggested the need to employ intelligent traffic management systems including smart traffic lights to reduce fuel wastage.

13.3 OPPORTUNITIES

Action is required in all countries, but opportunities are greatest for developing countries, where future patterns of transport can be shaped by the investment and planning decisions made today. Investing in green transport will enable such countries to leapfrog towards a sustainable path, rather than reproducing the mistakes made by industrialised countries.

A. Avoid, Shift and Improve strategy

Making a decisive shift to green transport arguably requires a holistic strategy that combines the following three elements:

1. Avoiding or reducing the number of journeys taken

This can be achieved by integrating land use and transport planning; designing denser, more compact settlements; harnessing telecommunication technologies such as teleconferencing and localising production and consumption. Demand for freight transport can be reduced by localising production and consumption and by optimising logistics to reduce empty runs and ensure a high load factor.

2. Shifting to more environmentally efficient forms of transport

This involves promoting public transport as well as walking and cycling, which usually requires substantial investment in infrastructure. For public transport to rival the private car it needs to be frequent, reliable, affordable and comfortable. Railways and waterways are generally greener methods of transporting freight, and shifting to them frees up road space.

3. Improving vehicle and fuel technology to reduce adverse environmental effects such as pollution and resource depletion

Enhancing the fuel economy of conventional engines; reducing the weight of vehicles and developing alternatives such as electric and hybrid vehicles, biofuels, and hydrogen fuel technologies are all examples of this strategy. Further efficiency gains can be achieved through an improvement in the occupancy rate of vehicles, or through better driving (eco-driving).

Given that transport systems vary greatly around the world, it is important that the above three strategies are applied in ways which fully consider the context and main problems facing each region. Many developing countries are heavily reliant upon non-motorised transport and therefore present opportunities for creating more sustainable transport systems than those in developed nations (Table 13.1).

Enacting the Avoid, Shift and Improve strategy requires adequate investment in the research, development, production and operation/management of:

- *Transport Infrastructure* such as tracks for buses and rail, pavements and cycle routes and park-and-ride facilities;
- *Greener vehicles and transport modes* (including bicycles, public transport vehicles and low emission vehicles, utilising technologies;

Table 13.1: Avoid, Shift and Improve strategy

Strategy	Developed Countries	Developing Countries
Avoid	Reduce vehicle kilometres (VKM) through Transport Demand Management (TDM), land use planning, localised production, and shorter supply chains.	Avoid unnecessary generation ofVKM through land use and transport planning.
Shift	Shift from private vehicles to Non-Motorised Transport (NMT) and Public Transport (PT) and from aviation to rail/ PT. Transfer freight from road to rail and water transport.	Enable conditions for the lowest-emitting modes (both freight and passenger). Prevent shift from NMT and PT to private vehicles by ensuring that attractive alternatives to private vehicles exist.
Improve	Improve existing vehicles. Down-scale vehicle engine size. Increase penetration of electric vehicles and carbon-neutral liquid fuels. Electrify rail (for both freight and passengers).	Ensure future vehicles/fuels are cleaner, encouraging small efficient cars. Design innovations for traditional NMT such as cycle rickshaws.

- *Cleaner fuels;*
- *Telecommunication technology* to substitute conventional transport, *e.g.* telework/ teleconferencing; and
- *Technologies* to enact green transport, *e.g.* GPS systems, Intelligent Transport Systems, green logistics, etc.

The above would need to be supported by appropriate enabling conditions. It is important that the generation of electricity, production of hydrogen and biofuels are all conducted in a sustainable manner.

Box 13.1: The Role of Transport in Reducing Rural Poverty

There is a large body of empirical evidence that shows a positive correlation between transport investment and economic outputs. Rural road investment directly contributes to the growth of agricultural output, increased use of fertilisers, commercial bank expansion and overall improvements in the socio-economic conditions of rural villages in India. Rural road investments in Bangladesh reduced poverty significantly through higher agricultural production, higher wages, lower input and transportation costs and higher output prices. Rural roads were also found to lead to higher rates of school attendance for both girls and boys and to be pro-poor. However, together with rural road infrastructure, investments also need to be made in facilities for public transport, NMT and a multimodal transport infrastructure. For those without access to private motor vehicles, this would lead to their increased mobility. This is especially the case when connecting urban centres with rural areas. Van de Walle (2002), in her work for the World Bank, argues that failing to consider the equity objective alongside the efficiency one can bias investment against poorer areas and poor people. This is particularly true in Asian transition economies where roads are one of many constraints on development. Their economic, social and environmental benefits will be dependent on other factors such as whether or not affordable transport services follow the road investment.

Investment in transport is often justified on the grounds that the movement of goods, services and workers is the vital fuel of the economic engine.

Freight transport volumes have traditionally been thought to strongly correlate with economic growth on the supply side and passenger car use to be driven by economic growth on the demand side. There is evidence, however, to suggest that high levels of GDP can be accompanied by transportation systems that rely less on the private car.

Of the various channels through which investment can flow into green transport, investment in infrastructure offers the greatest potential for economic growth by encouraging government investment and stimulating new business opportunities. Investment in green transport technology is also likely to benefit the overall economy, particularly through its potential to stimulate government investment.

Green transport business in the Avoid, Shift and Improve groups in Table 13.2 are tabulated below:

Table 13.2: Green transport businesses in the Avoid, Shift and Improve groups

Avoid, Shift, Improve	Sustainable business	Emissions reduction potential	Examples
Avoid	Telecommunication technology and services	Medium—Provides alternatives to physical travel	Teleconferencing and teleworking by major companies in Europe, US, etc.
	Parking providers	High—by providing formal parking space and replacing informal parking	Private parking operators in Tokyo
Avoid and Shift	Shared vehicle systems	High—by encouraging less private car usage	Car-sharing integrated with rail and public transport in Switzerland; Bicycle sharing such as: JCDecaux/Cyclocity, Paris, Clear Channel/ SmartBike, Barcelona
	Public transport operations (including fare collection, depot/fleet management, station management, security)	High—by increasing the quality of service and making transit systems more attractive	Bus Rapid Transit systems in Bogota, Pereira, Curitiba, Ahmedabad, Guayaquil, Mexico, Leon, Guadalajara, Guatemala; Bus systems in Santiago, Sao Paulo (and most Brazilian cities); Metro rail systems in Singapore, etc.
	Taxis and paratransit operations	Medium—by providing door-to-door alternative to private cars (depends on fuel type and operational efficiency)	Auto-rickshaws in India, Pakistan
Shift	Non-motorised transport (NMT) services	High—particularly when coupled with land use patterns that support shorter journeys achievable by NMT.	Bicycle rickshaws in India, New York City, San Francisco; Bicycle stations in Germany; Bicycle rentals in Amsterdam; Walking tours in Boston

	Intelligent Transportation Systems	Medium—optimising transportation system performance to minimising vehicle delays and making public transport attractive	Technology providers in Santiago, Guayaquil
	Commercial enterprises in public spaces, advertising and street furniture	Medium—improves the user experience of transit/non-motorised transport oriented cities	Barcelona, Buenos Aires, Guayaquil
	Low carbon vehicles	Highvby allowing better energy efficiency	Small, lightweight vehicles, ultra low emission engines, hybrid vehicles, plug-in hybrids linked with sustainable generation of electricity
Improve	Alternative fuels	High—by allowing lower CO_2 per unit of energy	Biofuels, conforming to stringent and comprehensive sustainability criteria
	Vehicle Maintenance	Medium—proper vehicle maintenance can reduce emissions and GHG	Annual vehicle checks in *e.g.* Indonesia

Transport is a complex sector, which is shaped over a long period of time, and by various external sectors and factors. Therefore, a combination of strategic approaches and policy instruments is required to green the transport sector.

A wide range of policies could support the Avoid, Shift and Improve strategies for green transport. These are tabulated in Table 13.3.

Table 13.3: Overview of instruments to support Avoid, Shift and Improve strategies

Type	Avoid	Shift	Improve
Planning	High density mixed land use development. Parking standards.	Integrated public transport planning. Land use planning.	Planning of smart grids. Planning of decarbonised electricity sources.
Regulatory	Traffic restrictions and travel bans (*e.g.* in city centres).	Parking restrictions. Road space allocations. Restrictions on the type of vehicles.	Vehicle standards (on *e.g.* emissions). Speed limits. Regulation of production processes.
Information	Increase awareness of the real costs of travel by various modes. Mobility management and marketing.	Increase awareness of alternatives. Mobility management and marketing. Co-operative schemes.	Eco-driving Public awareness campaigns. Labelling of the environmental performance of vehicles.
Economic	National subsidies for low carbon transport city design and planning.	Public-private partnerships for public transport systems (esp. BRT and lightrail). Removal of fuel subsidies/ taxing of fuels. Allocating fixed percentage of road infrastructure for NMT.	Fiscal incentives for cleaner and more efficient vehicles. "Cash for clunkers" programmes (buyout of old/ polluting vehicles). Fiscal incentives for cleaner fuels.

■ ■ ■

14

Guidelines for Green Tourism

14.1 INTRODUCTION

A growing body of evidence shows that greening tourism can lead to broad economic, social and environmental benefits for host countries and their communities. Tourism's potential for creating employment, supporting livelihoods and enabling sustainable development is huge, given that it is one of the main sources of foreign-exchange income—the principal source for one-third of developing countries and one-half of the world's LDCs according to the UN Conference on Trade and Development.

14.2 TOURISM IN A GREEN ECONOMY

Tourism in a green economy refers to tourism activities that can be maintained, or sustained, indefinitely in their social, economic, cultural and environmental contexts: "sustainable tourism." Sustainable tourism is not a special form of tourism; rather, all forms of tourism may strive to be more sustainable. A clear distinction should be made between the concepts of ecotourism and sustainable tourism: "the term ecotourism itself refers to a segment within the tourism sector with focus on environmental sustainability, while the sustainability principles should apply to all types of tourism activities, operations, establishments and projects, including conventional and alternative forms."

Sustainable tourism describes policies, practices and programmes that take into account not only the expectations of tourists regarding responsible natural-resource management (demand), but also the needs of communities that support or are affected by tourism projects and the environment (supply). Thus, sustainable tourism aspires to be more energy efficient and climate sound (*e.g.* by using renewable energy); consume less water; minimise waste; conserve biodiversity, cultural heritage and traditional values; support intercultural understanding and tolerance; generate local income and integrate local communities with a view to improving livelihoods and reducing poverty. Making tourism businesses more sustainable benefits local communities and raises awareness and support for the sustainable use of natural resources.

The movement toward more sustainable tourism promotes significant improvements in the performance of conventional tourism, as well as growth and improvements in smaller, niche areas centred on natural, cultural and community resources. The expansion of the latter, as a proportion of the industry as a whole, may have especially positive implications for biodiversity conservation and rural poverty reduction. However, the greening of conventional and mass tourism is likely to have its largest effects on resource use and management, as well as on increased economic spillovers and the inclusion of disadvantaged populations.

14.3 CHALLENGES

The tourism industry faces a multitude of significant sustainability-related challenges. Challenges that need to be resolved through the greening of the industry include (1) energy and GHG emissions; (2) water consumption; (3) waste management; (4) loss of biological diversity; and, (5) effective management of cultural heritage.

A. Energy and GHG Emissions

The tourism sector's growing consumption of energy, especially in travel and accommodation, and its dependence on fossil fuels has important implications for global GHG emissions and climate change as well as for future business growth. Several elements contribute to tourism's increasing energy consumption, including growth rates in international tourist arrivals and domestic travel; trends to travel further and over shorter periods of time; as well as preference given to energy-intense transportation (*e.g.* aircraft and car travel over train and bus, and flying first and business class instead of economy. The sustainability and competitiveness of tourism depends in part on energy efficiency (reductions in overall energy use) and a more intensive use of renewable resources.

After transport, accommodation is the most energyintensive component of the tourism industry, through its demand for heating or cooling, lighting, cooking (in restaurants), cleaning, pools and, in tropical or arid regions, the desalination of seawater. A general rule is that the more luxurious the accommodation, the more energy will be used.

B. Water Consumption

While water use by tourism, on a global basis, is far less important than agriculture, industry, or urban domestic use, in some countries and regions, tourism can be the main factor in water consumption. In such areas, it can increase pressure on already diminished water resources and compete with other sectors as well as subsistence needs of local populations (Box 1). Tourism can also directly affect water quality, for example, through the discharge of untreated sewage or freshwater abstraction.

Box 14.1: Water Consumption for Tourism and Local Communities

Tourism development is concentrated in coastal areas and on small islands, where potable water is typically scarce. This scarcity can be caused by either a physical absence of freshwater, or because the necessary infrastructure or resources are lacking. A tourism-thirsty industry can secure its water needs wherever it operates, although this can create situations of stark water inequity between tourists and neighbouring communities. Tourism's water demands can even lead to the appropriation of supply to the detriment of local domestic and agricultural needs, caused by the overexploitation of aquifers and reservoirs and the lowering of groundwater tables.

In a popular resort area of one South Asian country, for example, privately-owned water tankers buy water from villages through local elites and transport it to supply nearby hotels. This leaves villagers with water supply to their communal standpipes for a few hours a day only. Luxury resorts on an East African island are estimated to use up to 2,000 litres of water per tourist per day, almost 70 times more than the average daily domestic consumption of local people.

Golf tourism is rapidly expanding. An estimated 9.5 billion litres of water are used to irrigate the world's golf courses per day, equivalent to the daily needs of 80 per cent of the global population. One Mediterranean island, where water is so scarce it must sometimes be shipped in, is planning to increase its golf courses from three to 17, with tourism cited as the principal driver. This will involve building over agricultural land and constructing several desalination plants to ensure continual supply.

Global direct water consumption by international tourism (accommodation only) is estimated to be 1.3 km per year. Available data suggests that direct water use in tourism varies between 100 and 2,000 litres per guest night, with a tendency for larger, resort-style hotels to use significantly more water than smaller, pension-like establishments or campsites.

C. Waste Management

Waste management is another increasing and well- recognised challenge in the industry. Impacts are also considerable for wastewater management, even in high-income countries. In the Mediterranean region, for instance, it is commonplace for hotels to discharge untreated sewage directly into the sea, with 60 per cent of water used in tourism resulting in sewage in need of disposal. In the European Mediterranean, only 30 per cent of municipal wastewater from coastal towns receives any treatment before discharge. Anecdotal evidence suggests that this is also the case in many other countries outside the European Union.

D. Loss of Biological Diversity

There are many examples where large-scale tourism has had detrimental effects on biodiversity, including coral reefs, coastal wetlands, rainforests, arid and semi-arid ecosystems and mountainous areas. Biodiversity will be greatly affected by the way in which tourism grows and develops, especially in developing countries. Moreover, failure to incorporate biodiversity concerns in destination planning and investment will have detrimental effects on the natural environment, increase conflict with local communities, and lead to reduced value-creation potential for both the destination and investors (notably as interest in nature- based tourism is growing rapidly around the world and represents a strategic argument for maintaining biodiverse environments, which are often tourist destinations in developing countries).

E. Management of Cultural Heritage

Interest in unique cultures by tourists can result in adverse impacts and severe disruption for communities. There are examples of communities overrun by large numbers of visitors, commercialisation of traditions and threats to cultural survival from unplanned and unmanaged tourism. Tourism destinations are occasionally built by outsiders (usually with government approval) in areas that indigenous or traditional communities consider to be theirs, and where the development was neither desired nor locally validated. These situations lead to conflicts that make cooperation and mutual benefits nearly impossible to achieve, and instil animosities that negatively affect the local communities and the tourism destination. Frequently, the cultural issues overlap and are aggravated by environmental issues such as access to water, coastal resources and wildlife. Over the last two decades, with the growth in ecotourism and alternative travel, tourism impacts on vulnerable cultures has begun to be taken seriously by the tourism industry, governments, non-governmental organisations and the cultural groups involved.

14.4 OPPORTUNITIES

The following trends and developments provide a particularly promising space for greening tourism: (1) sizing and growth of the sector; (2) changing consumer patterns; and (3) maximising potential for addressing local development and poverty reduction.

A. Sizing and Growth of the Tourism Sector

Tourism is one of the most promising drivers of growth for the world economy. The sheer size and reach of the sector makes it critically important from a global resource perspective. Even small changes toward greening can have important impacts. Furthermore, the sector's connection to numerous sectors at destination and international levels means that changes in practices can stimulate changes in many different public and private actors.

The tourism economy represents 5 per cent of global GDP, while it contributes to about 8 per cent of total employment. International tourism ranks fourth (after fuels, chemicals and automotive products) in global exports, with an industry value of US$ 1 trillion a year, accounting for 30 per cent of the world's exports of commercial services or 6 per cent of total exports. Tourist arrivals have shown continuous yearly growth over the last six decades, with an average 4 per cent annual increase during 2009 and 2010. This trend has held in spite of occasional short drops from international crises, such as pandemics, recessions and terrorism.

B. Changing Consumer Patterns

Tourist choices are increasingly influenced by sustainability considerations. For instance, in TripAdvisor surveyed travellers worldwide and 38 per cent said that environmentally-friendly tourism was a consideration when travelling, 38 per cent had stayed at an environmentally-friendly hotel and 9 per cent specifically sought such hotels, while 34 per cent were willing to pay more to stay in environmentally-friendly hotels. The Center on Ecotourism and Sustainable Development (CESD) and International Ecotourism Society (TIES) found that a majority of international tourists are interested in the social, cultural and environmental issues relevant to the destinations they visit and are interested in patronising hotels that are committed to protecting the local environment. Increasingly, they view local environmental and social stewardship as a responsibility of the businesses they support.

C. Potential for Local Development and Poverty Reduction

Making tourism more sustainable can create stronger linkages with the local economy, increasing local development potential. Of particular and recognised importance are: purchasing directly from local businesses, recruiting and training local unskilled and semi-skilled staff, entering into neighbourhood partnerships to make the local social environment a better place to live, work and visit for all, and ability to improve the local natural environment within its areas of direct and indirect influence. The move toward more sustainable tourism has been shown in a number of destinations to enhance this local development potential through several means:

1. Its ability to harness biodiversity, landscape and cultural heritage available in developing countries can play a major role in enhancing incomes and employment opportunities;

2. Tourism is a relatively labour-intensive sector traditionally dominated by micro and small enterprises with activities particularly suited for women and disadvantaged groups;
3. A tourism product is a combination of different activities and inputs produced by many sectors: enhanced spending by tourists can benefit agriculture, handicrafts, transport, water and waste management, energy efficiency and other services;
4. As tourism development at destinations requires investment in facilities such as roads, water supply, and energy, it improves the basic common infrastructure facilities required for development of other sectors and improvement of quality of life (Bata 2010); and
5. Tourism employs more women and young people than most other sectors: providing economic benefits and independence to women is very important in terms of supporting child development and breaking the cycle of poverty.

Box 14.2: Wetland Shield

World Wetlands Day was celebrated in style at Yamuna Biodiversity Park (YBP), where experts and students discussed how to conserve wetlands in the national capital. Observed every year on February 2, this year's theme was—Wetlands for disaster risk reduction.

Several participants pointed out that wetlands play a critical role in reducing the effects of floods and cyclones by acting as a buffer.Faiyaz Khudsar, top scientist at YBP, said successful revival and restoration of wetlands is the need of the hour in Delhi with two such waterbodies restored through natural means at the park.

YBP developed a wetland over seven acres in Phase II, which is now attracting migratory birds like red-crested pochard, northern shoveler, northern pintails, great cormorants, painted stork, among several other species.The number of red-crested pochard spotted in 2016 at the park rose to 103, up from 52 in the previous year.

14.5 OVERCOMING BARRIERS: ENABLING CONDITIONS

Tourism can have positive or negative impacts depending on how it is planned, developed and managed. A set of enabling conditions is required for tourism to become sustainable: to contribute to social and economic development within the carrying capacities of ecosystems and socio-cultural thresholds. This section presents recommendations to create the enabling environment for increased investment in sustainable tourism development, overcoming barriers in the areas of: (1) private-sector orientation; (2) destination planning and development; (3) fiscal and government investment policies; (4) finance and investment; (5) local investment generation. Among these, the following may be mentioned.

14.5.1 Private-sector Orientation

Tourism businesses are no different to other businesses when it comes to the criteria that must be considered in deciding whether to invest in them. However, there are some specific characteristics that will affect tourism business costs:

- Tourism businesses are relatively labour-intensive and therefore labour costs often make up the largest proportion of operating costs;
- The cost of inputs for capital investment and operation are higher for remote locations;
- The cost of capital will attract a premium if there is uncertainty about returns from investment in tourism;
- The price of land in tourist-desirable locations will be governed by competition with other land uses which may be able to pay more (due to higher returns);
- Project planning and approvals cost will be high if assessment is lengthy or complex; and
- Labour and land make up a high proportion of inputs and are subject to payroll tax and land tax.

The business case for investing in these areas is sound. At the private-sector level, hotel owners, tour operators, and transport services can play a key role in protecting the environment and influencing tourists to make sustainable choices. Increased public environmental awareness, including traveller awareness, has contributed to the development of a host of voluntary industry initiatives and the definition of environmental performance at the national, regional and international levels.

14.5.2 Enabling Conditions for Engaging the Industry

1. Tourism promotion organisations, resource management agencies and Destination Management Organizations (DMOs) should link tourism products (*i.e.* parks, protected areas and cultural sites) more closely with marketing positions. This will ensure a consistent and unique selling position in world tourism markets based on high-value experiences at natural and cultural sites in a compact geographical area.
2. Tourism industry associations and wider industry platforms play an important role in engaging tourism businesses in sustainability as well as developing practical tools to respond to many common challenges. As in most industries, the concept of corporate social responsibility (CSR) is increasingly recognised in the tourism sector and is being promoted by industry bodies, at the international as well as national levels. However, a formal response, including measures such as triple-bottom-line reporting, environmental management systems

and certification appears to be prevalent only within a selection of larger firms.

3. International development institutions, such as multilateral and bilateral cooperation agencies, and Development Finance Institutions (DFIs) should engage directly to inform, educate and work collaboratively with the tourism industry to integrate sustainability into policies and management practices, and secure their active participation in developing sustainable tourism. At the national level, government and civil-society engagement should be a critical part of these efforts to coordinate action.
4. The increased use of industry-oriented decision- support tools would help speed the adoption of green practices.
5. The promotion and widespread use of internationally recognised standards for sustainable tourism is necessary to monitor tourism operations and management. The private sector tends to perform best when clear criteria, objectives and targets can be identified and incorporated into their investment plans and business operations.
6. Economies of scale in the tourism sector could be achieved by means of *clustering*. A high environmental quality is a key input by those companies that pursue competitive advantages based on sound environmental management. In the case of tourism, the conservation of the natural capital of a country has a chainable effect and complementary influence on many firms. Clustering can strengthen backward and forward linkages in the tourism value chain and drive sustainability in the whole industry.

14.5.3 Destination Planning and Development

Advancing greening goals through tourism planning and destination development requires the ability and institutional capacity to integrate multiple policy areas; consider a variety of natural, human and cultural assets over an extended time frame; and put in place the necessary rules and institutional capacity. A destination cannot successfully implement a green tourism strategy without the right laws and regulations in place, or the right governance structure to oversee them. Legislation should protect the environment, limit potentially harmful development, control detrimental practices, and encourage healthy behaviour. Clear rules in these areas, based on the destination strategy and its unique asset base, determine the direction, scale and scope of government and private investment in more sustainable tourism.

14.5.4 Enabling Conditions for Greener Destination Planning

1. Higher-level government, community and private tourism authorities must establish mechanisms for coordinating with ministries

responsible for the environment, energy, labour, agriculture, transport, health, finance, security, and other relevant areas, as well as with local governments.

2. Organisations engaged in developing tourism strategies should make use of credible scientific methods and tools encompassing economic, environmental and social approaches and assessments for sustainable development that will help stakeholders related to different components of the value chain understand their environmental and socio-cultural impacts.
3. Tourism master plans or strategies provide a supply-side approach for developing a tourism destination. Environmental and social issues must be included in these plans in order to manage the critical assets and promote greener outcomes.

14.5.5 Fiscal Policies and Economic Instruments

Tourism investment from government should focus on business motivations for sustainable management as key targets. Incentives should be consistent with both environmental protection and value added creation. Market trends and competitive advantages need to be mutually reinforced. In this regard, policy coherence is a necessary condition. From a national perspective, sustainable tourism policy should address market failures (including externalities) in a consistent manner, avoiding the creation of additional distortions through government interventions.

14.5.6 Financing Green Tourism Investments

Environmental and social investments are relatively new, and remain outside the mainstream of financial markets (particularly in developing countries). In many cases, barriers are based on misperceptions or lack of knowledge. For example, for many green investments, payback periods and amounts are not clearly established (due to limited experience with them), creating uncertainty for banks or other investors that can jeopardise financing. Also, the return on many green investments includes easily measurable components (such as energy savings), combined with more difficult to measure components such as guest satisfaction which can make calculating returns tricky.

In other cases, framework conditions in destination countries limit investment. For example, higher interest rates in many countries make investments that are completely viable in wealthy countries, unviable in the local environment. Another frequently cited situation found in many developing countries is that the financial regulatory systems classify environmental investments as nonproductive assets, requiring banks to hold greater reserves, resulting in higher interest rates and less investment.

14.6 SUM UP

Tourism can have positive or negative impacts depending on how it is planned, developed and managed. Various enabling conditions are required for transforming tourism to contribute to social and economic development within the carrying capacities of ecosystems.

To promote sustainable tourism in a green economy, the national, regional, and local economy should first provide a good investment climate, featuring security and stability, regulation, taxation, finance, infrastructure, and labour. Various tourism stakeholders should collaborate and share knowledge and tools in order to understand the overall picture of environmental and socio-cultural impacts of tourism activities at destinations. There is also a need for policy coherence, which can include economic instruments and fiscal policy to reward sustainable investments and practices and discourage the most costly externalities associated with uncontrolled tourism expansion. In the case of tourism, government and private tourism authorities should coordinate with ministries responsible for the environment, energy, agriculture, transport, health, finance, security, and other relevant areas, as well as with local governments.

APPENDIX

India's New Tourism Policy

India will unveil a new tourism policy on 15 May, 2017, outlining clear-cut action plans and timelines, as it aims to attract more foreign travellers and boost domestic tourism.

The policy will create a new way of working in which a task is identified and a person assigned for it, said an official statement citing the tourism minister at a tourism industry workshop held in New Delhi.

The timelines will be fixed and implementation closely monitored, Mahesh Sharma, minister of state for tourism, culture and civil aviation, said at the workshop to seek inputs for National Tourism Policy.

The move to float a new policy, which will replace the one in place since 2002, is in line with the National Democratic Alliance (NDA) government's aim to make tourism one of the major drivers of economic growth.

The minister also appreciated the proposal to set up a National Tourism Board, which will create synergies between various ministries to promote tourism, the statement said.

The existing tourism policy focuses on positioning tourism as a major engine of economic growth, harnessing the direct and multiplier effects of tourism for employment generation, making domestic tourism a major growth driver, establishing India as a global brand and acknowledging the role of private sector, among other things.

"The potential of the existing policy of 2002 has not been used fully and the new policy will use the framework of the existing National Tourism Policy of 2002 to make it dynamic to meet the challenges that have emerged in today's world, including technological and global competitiveness," Sharma said.

"The new policy will contain a definite action plan with a clear-cut road map and timelines," added Lalit Panwar, secretary, tourism.

India is trying to raise the share of foreign travel arrivals to 1% of total global tourists over the next couple of years, from 0.64% at present.

In 2014, the annual foreign exchange earnings from tourist arrivals totalled $19.657 billion, compared with $18.445 billion in 2013.

The new policy will follow several steps to boost tourism, including extending visas-on-arrival facility for more countries as well as issuing electronic visas.

■ ■ ■

15

Guidelines for Green Cities

15.1 INTRODUCTION

A city is a social, ecological, and economic system within a defined geographic territory. It is characterised by a particular human settlement pattern that associates with its functional or administrative region, a critical mass and density of people, man-made structures and activities. Most commonly, cities are differentiated from other settlements by their population size and functional complexity.

15.2 GREEN CITIES

Green cities are defined as those that are environmentally friendly. Indicators measuring environmental performance can include: levels of pollution and carbon emission, energy and water consumption, water quality, energy mix, waste volumes and recycling rates, green-space ratios, primary forests, and agricultural land loss. Other indicators include the share of apartment living, motorisation rate, and modal share of urban transport. Another important measure of humanity's demand on nature is the Ecological Footprint. Defining green cities by their environmental performance does not mean social equity issues are ignored. In fact, greener living environments can play an important role in making cities more equitable for their residents.

15.3 CHALLENGES

15.3.1 The Rapid Pace of Urbanisation

In 2007, for the first time in human history, 50 per cent of the global population lived in urban areas. Only a century ago, this figure stood at 13 per cent but it is now predicted to reach 69 per cent by 2050. In some regions, cities are expanding rapidly, while in others, rural areas are becoming more urban. A significant part of this urbanisation is taking place in developing countries as a result of natural growth within cities and large numbers of rural-urban migrants in search of jobs and opportunities. Often this happens despite widespread anti-urbanisation policies, which aim to balance development and to sustain rural economies. However, such efforts have mostly been unsuccessful and risk that urban agglomerations are left unprepared for inevitable increase in population growth. Rapid urban growth tends to overwhelm cities where the struggle to develop infrastructure, mobilise and manage resources has negative consequences for the environment.

The scale of the problem comes into sharp focus in India and China. India's urban population grew from 290 million in 2001 to 340 million in 2008 and it is projected to reach 590 million by 2030. The country will have to build 700–900 million square metres of residential and commercial space a year to accommodate this growth, requiring an investment of US$ 1.2 trillion to build 350–400 kilometres of subway and up to 25,000 kilometres of new roads per year. Similarly, China's urban population is expected to increase from 636 million in 2010 to 905 million by 2030. It is predicted that by 2050 the country will need to invest 800-900 billion RMB per year to improve its urban infrastructure, about one-tenth of China's total GDP.

At the national level, urbanisation goes hand in hand with increasing resource consumption, more energy intensive food supply, and ever-increasing flows of goods and people.

15.3.2 The Social Implications of Traditional Urban Development

Patterns of urbanisation in many areas also raise important social challenges. The traditional business as-usual (BAU) model of urban development—typical of rapidly urbanising areas—is characterized by uncontrolled, often even incentivised, horizontal expansion. This leads to urban sprawl of affluent populations with lower development densities and increased dependency on the private car and to peripheralisation of the urban poor, decreasing their access to the city and its workplaces, services and infrastructure. Typical developments further include the emergence of socially divisive neighbourhoods in the form of gated communities, shopping centres and business districts and, a significant increase in the level of informal development with large swathes of slum housing with no access to basic services, infrastructure and sanitation. At a general level, the rapid growth of many cities coupled with insufficient resources and poor management compromises fresh water and electricity supply, waste treatment, transport, and other infrastructure provision, affecting the urban poor most.

15.4 OPPORTUNITIES

15.4.1 Structural Capacity

The environmental performance of cities is dependent on a combination of effective green strategies and physical structure - urban form, size, density and configuration. They can be designed, planned and managed to limit resource consumption and carbon emissions. Or, they can be allowed to become voracious, land-hungry, all-consuming systems that ultimately damage the delicate global energy equation.

More compact urban forms, reduced travel distances and investment in green transport modes lead to greater energy efficiency. Lower surface-to-volume ratios of denser building typologies can result in lower heating and cooling loads. Greater utilisation of energy efficient utilities can contribute to lower embedded energy demand for urban infrastructure. Cities can be structured to make use of green grid-based energy systems such as combined heat and power and micro-generation of energy as well as rainwater harvesting, access to clean water and efficient waste management. In short, effective urban planning and governance, as will be shown below, can have significant effects on sustainable urban lifestyles, making the most of urban critical mass and reducing individual patterns of consumption.

Despite a rich debate on the links between physical structure and energy use in cities, there is growing evidence that compact urban environments, with higher-density residential and commercial buildings (as opposed to low density, sprawl-like development) and a well distributed pattern of uses and an efficient, transport system based on public transport, walking and cycling reduce the energy footprint. Research has shown that the so-called "compact city"model has lower per-capita carbon emissions as long as good public transport is provided at the metropolitan level.

15.4.2 Technological Potential

Cities are incubators of innovation due to the close interaction of their residents and workers who benefit from the exchange of ideas and opportunities. In particular, they benefit from the concentration of diverse yet specialised skill-sets in research institutions, firms and service providers that can pilot and scale new technologies in an already highly networked environment.

An urban setting, which tends to support a diverse and compact pattern of production and consumption is further advantageous to advance the notion of "industrial ecology." By optimising and synergising different industrial sectors and resource flows, outputs of one sector that become the input of another create a circular economy. Principles of symbioses can also help minimise or recycle waste. These opportunities have led to intensified efforts in designing cross-sectoral green city strategies when developing new districts or eco-cities.

15.5 THE CASE FOR GREEN CITIES

The case for green cities can be made in terms of inter-linked economic, social, and environmental benefits. Economically, the benefits include agglomeration economies, lower infrastructure costs and reduced congestion cost while reducing carbon emissions and other environmental pressure. Socially, the benefits include employment creation, poverty reduction and improved equity, and quality of life including improved road safety and community cohesion, among others. Environmental benefits are embedded in most of the economic and social benefits. Additional environmental benefits include reduced pollution, which helps improve public health. Another environmental benefit is the potential for improving ecosystems within urban areas.

15.5.1 Economic Benefits

A. Agglomeration Economies

Larger, denser cities—which help lower per capita emissions—are good for economic growth. From an economic perspective, cities matter because they bring people and things closer together, help overcome information gaps, and enable idea flows. These agglomeration economies translate into productivity gains for firms, and higher wages and employment rates for workers. Agglomeration economies exist in both developed and developing countries.

B. Lower Infrastructure and Operating Costs

Densification reduces the capital and operating costs of infrastructure. Evidence suggests that linear infrastructure including streets, railways, water and sewage systems as well as other utilities come at considerably lower cost per unit the higher the urban density. A recent study of Tianjin concluded that infrastructure cost savings as a result of compact and densely clustered urban development reach 55 per cent compared with a dispersed scenario.

Table 15.1: Infrastructure costs for different development scenarios in Calgary

	Total cost (CA$ billion)			
	Dispersed scenario	**Recommended direction**	**Difference**	**Percent difference**
Road capital cost	17.6	11.2	6.4	–36
Transit capital	6.8	6.2	0.6	–9
Water and wastewater	5.5	2.5	3.0	–54
Fire stations	0.5	0.3	0.2	–46
Recreation centres	1.1	0.9	0.2	–19
Schools	3.0	2.2	0.9	–27
Total	34.5	23.3	11.2	–33

Private transport fuel expenditure and urban density of selected cities shows how urban density can be an essential measure for decreasing long-term operating costs.

C Reduced Congestion Costs

Bigger, more productive cities tend to suffer from crowding and congestion, as firms and households compete for space in the most popular locations. One proven method for controlling congestion is demand management via charging.

Many public transport projects around the world have brought about significantly reduced congestion costs, notably BRT systems such as in Bogota and successfully emulated in Lagos, Ahmadabad and Guangzhou and Johannesburg. A synergetic interplay of compact urban form and an efficient bus system has been observed in Curitiba, which boasts the highest rate of public transport use in Brazil (45 per cent). There, reduced congestion means much less fuel is wasted in traffic jams: only US$ 930,000, compared with an estimated US$ 13.4 million in Rio de Janeiro.

15.5.2 Social Benefits

A. Job Creation

Greening the cities can create jobs on a number of fronts: (1) urban and peri-urban green agriculture; (2) public transport; (3) renewable energy; (4) waste management and recycling; and (5) green construction. Green services will generally be more urban-orientated than green manufacturing or primary industry, although there will be some high-tech green manufacturing clusters in or close to urban cores, drawing on knowledge spillovers from universities and research labs.

At the same time, specific sectors and firms may combine remote or off-shored production with highly urbanised consumer/service/support markets. This means that there is potential for cities to grow both green tradable activity (high value, exportable) and develop greener non-tradable activities (lower value, goods and services for local consumption). Overall, a green economy cannot be expected to create or destroy net jobs in the long run; the supply and demand for labour tend to equate in accordance with labour market conditions. In the short run, with unemployed resources, the net employment creation effect is likely to be larger.

First, there is considerable policy interest in urban and peri-urban agriculture. Green urban agriculture can reuse municipal wastewater and solid waste, reduce transportation costs, preserve biodiversity and wetlands, and make productive use of green belts.

Second, transport activities typically make up a significant share of a city's employment (operationally and in infrastructure development).

Third, the International Labour Organisation research (UNEP *et. al.* 2008) indicates that shifting from conventional to renewable energy will result in small net job losses, but cities are well-placed to benefit from new opportunities.

As well as research and development activity, renewable energy systems may often involve decentralised production, which locates power generation close to urban consumer cores. Critically, installation and servicing activities are both labour-intensive and urban-orientated. These domestic or personal service activities will be an important source of green jobs in urban areas.

Fourth, waste and recycling activity is similarly labourintensive. A recent estimate reveals that up to 15 million people are engaged in waste collection for their livelihood in developing countries.

Box 15.1: Green Jobs in the Urban Economy

The process of making the world's cities and urban fabric greener and maintaining them in a sustainable way will bring considerable employment opportunities. Upgrading to greener infrastructure generates jobs, whether by improving roads and buildings, establishing public transport networks, repairing and enhancing drainage and sewerage systems or creating and managing efficient recycling services. Many of these jobs will require knowledge of new technologies or working practices, for example, in constructing, installing and maintaining local hydrogen fuel-cell power stations or a network of charging points for electric vehicles. Providing training and support is fundamental to the process, within local authorities and for private companies, particularly small enterprises.

In creating the jobs that will enable cities to be greener, there is a great opportunity to address urban poverty, which is widespread (and in many places increasing at a faster rate than rural poverty), particularly in developing countries. Providing job opportunities where there are few is clearly important, but to make real inroads into poverty, employment must also encompass workers' rights, their social protection and social dialogue. The burgeoning international movement on "the right to the city" promotes community and consumers' rights but workers' rights are increasingly being recognized. Coalitions of urban workers in Brazil, for example, are helping to draw attention to and reduce informal, casualised labour. Inappropriate working and living conditions expose many urban workers to risk on a daily basis, while many do not have access to an adequate system of health care, pay for holidays and protection against loss of pay when they are unable to work. Several ILO initiatives provide a sound basis for action on improving social protection, and other efforts of communities to organise their own risk protection should be supported.

Fifth, many developed nations have also started looking at green construction as the largest possible employment provider. Retrofitting existing building stocks will provide a massive employment opportunity for many mature cities, since work is undertaken on site. Higher environmental standards for construction and fittings also create employment potential. Most excitingly, green construction has also the potential of making buildings to go from being exclusively consumers of resources to becoming producers—in esources like water, energy, food and materials, or even green space.

B. Poverty Reduction and Social Equity

The World Development Report describes increasing economic density—one of the main features of a green city—as "a pathway out of poverty."

Innovative approaches to urban planning and management can make urbanisation inclusive, propoor and responsive to threats posed by environmental degradation and global warming.

There are other examples of how greening cities can address poverty and equity concerns. Improving sanitation and fresh water supply can reduce persistent poverty and the adverse impacts of water-borne disease. Retrofitting older buildings in lower-income neighbourhoods can improve energy efficiency and resilience, reducing the vulnerability of poorer communities when energy prices rise. Upgrading infrastructure in slum areas offers both health benefits and fewer adverse impacts on the environment.

C. Improvement in Quality of Life

Community cohesion is one aspect of quality of life and affects individuals, families and social groups at the neighbourhood and district level. Social relationships not only have particularly positive impacts on physical and mental health but also on economic resilience and productivity. This is especially the case for disadvantaged people, as community cohesion and social inclusion are linked.

Improving the urban environment by measures such as traffic calming and promoting walkability can help foster a sense of community. Such changes are often designed to counteract instances of community severance, as identified:

- *Physical barriers* whereby either spatial structures themselves prohibit interaction or certain activities cause disruption, as in the case of road traffic;
- *Psychological barriers* that are related to the perception of certain areas determined by traffic noise and pollution or perceived danger; and
- *Long-term social barriers* where residents change behaviour following initial disruptions and create a more sustained form of being disconnected from certain people and areas close-by.

15.5.3 Environmental and Health Benefits

A. Reducing Pollution and Improving Public Health

Air pollution in cities remains a major public health burden, particularly in the developing world. Many cities have already taken decisive action and significantly improved the situation. There is a broader set of public health issues around healthier lifestyles in cities.

It is no coincidence that cities with a long tradition of applying land-use planning, public transport strategies and a focus on public green space are among the healthiest cities in the world.

B. Ecosystem Services and Risk Reduction

Urban greenery and vegetation represent a range of ecosystem services with significant wider welfare effects. Ecosystem services further play a critical role in risk reduction measures.

Restoration of urban ecosystems is part of the city greening effort, which can reduce the impact of freak weather conditions. Coastal regions in particular can benefit both in terms of lives and money.

Safeguarding natural ecosystems in cities' hinterlands is also important in reducing their exposure to risk. This is of particular relevance to fresh water supply and food security. As they have expanded, many cities have exhausted local fresh water sources and rely on importing water from their wider region. Such requirement to import water is already associated with enormous costs for cities.

15.6 GREENING URBAN SECTORS

15.6.1 Transport

Most green transport policies that follow the "avoid-shift-improve" paradigm can be found in cities. While "avoiding transport" is mostly covered by structural adjustments to the shape of cities introduced earlier, classic green transport strategies in cities primarily focus on reducing car use or at least slowing its growth.

In recent years, some cities have led efforts to electrify road-based transport, even though walking and cycling are still the greenest forms of transport.

15.6.2 Buildings

Tackling the energy demand of existing building stock is a priority for cities, and urban green building strategies also include more efficient use of other resources such as water and materials. Particularly in a developing world context, passive design solutions to improve environmental performance are by far the most cost-effective approaches.

Stringent building codes, mandatory energy certificates, tax incentives and loans, have had a measurable impact on energy demand in a number of cities.

15.6.3 Energy

Cities uniquely concentrate energy demand and rely on energy sources beyond their boundaries. But cities have the potential to either dissipate the distribution of energy or optimise their efficiency by reducing energy consumption and adopting green energy systems including renewable micro-generation, district heating, and combined heat and energy plants (CHP).

15.6.4 Vegetation and Landscape

While cities are principally made up of buildings and infrastructure, they can contain a significant proportion of open space. Despite sustained growth, cities like Johannesburg, London and Delhi have maintained high levels of green open space (parks, public and private gardens), while others like Cairo, Tokyo or Mexico City have far lower levels of green space. Parks, protected green space and gardens, street trees and landscaping provide vital ecosystem services, acting as green lungs absorbing and filtering air pollution or as acting as filters for waste water. They also provide a habitat for wildlife and offer recreational benefits to city dwellers.

In addition, the presence of green landscaped areas helps regulate natural processes, including the mitigation of local temperature extremes: a ten per cent increase in tree cover reduces cooling and heating energy use by between five per cent and ten per cent.

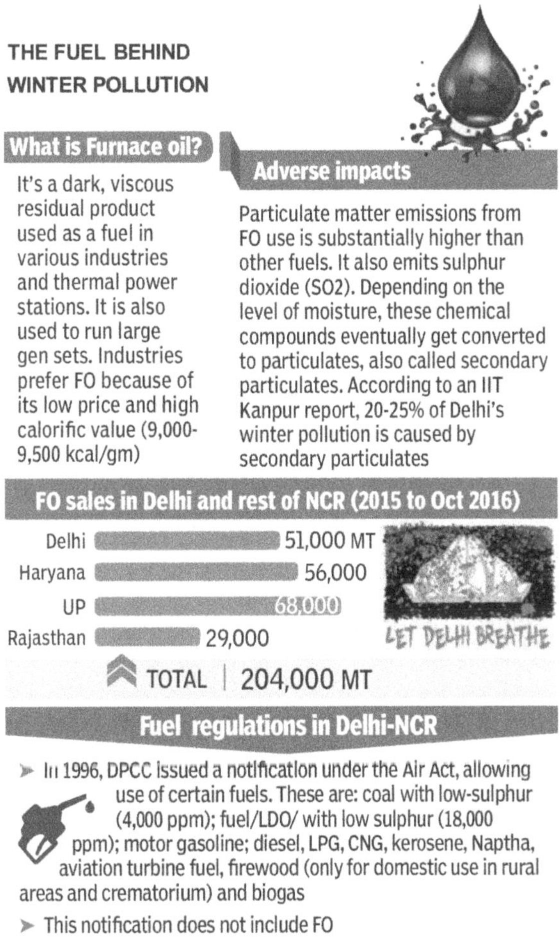

15.6.5 Water

Cities require significant transfers of water from rural to urban areas with water leakage being a major concern. Upgrading and replacement of pipes has contributed to net savings of 20 per cent of potable water in many industrialised cities. Many cities are introducing water meters and are shifting away from simple water access fees. A measure to maximise utility of fresh water is the cascading of water use where the waste water generated by one process can be used in another with a lesser quality requirement.

To further reduce water consumption and provide alternatives to piped water supply, rain can be harvested and used as drinking and non-drinking water. Such services can only be implemented in cities where there is a greater willingness to pay for water than in rural areas. To counter severe water shortages in Delhi, the Municipal Corporation made rainwater harvesting a requirement for all buildings with a roof area above 100 square metres and a plot area greater than 1,000 square metres. It is estimated that 76,500 million litres of water per year will be made available for groundwater recharge.

15.6.6 Food

The food footprint of a city has significant impacts on its green credentials, especially if one takes into account the energy use generated by transporting food from remote locations to urban marketplaces. More broadly, urbanisation is usually accompanied by a loss of nearby arable land and a rise in demand for processed foods by urban consumers. While there is some way to go to see a substantial reduction in the food footprints of highly consumptive cities there is evidence that farmers' markets are successfully re-establishing links between inner cities and regional agriculture.

15.6.7 Waste

By concentrating people and activities, cities have become centres of the waste economy, which plays a dominant role in a city's ecological footprint. Yet, cities have demonstrated considerable resilience in finding green solutions that reduce overall waste, increase recycling and pioneering new forms of environmentally friendly treatment of unavoidable waste. In developing world cities which typically suffer from insufficient formal waste collection, it is a large workforce of mostly informal recyclers and reclaimers. Green waste strategies in these contexts often fail to recognise the potential role of these actors and implement expensive, technology-driven recycling models.

15.6.8 Infrastructure and Digital Technology

A growing body of evidence suggests that cities are the natural sites of investment in smart infrastructure to deliver more sustainable environments. Cities provide a critical mass of potential users for a wide range of IT-based services which build upon complex physical infrastructure (such as roads, rail, cabling and distribution systems). The digital infrastructure of the internet and data centres create an intelligent infrastructure that connects people to people, people to city systems and city systems to each other, allowing cities and their residents to respond to changing circumstances by adapting in near real-time and to recognise patterns to help make informed decisions.

In addition, smart transport systems are being used to tackle congestion, facilitate road user charges or supply real-time information on traffic problems.

15.7 BARRIERS AND CONSTRAINTS

Today, most cities are adopting fundamentally non-sustainable practices as a result of a combination of the following barriers and constraints, which vary in significance according to geographical location and position with the economic and political development cycle:

- *Fragmented governance*—lack of coordination between policy frameworks that promote green economy measures at supra-national, national, regional and metropolitan levels;
- *Affordability*—even cost-effective green measures may be out of the reach of poorer cities, leaving them saddled with more wasteful urban infrastructure;
- *Lack of investment*—despite wider acceptance of the relevance of the green economy to well-being, the private and public sector have not prioritised green investment in basic city infrastructure (such as green planning, public transport and housing strategies);
- *Negative tradeoffs*—without effective policy intervention and infrastructure investment, (which promote productivity and resource efficiency) green city strategies can lead to greater congestion (of people and traffic), higher land values and costs of living;
- *Consumer preferences*—when given a choice consumers may not be willing to adopt new models of urban living that require changes in individual and collective patterns of consumption (e.g. high-density apartment living, public transport use);
- *Switching costs*—high short-term transition (welfare and capital) costs for businesses that shift from brown to green, leave many companies without adequate compensation to make the investment;
- *Vested business interests*—industry dynamics in construction, road-building and infrastructure are resistant to change that challenges existing business models and threatens the potential of short-term return on investment;
- *Risk aversion*—individuals, corporate and government organisations are resistant to any change that does not demonstrate immediate improvement in economic well-being, quality of life or enhanced status within the community;
- *Perverse policies*—these produce underpriced goods and services, thereby encouraging overconsumption. Such policies include subsidised

road infrastructure; the failure to charge developers fully for the cost of services and infrastructure new developments require; various tax abatements encouraging home ownership and other public policy measures that facilitate urban sprawl and the use of the private car as a dominant means of transport; and

- *Behavioural response and the rebound effect*—consumers may respond to reduced energy costs (generated by energy efficiency measures) by either increasing per capita energy consumption or by spending savings and increasing overall consumption per head.

■ ■ ■

16

Success Stories in Green Economy

16.1 INTRODUCTION

A Green Economy is characterized by substantially increased investments in economic sectors that build on and enhance the earth's natural capital or reduce ecological scarcities and environmental risks. These sectors include renewable energy, low-carbon transport, energy-efficient buildings, clean technologies, improved waste management, improved freshwater provision, sustainable agriculture, forestry, and fisheries. These investments are driven by, or supported by, national policy reforms and the development of international policy and market infrastructure.

These investments and policy reforms provide the mechanisms and the financing for the reconfiguration of businesses, infrastructure and institutions. Such reconfiguration could lead to a higher share of green sectors in GDP, greener jobs, lower energy and resource-intensive production, lower waste and pollution, and significantly lower greenhouse gas emissions. It can also assist in the reduction of persistent poverty through targeted wealth transfers, new employment, as well as improved access to the flow of ecosystem goods and services to the bottom of the economic pyramid.

The UNEP-led Green Economy Initiative, launched in late 2008, consists of several components whose collective overall objective is to provide the analytics and policy support for investing in green sectors and in greening brown sectors. The Initiative will assess how sectors—such as renewable energies, clean and efficient technologies, water services, and sustainable agriculture—can contribute to economic growth, creation of decent jobs, social equity, and poverty reduction while addressing climate and other ecological challenges. Within UNEP, the Green Economy Initiative includes three sets of activities:

1. Producing a Green Economy Report and related research materials, which will analyse the macroeconomic, sustainability, and poverty-reduction implications of green investment in a range of sectors from renewable energy to sustainable agriculture and providing guidance on policies that can catalyse increased investment in these sectors.
2. Providing advisory services on ways to move towards a green economy in specific countries.
3. Engaging a wide range of research, non-governmental organisations, business and UN partners in implementing the Green Economy Initiative.

Beyond UNEP, the Green Economy Initiative is one of the nine UN-wide Joint Crisis Initiatives (JCI) launched by the UN System's Chief Executives Board in early 2009. In this context, the Initiative includes a wide range of research activities and capacity building events from more than 20 UN agencies including the Bretton Woods Institutions, as well as an Issue Management Group (IMG) on Green Economy, launched in Washington, DC, in March, 2010.

16.2 SUCCESS STORIES IN GREEN INITIATIVE

The economic analysis in the Green Economy Report builds in part on the encouraging signs and results of many initiatives around the world. A number of these come from developing countries, including emerging economies, and illustrate a positive benefit stream from specific green investments and policies, that if scaled up and integrated into a comprehensive strategy, could offer an alternative development pathway, one that is pro-growth, projobs and pro-poor.

Eight of these examples, a limited selection from a growing range of experiences in different sectors, are summarized below, highlighting their economic, social and environmental benefits. While some represent established broad-based policies and investment programmes, others are newly initiated pilot projects or local ventures. In this sense the collection underlines that a green economy strategy is not limited to national or other government policy levels but can take root wherever there is the leadership and vision to make this transformation. Indeed, there is a growing body of evidence illustrating the growing interest among developing countries to seize opportunities to move to a Green Economy.

16.2.1 Energy Infrastructure in China

China is taking considerable steps to shift to a low-carbon growth strategy based on the development of renewable energy sources. The outline of 11th Five-year Plan (2006–2010) allocated a significant share of investments to green sectors, with an emphasis on renewable energy and energy efficiency.

The Plan projects that the per-unit GDP energy consumption by 2010 should have decreased by 20 per cent compared to 2005. In addition, the Chinese government has committed itself to producing 16 per cent of its primary energy from renewable sources by 2020.

Passed in 2005, China's Renewable Energy Law serves as the principal framework for development of the sector. The law offers a variety of financial incentives, such as a national fund to foster renewable energy development. discounted lending and tax preferences for renewable energy projects, and a requirement that power grid operators purchase resources from registered renewable energy producers. The combination of investments and policy incentives has encouraged major advances in the development of both wind power and solar power.

A. Wind Power

The additional generating capacity from wind power has exhibited an annual growth rate of more than 100 per cent from 2005 to 2009. With new installations of 13.8 GW coming on line in 2009, China led the world in added capacity, and is second in terms of installed capacity, after the U.S. To reflect increasing ambition in the industry, the government has indicated its intention to increase its previous target of 30 GW of installed capacity by 2020 to 100 GW.

To directly encourage local wind turbine manufacturing, China has implemented policies to encourage joint-ventures and technology transfers in large wind turbine technology and mandated the use of locally made wind turbines. The Ministry of Science and Technology has subsidized wind energy R&D expenditures at varied levels over time, beginning most notably in 1996 with the establishment of a renewable energy fund. Domestic wind turbine makers, such as Sinovel Wind, Goldwind Science and Technology, and Dongfang Electric, have contributed an increasing share of total new installations. Together they accounted for at least half of a market dominated by foreign firms until 2008.

China's National Development and Reform Commission issued the Interim Management Measures for Renewable Power Tariff and Cost Allocation in 2006, and the Interim Measures on Renewable Power Surcharge Collection and Allocation in 2007. Together with the Renewable Energy Law, the regulations aim to encourage a reduction in the price of wind power by stipulating that a competitive pricing bidding model be used for the majority of wind power development in China.

B. Solar Power

Being the largest Solar PV manufacturer in the world, China produced 45 per cent of global solar PV in 2009. The domestic solar market has started developing more recently, with about 160 MW solar PV installed and connected to grid in 2009. But with more than 12 GW of large projects in the pipeline, it could rapidly become a major market in Asia and the world. For solar PV, the government has also indicated that the target for installed capacity in 2020 could be increased from 1.8 GW to 20 GW.

China is now the world's largest market for solar hot water, with nearly two-thirds of global capacity. More than 10 per cent of Chinese households rely on the sun to heat their water with more than 160 million square metres as total collector area. The rapid development of the solar water heater sector is due to its basic profitability for both business manufacturing the units and households that install them. There are also considerable health and sanitation benefits afforded by the improved availability of hot water, made more feasible and economic with solar water heater systems. Within the context of the Eleventh Five-Year Plan for New and Renewable Energy, an Implementation Plan on Promoting Solar Thermal Utilization in China was adopted in 2007. Under this national policy, the installation of SWH systems is given priority for major hot water consumers, such as hospitals, schools, restaurants and swimming pools.

C. Job Creation

The energy sector as a whole generates output worth US$17 billion and employed an estimated 1.5 million at the end of 2009, of which 600,000 were in the solar thermal industry, 266,000 in biomass generation, 55,000 in solar photovoltaics and 22,200 in wind power. In 2009 alone, an estimated 300,000 jobs were created.

China's experience provides an example of policy- led growth in renewable energy that has created jobs, income and revenue streams for nascent low carbon industries.

16.2.2 Feed-in Tariffs in Kenya

Kenya's energy profile is characterized by a predominance of traditional biomass energy to meet the energy needs of the rural households and a heavy dependence on imported petroleum for the modern economic sector needs. As a result, the country faces challenges related to unsustainable use of traditional forms of biomass and exposure to high and unstable oil import prices. In March 2008, Kenya's Ministry of Energy adopted a Feed-in Tariff, based on the realization that "Renewable Energy Sources (RES) including solar, wind, small-hydro, biogas and municipal waste energy have potential for income and employment generation, over and above contributing to the supply and diversification of electricity generation sources".

A Feed-in Tariff (FIT) is a policy instrument that makes it mandatory for energy companies or "utilities" responsible for operating the national grid to purchase electricity from renewable energy sources at a pre-determined price that is sufficiently attractive to stimulate new investment in the renewable sector. This, in turn, ensures that those who produce electricity from identified renewable energy sources such as solar, wind and other renewable sources have a guaranteed market and an attractive return on investment for the electricity they produce. Aspects of an FIT include access to the grid, long-term power purchase agreements and a set price per kilowatt hour (kWh).

Kenya's FIT policy has as its objectives to:

(*a*) facilitate resource mobilization by providing investment security and market stability for investors in Renewable Energy Sources (RES) electricity generation,

(*b*) reduce transaction and administrative costs by eliminating the conventional bidding processes, and

(*c*) encourage private investors to operate the power plant prudently and efficiently so as to maximize its returns. By taking a long-term commitment to the development of renewable sources of energy and stipulating a long-term power purchase agreements of a minimum of 20 years, the Kenya Government has taken a critically important step in the development of the country's significant potential for renewable energy generation, while pursuing equally important economic, environmental and social policy objectives.

In January 2010, Kenya revised the FIT policy, which resulted in the addition of three renewable energy sources: geothermal, biogas, and solar energy resource generated electricity. In addition, the revised policy extended the period of the power purchase agreements from 15 to 20 years and increased the fixed tariffs per kilowatt-hour for pre-existing wind, biomass and small-hydro power under the FIT.

A. Expected Benefits

The advantages of this policy include:

(*a*) environmental integrity including the reduction of greenhouse gas emissions;

(*b*) enhancing energy supply security, reducing the country's dependence on imported fuels; and coping with the global scarcity of fossil fuels and its attendant price volatility; and

(*c*) enhancing economic competitiveness and job creation. Initially covering wind, biomass and small-hydro, the policy is planned to include geothermal sources of energy.

It is expected that the FIT policy in Kenya could stimulate about 1300 MW of electricity generation capacity. If the projected generation capacity is realized, this could contribute significantly to ensuring security of electricity supply in the country by increasing the reserve margin. Furthermore, since the resources used consist of relatively low- cost local fuels, it is likely to reduce costs for the consumer. Benefits targeted are a "triple-win" of additional renewables-based generation capacity to the country; enhancing employment and poverty alleviation in the rural areas; and increasing income opportunities for business development.

As Kenya's greatest renewable energy potential is in rural areas, the effects of the FIT policy are expected to trickle down and stimulate rural employment. This can happen through the construction of power plants, but also in the context of agro-industries, in particular sugarcane, which is predominant in the country. It is estimated that the sugar factories have directly and indirectly contributed to job creation by supporting about 200,000 small-scale farmers within the sugar belt in western Kenya, and that between five and six million people either directly or indirectly benefit from the sugar factories.

B. Observed Impacts

Since the announcement of the FIT policy, some sugar companies have planned to upgrade their biomass-based cogeneration potential in order to benefit from the FIT policy. While full effects have not been felt yet, Kenya provides an example of how forward-looking energy policy could contribute to matrix diversification, improved benefit streams to small rural producers, and enhanced local development.

16.2.3 Organic Agriculture in Uganda

Uganda has taken important steps in transforming conventional agricultural production into an organic farming system, with significant benefits for its economy, society and the environment. Organic Agriculture (OA) is defined by the Codex Alimentarius Commission as a holistic production management system, which promotes and enhances agro-ecosystem health, including biodiversity, biological cycles and soil biological activity. It prohibits the use of synthetic inputs, such as drugs, fertilizers and pesticides.

Uganda uses among the world's lowest amount of artificial fertilizers, at less than 2 per cent (or Ikg/ha) of the already very low continent-wide average of 9kg/ha in Sub Saharan Africa. The widespread lack of fertilizer use has been harnessed as a real opportunity to pursue organic forms of agricultural production, a policy direction widely embraced by Uganda.

A. Economic, Social and Environmental Benefits of Organic Farming in Uganda

In Uganda 85 per cent of the population was engaged in agriculture production, contributing to 42 percent of the national GDP and 80 percent of the exports earnings in 2005/06. As early as 1994 a few commercial companies began deliberately engaging in organic agriculture. At the same time in Uganda, there was a general movement in the agricultural sector towards developing sustainable agriculture as a means of improving people's livelihoods.

By 2003, Uganda had the world's 13th-largest land area under organic agriculture production and the most in Africa. By 2004, Uganda had around 185,000 ha of land under organic farming covering more than 2 per cent of agricultural land, with 45,000 certified farmers. By 2007,296,203 hectares of land were under organic agricultural production with 206,803 certified farmers.This constitutes an increase of 359 per cent in terms of number of farmers and 60 per cent in terms of acreage, respectively, from 2002 to 2007.

According to IFOAM, the global market for organic foods and drinks is estimated to be around US$50 billion, and increased by 10–20 per cent annually between 2000 and 2007. This sub-sector provides a unique export opportunity for many developing countries, owing to the fact that 97 per cent of the revenues are generated in the OECD countries, while 80 per cent of the producers are found in developing countries of Africa, Asia and Latin America.

As a significant producer of organic products, Uganda benefits from an important source of export earnings and revenue for farmers. Certified organic exports increased from US$ 3.7 million in 2003/4, to US$ 6.2 million in 2004/5, before jumping to US$ 22.8 million in 2007/8. In terms of price premiums and income for farmers, studies commissioned by UNEP and UNCTAD indicate that in 2006 the farm-gate prices of organic pineapple, ginger and vanilla were 300 per cent, 185 per cent, and 150 per cent higher, respectively than conventional products. Through organic farming, Uganda not only gains economically, it also contributes to mitigating climate change, as GHG emissions per ha are estimated to be on average 64 per cent lower than emissions from conventional farms. Various studies have shown that organic fields sequester 3-8 tonnes more carbon per ha than conventional agriculture.

B. Policy and Systemic Change to Transform Uganda's Agricultural Sector

On the policy side, in 2004 the Uganda Organic Standard was adopted, while in 2007, as part of the East African Community, Uganda adopted the regional standard, the East African Organic Products Standards (EAOPS) developed under a joint UNEP-UNCTAD initiative. In July 2009, the government released a Draft Uganda Organic Agriculture Policy. The draft policy describes the vision, mission, objectives and strategies to support the development of organic agriculture as "one of the avenues for delivering self- sustaining growth as it provides mechanisms for individual farmers to improve productivity, add value and access markets which are keys to achievement of the Poverty Eradication Action Plan objectives".

The strategy put in place to implement the policy is based on interventions in nine policy areas: the promotion of organic agriculture as a complementary agricultural production system; the development of a system of standards, certification and accreditation; the promotion of research, to enable technology development and dissemination; support to the development of local, regional and international markets for organic products; the generation of information, knowledge and skills through education and training; the improvement of post-harvest handling practices, preservation, storage and value addition; the sustainable use of natural resources; and participation of the special interest groups such as women, youth, and the poor and vulnerable.

In sum, Uganda has taken an apparent liability—limited access to chemical inputs—and turned this into a comparative advantage by growing its organic agriculture base, generating revenue and income for smallholder farmers.

16.2.4 Sustainable Urban Planning in Brazil

Rapid growth of urban areas presents both environmental and socio-economic challenges to residents, businesses and municipalities. With inadequate planning and limited finances accommodating the increasing urban populations often results in expansion of informal housing in cities or suburban developments requiring high use of private transport.

Brazil has the fourth-largest urban population after China, India, and the US, with an annual urban growth rate of 1.8 per cent between 2005 and 2010. The city of Curitiba, capital of Parana State in Brazil has successfully addressed this challenge by implementing innovative systems over the last decades that have inspired other cities in Brazil, and beyond. Particularly known for its Bus Rapid Transit system, Curitiba also provides an example of integrated urban and industrial planning that enabled the location of new industries and the creation of jobs.

A. Sustainable Planning for Growth in Curitiba

Through its innovative approaches in urban planning, city management, and transport planning going back as far as the 1960s, Curitiba has been able to grow in population from 361,000 (in 1960) to 1.828 million (in 2008), without experiencing typical drawbacks from congestion, pollution and reduction of public space. The population density in the city has increased three-fold from 1970 to 2008. At the same time, the average green area per person increased from 1 km^2 to over 50 km^2.

One of the key elements of urban planning was a choice for growth in a 'radial linear-branching pattern', which served to protect both density and green areas. This pattern encouraged—through a combination of land-use zoning and provision of public transport infrastructure—a diversion of traffic from the city centre and the development of housing, services and industrial locations along the radial axes.

B. Economic and Environmental Benefits

Cities offer considerable opportunities to reduce $C0_2$ emissions when applying coordinated approaches to emission reductions in transport and buildings, which are the two of largest sources. As a result of integrated urban planning, Curitiba has the highest rate of public transport use in Brazil (45 per cent of journeys), and one of the country's lowest rates of urban air pollution.

The economic and resource efficiency benefits of such initiatives are considerable. Curitiba's fuel usage is 30 per cent lower than in Brazil's other major cities. Excessive fuel use due to severe traffic congestion - estimated at a value of US$1 million for Curitiba in 2002 - was about 13 times and 4 times less in per capita terms than those in Sao Paulo and Rio de Janeiro, respectively. The per capita loss due to time spent in severe congestion in Curitiba is approximately 11 and 7 times less than in those two cities, respectively.

C. Policy and City Planning for Ecological Infrastructure and Industrial Activity

By turning areas vulnerable to flooding into parks planted with many trees, and creating artificial lakes to hold floodwaters, Curitiba has managed to address its potentially costly flooding problem, in terms of flood control and drainage. The cost of this strategy, including the relocation costs of slum dwellers, is estimated to be 5 times less than building concrete canals. Also, as a result, the property values of neighbouring areas appreciated, and tax revenues increased.

The local government established the Curitiba Industrial City (CIC) on the city's west side, taking into account wind direction to avoid polluting the central city. The CIC has strict environmental regulations and "polluting" industries are not allowed. After three decades, the CIC today accommodates more than 700 companies, including an automaker producing BRT buses and information technology companies. The CIC has already created about 50,000 direct jobs and 150,000 indirect jobs, and about 20 per cent of the state's exports are from the CIC.

Curitiba has also promoted waste management infrastructure and public awareness on waste separation and recycling. With 70 per cent of the city's residents actively recycling, 13 per cent of solid waste is recycled in Curitiba, as compared to only 1 per cent in Sao Paulo.

All in all, Curitiba presents a case study of how smart urban planning can avoid significant costs in the future and improve efficiency, productivity and quality of life for its inhabitants.

16.2.5 Rural Econological infrastructure in India

India's National Rural Employment Guarantee Act, 2005 (NREGA) is a guaranteed wage employment programme that enhances the livelihood security of marginalized households in rural areas. Implemented by the Ministry of Rural Development, NREGA directly touches the lives of the poor, promotes inclusive growth, and also contributes to the restoration and maintenance of ecological infrastructure.

In its first two-and-a-half years of operation, from 2006 to 2008, NREGA generated more than 3.5 billion days of work reaching on average 30 million families per year. The programme is implemented in all 615 rural districts of the country, with women representing roughly half the employed workforce. The emphasis is placed on labour-intensive work, prohibiting the use of contractors and machinery.

A. Investing in Ecological Infrastructure

In addition to supplementing wage employment, NREGA's secondary objective is to strengthen rural natural resource management. This is achieved by financing rural works that address causes of drought, deforestation and soil erosion, thus restoring the natural capital base on which rural livelihoods depend.

As one of the world's leading crop producers, India has seen a surge in water consumption in the agriculture sector over the years. With industrial water consumption expected to quadruple between 2000 and 2050 as well, groundwater aquifers in the four major river basins may be depleted by half by 2050. India is moving towards water scarcity due to increasing demand and a drop in supply of clean water as well, with forecasts of even a 50 per cent deficit by 2030. Increasing water scarcity is a major constraint to agriculture and livestock production affecting livelihoods and food security.

Water conservation accounts for about half of the total projects supported under NREGA, with 850,000 water conservation works funded and completed from 2006 to 2008. For example, in the District of Jalaun (Uttar Pradesh), NREGA provided training and jobs for villagers to develop solutions to their heavily silted water harvesting infrastructure, alleviating their water shortage. In 2007–2008, more than 3,000 new soak pits, together with hand pumps were constructed. This has helped conserve an estimated 5 million litres of water.

Similarly in Andhra Pradesh, NREGA supported the restoration of a network of water storage tanks dating back over 500 years in the principal arid zone. Repairs to the gates of the tanks, as well as works to desilt the channels feeding them, has restored to full capacity. This not only boosts crop and livestock production but has contributed to groundwater replenishment.

In its short history, NREGA has produced many such success stories across the country. In promoting to inclusive growth and the restoration of ecological infrastructure, the programme also has a strong impact on empowerment of poor or marginalized groups. It has contributed to boosting the average wage for agricultural labourers by more than 25 per cent over its three-year history.

16.2.6 Forest Management in Nepal

Community forestry has contributed to restoring forest resources in Nepal. Forests account for almost 40 per cent of the land in the country. Although this area was decreasing at an annual rate of 1.9 per cent during the 1990s, this decline was reversed, leading to an annual increase of 1.35 per cent over the period 2000 to 2005.

Community forestry occupies a central place in forest management in Nepal. In this approach, local users organised as Community Forest User Groups (CFUGs) take the lead and manage resources, while the government plays the role of supporter or facilitator. Forest management is a community effort and entails little financial or other involvement on the part of the government.

Since 1980, about 14,000 CFUGs have been formed. About one-fourth of Nepal's national forest is now managed by more than 35 per cent of the total population. Community forestry is now the second-largest forest management regime after government-managed forests.

Forest user groups develop their own operational plans, set harvesting rules, set rates and prices for products, and determine how surplus income is distributed or spent. There is evidence of significant improvement in the conservation of forests (both increased area and improved density) and enhanced soil and water management.

A. Benefits from Community Forestry Management

The advantages of community forestry include employment and income generation from forest protection, tree felling and log extraction, as well as non-timber forest products. Additional economic benefits are in the form of sustained wood fuel sources, which contribute more than three-quarters of energy needs to households. Improved forest management and cover also contribute to nature conservation.

Community forestry promotes inclusive growth. Some community forest activities have initiated a scholarship programme for low income people, as well as savings and credit operations among members, including loans to finance income generation activities. Community forestry also empowers CFUGs with greater influence over decision making through participation in planning and management.

B. Policies to Protect Deforestation

The Ministry of Forest and Soil Conservation (MFSC), as the lead institution for creating an enabling environment for forest resource management, has implemented two policies, including Community Forestry and the Leasehold Forestry Policy, that have helped the government of Nepal achieve significant success in forest conservation and environmental protection.

The Master Plan for the Forestry Sector prepared in 1989, the Forest Act of 1993, the Forest Regulations of 1995, and the Forestry Sector Policy of 2000 were developed and implemented to support the community forestry programme, after earlier efforts at centralized control. The Forest Act and Forest Rules accelerated the transfer of forests to forest user groups; they provide the legal basis for the implementation of community forestry and recognize CFUGs as "self-governing autonomous corporate bodies for managing and using community forests".

The remarkable turnaround in forest management in Nepal is directly attributable to the benefits generated for community groups, in the form of environmental goods, services and welfare enhancements.

16.2.7 Ecosystem Services in Ecuador

The city of Quito offersa leading example of the potential for developing markets that channel economic demand for water to upstream areas from which it is supplied. In Quito's case, the availability of water for the city of about 1.5 million inhabitants and surrounding areas depends on the conservation of protected areas upstream, with 80 per cent of the water supply originating in two ecological reserves, the Cayambe-Coca (400,000 ha) and the Antisana (120,000 ha).

The Fund for the Protection of Water—FONAG—was established in 2000 by the municipal government, together with a non-governmental organisation, as a trust fund to which water users in Quito contribute. FONAG uses the proceeds to finance critical ecosystem services, including land acquisition for key hydrological functions.

A. Demand and Supply

The users include farmers, hydropower companies, industries and households, who pay differentiated rates depending on use. The largest share of payments comes from the Quito water utility (Metropolitan Enterprise of Water and Sewer Systems in Quito—EMMAP-Q) which contributes 1 per cent of monthly water sales. Hydropower companies make fixed annual payments, as does the Cerveceria Andina brewery. Farmers drawing water through irrigation also contribute. The fund held more than US$ 7 million at the end of 2009 (up from US$ 3 million in 2005) and invested about US$ 0.8

million in 2008. Administration costs are limited to between 10 and 20 per cent of total expenditures.

FONAG finances both watershed management projects in micro river valleys and longer-term programmes (at least 20 years in duration) oriented towards communication, environmental education, forestry, and the river basin management training. These projects and programmes are undertaken with the participation of different community actors, local authorities, educational institutions, and governmental and non-governmental organisations.

B. Benefits

FONAG is contributing to securing present and future water supplies for Quito. Through the fund, more than 65,000 ha of watersheds are now under improved management. Upstream farmers receive support for watershed protection programmes, as opposed to cash payments. More than 1800 people are estimated to receive increased economic benefits associated with watershed management and conservation.

FONAG has served to inspire the development of similar schemes elsewhere in Latin America and beyond. For example, in South Africa, where water forms one of the greatest constraints on development, a recently-launched initiative in the Maloti Drakensberg Mountains aims to implement a payments for watershed services programme, with support from UNEP and the BASF Social Foundation. This initiative will use payments from downstream users to support the restoration of dongas, and the improvement of grazing and veld fire management regimes in order to reduce sedimentation and increase the quality and quantity of water flows. In so doing, employment will be generated for local households and the productive potential of agricultural activities should increase.

In sum, FONAG is one of a growing number of initiatives that bring together different stakeholders in the creation of innovative mechanisms to channel economic demand for water to areas critical for the supply, in the process improving livelihoods and generating employment, while investing in key ecosystem services.

16.2.8 Solar Energy in Tunisia

To reduce the country's dependence on oil and gas, Tunisia's government has undertaken steps to promote the development and use of renewable energy.

A law establishing an "energy conservation system" on energy management in 2005 was immediately followed with the creation of a funding mechanism—the National Fund for Energy Management - to support increased capacity in renewable energy technologies and also improved energy efficiency. The replenishment of this Fund is based on a duty levied on the first registration of private, petrol-powered and diesel powered cars and on import duty or local production duty of air-conditioning equipments with the exclusion of those produced for exports.

Between 2005 to 2008, clean energy plans have already allowed the government to save $1.1 billion in energy bills, relative to initial investments of $200 million in clean energy infrastructure. Primary energy consumption from renewables, together with savings from energy efficiency, are expected to reach 20 per cent of total energy consumption in 2011.

In December, 2009, the government presented the first national Solar Energy Plan and other complementary plans with the objective of increasing the share of renewable energy sources from just under 1 per cent to 4.3 per cent in 2014. The plan includes the use of solar photovoltaic systems, solar water heating systems and solar concentrated power units for electricity generation.

Total financial resources to implement the plan have been estimated at $ 2.5 billion,including $ 175 million from the National Fund, $ 530 million from the public sector, $ 1,660 million from private sector funds, and $ 24 million from international cooperation, all to be spent by 2016 on 40 renewable energy projects. Approximately 40 per cent of the resources are devoted to the development of energy export infrastructure.

The energy savings expected to result from the Solar Energy Plan could reach 22 per cent for 2016, with a reduction of 1.3 million tonnes per year of CO_2.

A. Solar Water Heating Systems—The PROSOL Programme

The Tunisian Solar Programme (PROSOL)—a joint initiative of the Tunisian National Agency for Energy Conservation (ANME), the state utility Societe Tunisienne de I'Electricite et de Gaz (STEG), the United Nations Environment Programme and the Italian Ministry for the Environment, Land and Sea—provides an example of solar thermal market development.

Financial and fiscal support combines a capital grant qualifying for a VAT exemption, customs duty reduction and a bank loan with a reduced interest rate. Repayment of the loan is organized through the regular utility bill of the state electric utility STEG, with local banks receiving support that allows them to finance SWH projects with reduced interest rates.

This arrangement has generated direct financial benefits for the end users, when comparing the size of the monthly instalments for a SWH system to the earlier electricity bills. A complementary interest rate subsidy was available during the first 2 years (2005–2006) of the programme, reducing the interest rate of the loan to 0 per cent to the final end user. This support was removed in 2007 and annual interest rates for loan repayment have been 6.5 per cent.

The government provides a subsidy of 20 per cent of the system cost or $ 75 per square meter, while customers are expected to finance a minimum of 10 per cent of the purchase and installation costs.

Over 50,000 Tunisian families now get their hot water from the sun based on loans amounting to more than \$ 5 million in 2005 and \$ 7.8 million in 2006—a substantial leverage to PROSOL's initial cost of \$ 2.5 million. With installed surface of the programme reaching 400 000 m^2, the government has now set a more ambitious target of 750,000 m^2 for the period 2010–2014, a level comparable to much larger countries such as Spain or Italy. As of 2008, PROSOL helped avoid 214,000 tonnes of cumulative CO_2 emissions. Jobs have been created as 42 technology suppliers were officially registered and at least 1000 companies installed the systems.

In conclusion, the experience in Tunisia demonstrates the potential returns on investing in renewable energy, creating new jobs, and reducing dependency on fuel imports.

■ ■ ■

17

International Initiatives for Green Economy

17.1 INTRODUCTION

Until the 1970s, there was little in the way of international institutions or international law relating to environmental protection. Environmental degradation was regarded as a local or national problem of little concern to the international community. This changed in 1972 with the meeting in Stockholm of the U.N. Conference on the Human Environment. The U.N. General Assembly resolution convening this conference recognized "an urgent need for intensified action at national and international levels to limit and, where possible, to eliminate the impairment of the human environment."

The Stockholm Conference, as it is now called, was a grand success; four major initiatives resulted from it. First, the Stockholm Declaration of Principles was adopted—this proved to be a seminal document in the development of international environmental law. Second, the U.N. Environment gram (UNEP) was established as a permanent new international organisation within the United Nations to foster protection of the global environment. The decision has taken to create a "program" and not a full-fledged organisation because of the related work of many existing U.N. organisations, such as the Food and Agriculture Organisation, the International Maritime Organisation and the World Health Organisation. Third, the Stockholm Conference adopted an action plan for the development of environmental policy to be administered by UNEP, and fourth an Environmental Fund was created to receive voluntary contributions.

In the years following the Stockholm Conference, discussion about international environmental protection measures became entangled with the increasing influence of developing countries and their emphasis on development and a

New International Economic Order. Developing countries came to resent calls for environmental measures that would inhibit their right to develop. The stand-off remained until the World Commission on Environment and Development published a report, *Our Common Future* (called the Brundtland Report after its leader, Gro Harlem Brundtland, later Prime Minister of Norway), calling for a new approach—*sustainable development*—combining the right to develop with environmental concerns.

17.2 UN CONFERENCE ON ENVIRONMENT AND DEVELOPMENT

The Brundtland Report opened the way for the U.N. General Assembly to convene a new conference, which was held amid great fanfare in Rio de Janeiro in 1992 and attended by 176 states and 103 heads of state or government. This U.N. Conference on Environment and Development (UNCED) was a major event that built upon the Stockholm meeting, but went beyond it in scope and depth. Five important international instruments were adopted at it:

1. *The Rio Declaration on Environment and Development*—a set of twenty-seven nonbinding principles setting out the recommended contours of sustainable development;
2. *Agenda 21*—an action program containing recommended strategies to be followed by all states in order to deal with environmental issues and resources. Every conceivable environmental problem is addressed in its forty chapters, and the interconnections among economic, developmental, and environmental issues are explored in depth. The financial chapter recommends a doubling of financial aid to developing countries to assist them in carrying out the recommended goals;
3. *The U.N. Framework Convention on Climate Change*—the beginning of a new international regulatory regime to deal with the consequences of rampant use of fossil fuels for the production of energy;
4. *The U.N. Convention on Biological Diversity*—a worldwide initiative to combat the loss of biological diversity through overexploitation of natural resources; and
5. *The Non-Binding Authoritative Statement of Principles on the Management, Conservation, and Sustainable Development of All Types of Forests*—an attempt to forge a consensus leading to reduced exploitation of forests. Developing countries blocked both a legally binding treaty and a convention on tropical forests alone. Many developed countries, led by the United States, opposed any treaty that would place any international constraints on harvesting temperate zone forests.

All these measures are being ignored or are in severe difficulty. The United States not only played no role in organizing these U.N. initiatives but also consistently obstructed them.

In 2002, under a mandate from the U.N. General Assembly, the World Summit on Sustainable Development was held in Johannesburg, South Africa. At this summit there was a ten-year review of the progress made toward achieving the sustainable development goals agreed upon at Rio de Janeiro. The news was not good, and the Johannesburg Summit produced only a political declaration and laments on the lack of progress and inability to agree on future actions. The United States came in for particular criticism, and the U.S. delegation led by Health and Human Sendees Secretary Tommy Thompson walked out and went home early.

17.3 INTERNATIONAL ENVIRONMENTAL GOVERNANCE

A multiplicity of international regimes have responsibility for the global governance of environmental matters. All are weak, especially compared with international economic institutions. The universal rhetoric of states that the global environment is a concern has not translated into effective action. The organisational structure of environmental protection is fragmented. There is an urgent need for a *World Environment Organization* similar to the World Trade Organization that would centralize and coordinate the international regimes that deal with global environmental issues.

At present the United Nations Environment Program (UNEP) plays the central role in international environmental governance, but its charter authorizes it only "to promote international cooperation in the field of the environment and to recommend, as appropriate, policies to this end"—not a very strong mandate. Fifty-eight member states are elected triennially to UNEP's Council by the General Assembly, and there is an executive director and a permanent secretariat based in Nairobi, Kenya. Despite little funding and a narrow mandate, UNEP has a solid record of accomplishment. It has served as a forum for the conclusion of important international agreements, developed new international legal principles and guidelines, and assisted developing countries in the reform and administration of their environmental laws. In 1997 the General Assembly of the United Nations established a Global Ministerial Forum to confirm UNEP's role as the setter of the global environmental agenda, the coordinator of the UN's environmental protection efforts, and the advocate on behalf of the global environment. UNEP now has responsibility for the coordination of U.N. environmental treaties and their implementation.

17.4 THE UNITED NATIONS ENVIRONMENT PROGRAMME

The United Nations Environment Programme (UNEP) was established by U.N. General Assembly Resolution 2997 on December 15, 1972. UNEP has a Governing Council composed of fifty-eight members elected by the U.N. General Assembly to serve four-year terms. The day-to-day functions of UNEP are the responsibility of its Executive Director.

The functions and responsibilities of the Governing Council are as follows:

- to promote international cooperation in the field of environment and to recommend, as appropriate, policies to this end,
- to provide general policy guidance for the direction and coordination of environmental programs within the U.N. system,
- to receive and review the periodic reports of the executive director of UNEP on the implementation of environmental programs within the U.N. system,
- to keep under review the world environmental situation in order to ensure that emerging environmental problems of wide international significance receive appropriate and adequate consideration by governments,
- to promote the contribution of the relevant international scientific and other professional communities to the acquisition, assessment, and exchange of environmental knowledge and information and, as appropriate, to the technical aspects of the formulation and implementation of environmental programs within the U.N. system,
- to maintain under continuing review the impact of national and international environmental policies and measures on developing countries,
- to review and approve the program of utilization of the Environment Fund.

A second U.N. environmental body, the Commission on Sustainable Development (CSD), was created after the 1992 UNCED Conference to review the implementation of Agenda 21 and other Rio documents. The CSD comprises of representatives of fifty-three states elected by the U.N. Economic and Social Council for three-year terms. Meeting only annually, the CSD has confined its work to making policy recommendations, such as the Program for the Further Implementation of Agenda 21, which was adopted in 1997. Agenda 21 has languished as a result.

Many other U.N. specialized agencies are involved in some aspect of the environment or resources. For example, the *Food and Agriculture Organisation* (FAO), which was created to improve efficiency in the production of food and agricultural produce, promotes sustainable approaches to fishing and water resource management and raises awareness of the harmful impacts of chemical pesticides and fertilizers used in agriculture. The FAO played a key role in the negotiation of regional fisheries agreements and the 1995 Agreement on Straddling and Highly Migratory Fish Stocks. The FAO also collaborated w ith UNEP in the negotiation of the 1998 Convention on the Prior Informed Consent Procedure for Certain Hazardous Chemicals and Pesticides in International Trade.

The *International Maritime Organisation* (IMO) in London has responsibility for the environmental protection of the oceans. Beginning in the 1950s, the IMO convened international conferences, which led to international conventions preventing and controlling pollution from ships. As a result, oil pollution from shipping discharges and accidents has greatly decreased, although the transport of oil by water has increased. The IMO established a Marine Environmental Protection Committee after the Stockholm Conference of 1972, and there is also a legal committee. Agenda 21 called upon the IMO to address broader environmental issues; it responded with regulations protecting sensitive sea areas and the regulation of the discharge of air and ballast-water, garbage, and plastics pollution from ships. To be effective, however, these rules must be implemented by member states.

The *U.N. Economic, Scientific, and Cultural Organization* (UNESCO) administers the 1972 World Heritage Convention, which calls upon states to designate areas within their territory that deserve special recognition and preservation. UNESCO's Man and the Biosphere Program establishes a worldwide system of natural areas designated as "biosphere reserves." Designation is voluntary, however, and reserves remain under state sovereignty.

Additional international environmental governance is exercised by scores of treaty regimes that have been established to deal with particular environmental or resource problems—from the Amazon to wetlands and whales and everything in between. Because there is no central, coordinating environmental organization, there is great need to upgrade UNEP's status to become the operating agency or the coordinating organization for all international environmental agreements and understandings.

Non-governmental organisations (NGOs) have played a leading role in international environmental efforts. They range from advocacy groups, such as Greenpeace International, to scientific bodies, such as the World Wide Fund for Nature, to political NGOs, such as the Water Resources Institute. Perhaps the leading NGO is the International Union for the Conservation of Nature (IUCN, also known as the World Conservation Union) in Gland, Switzerland, whose membership includes 79 states, 149 government agencies, 690 national NGOs, and 68 international NGOs. The IUCN passes resolutions and offers advice and expertise to governments and has played a key role in preparing international conventions and documents, such as the World Charter for Nature, which was adopted by the U.N. General Assembly in 1982.

Environmental governance is exercised by many international bodies, but despite their individual competence, there is no clear leadership or direction to the effort to deal with global environmental problems.

17.5 DISTINCTIVE DOCTRINES OF INTERNATIONAL ENVIRONMENTAL LAW

17.5.1 "Soft" Law and Framework Treaties

A distinctive feature of international environmental law is the development of legal doctrines and principles, which serve as policy guides for more particular, concrete standards and legal norms. These policy guides are too general to constrain conduct and so are regarded as "soft" law, which appears to be a contradiction, because law denotes social rules that are compulsory and are therefore "hard." Nevertheless, because of its novelty, international environmental law has made use of halfway, general principles that attract wide assent, avoiding details and standards that bite or reserving them for later. Soft law can thus be regarded as the essential beginning of a law-making process. Soft law may also be aspirational, preparing the way for future progress. Soft law also denotes general, commonly agreed-upon principles that provide the basis for international negotiations.

International environmental law also makes use of a distinctive method of law making known as the *framework treaty process.* Typically this process involves concluding an initial multilateral treaty at which states agree only that there is an international environmental problem that requires action, and suggests practicable measures.' Although such a treaty has little immediate impact, it may be very effective in providing a framework for future action. To accomplish this the treaty must establish a permanent secretariat and a conference of the parties that meets periodically. This process will result in the adoption of protocols, annexes, or action plans that set out precise measures to be taken by the parties. Environmental treaties have a dynamic character and flexibility that, if taken advantage of, enable the parties to respond to changed conditions and circumstances. An example is the 1992 Climate Change Convention, which was followed by such agreements as the Kyoto Protocol.

17.5.2 Sustainable Development

Sustainable development is an overarching concept of international environmental law. The Rio Declaration and most treaties and other international understandings state sustainable development as their general goal. The origin of the idea of sustainable development is that environmental protection should not prevent economic growth, especially in the developing world. But what does this mean? Sustainable development is a complex issue containing both substantive and procedural elements. Its substantive component means that, although there is a right to development, resources should be used wisely and efficiently, pollution should be prevented or controlled, and environmental impacts should be minimized. Resources should be used so as not to endanger future needs, and recycling and reuse should be the norm. Procedurally, sustainable development means that environmental considerations will be integral to developmental decision making; there should be full disclosure of information and public participation in significant decisions. Nevertheless, there is an inherent ambiguity in the concept of sustainable development. The product of political compromise that seeks to integrate environmental protection and economic growth, sustainable development contains an undeniable tension that is difficult to resolve.

17.5.3 Intergenerational Equity

The 1987 report of the World Commission on Environment and Development, *Our Common Future,* defined sustainable development as meeting the needs of the present "without compromising the ability of future generations to meet their own needs." Thus, intergenerational equity is an important concept in conservation and preservation efforts. Principle 3 of the Rio Declaration echoes this concern.

Few would argue with intergenerational equity as an abstract principle, but how is it to be realized? There is no accepted way to measure whether any action will harm any future generation, and there is no recognized advocate for future generation rights. This principle would seem to have limited utility. Intergenerational equity is, however, stated as a reason for the conclusion of many environmental treaties, such as, for example, the International Whaling Convention (1946). The Supreme Court of the Philippines has also cited this principle as a reason for allowing citizens to sue the government over excessive timber logging.

17.5.4 Common but Differentiated Responsibility

A fundamental but controversial principle of international environmental law is that, although all states have common responsibility for environmental, amelioration, developed countries bear the greatest burden. This is based on the idea that the developed world has and continues to pollute more than developing countries. No specific differentiation of responsibility is defined by this principle, but its realisation may be embedded in various ways in most environmental treaties. This principle has three aspects.

First, developing countries will not be called upon to bear the same corrective burden as developed countries. Rio Principle 11 makes this explicit, stating, "[Environmental] Standards applied by some countries may be inappropriate and of unwarranted economic and social cost to other countries, in particular developing countries."

Second, developing countries have a right to receive technical and financial help to comply with environmental obligations. This principle was partially implemented with the establishment of the Global Environmental Facility (GEF) in 1991, which provides funds for agreed-upon incremental costs of environmental measures in developing countries. Although the World Bank acts as trustee, GEF is operated by its own council; composed of thirty-two states balanced between donor and recipient countries, this council the manages GEF. Decisions require a double majority of 60 percent of all members and 60 percent (by contribution) of donors.

Third, Rio Principle 5 injects the idea of intragenerational equity, that there is an obligation to "decrease the disparities in standards of living and better meet the needs of the majority of the people of the world." This statement connects the common but differentiated responsibility idea to the right to development. Under Rio Principle 9, developed states also "should cooperate to strengthen capacity building of developing states for sustainable development through the transfer of technology and scientific knowledge."

17.5.5 Precautionary Principle

Article 15 of the Rio Declaration refers to the "precautionary approach" to protecting the global environment. This is the idea that "where there are threats of serious or irreversible damage, lack of full scientific certainty shall not be used as a reason for postponing cost-effective measures to prevent environmental degradation."

The precautionary principle is a feature of most international environmental treaties that deal with resources or environmental degradation. As a principle it is inherently flexible—it does not define the kind of threat that triggers its application, nor is it specific about measures to be taken. As a principle it can be adapted to specific treaties and particular circumstances.

All versions of the principle justify action when there is some evidence but not proof that a human practice is harming the environment. The central concept here is *causation* and the evidence necessary to establish it. First, it can justify taking action in the form of environmentally protective measures; second, it can remove a reason to avoid actions-scientific uncertainty cannot be used to avoid action—so that decisions can be made on other, clearer grounds; and third, it can justify a reversal of the burden of proof—instead of requiring proof bv an environmental interest, the burden can be placed on the side that believes an environmental measure to be unnecessary.

The precautionary principle also makes possible a focus on risk and risk assessment. Scientific findings may be expressed in terms of risk either because limitations of inherent in data or because of the inability to conduct experimentation on human beings. The focus on risk makes possible a decisionmaking process that takes into consideration the degree and magnitude of the potential harm. The greater the risk, the greater role there is for precaution and for precautionary action. But the principle does not state what level of risk is socially acceptable or how to control any particular risk. Thus, invocation of the precautionary principle cannot Ire determinative.

17.5.6 State Responsibility for Environmental Harm

The *Trail Smelter case* promulgated the principle that states are responsible for significant transboundary environmental harm. International law accordingly does not allow states to either conduct or permit activities within their territories that harm other states or the environment. This doctrine, called the *harm prevention principle,* is derived from the Latin maxim, *sic utere tuo, ut non laedas*—a principle of "good neighborliness" that is now a duty of customary international law. In its *Advisory Opinion on the Legality of the Threat or Use of Nuclear Weapons,* the International Court of Justice affirmed that.

The existence of the general obligation of states to ensure that activities within their jurisdiction and control respect the environment of other states and of areas beyond national control is now part of the corpus of international law relating to the environment."

The harm prevention principle is now recognized by state practice and was restated in both the 1972 Stockholm Declaration (Principle 21) and the 1992 Rio Declaration (Principle 2). The U.N. International Law Commission has elaborated draft articles creating liability under international law for transboundary environmental damage.

State responsibility under the harm prevention principle is now recognized as applying not only to transboundary harm but also to the global environment and to areas beyond the jurisdiction of all states, such as the oceans. Thus, the right to development is qualified, and integration of development and environmental protection is required.

Three aspects of the harm prevention principle are apparent: First is the duty to prevent harm, second is the duty to compensate for harm that has occurred, and third is the duty to cease or modify an activity that has caused harm.

However, uncertainty obscures many aspects of the harm principle. For example, is there state responsibility without fault? What degree of damage is necessary before responsibility arises? These ambiguities make the harm prevention principle difficult to apply in particular circumstances. Thus, the development of specific international regimes through appropriate international agreements on a global, regional, and bilateral basis is essential.

17.5.7 Cooperation

States are obliged to cooperate with each other and with relevant international institutions to deal with international environmental problems. This duty was first recognized in the case of shared natural resources. In the *Lac Lanoux* arbitration case, the tribunal posited a duty of notification and good faith negotiation with respect to a proposed diversion of water by France from a watercourse shared with Spain. Conflicting state interests must be reconciled through negotiations and mutual concessions. This duty of cooperation has been extended by state practice and treaties to include the management of transboundary risks posed by hazardous or potentially harmful activities, such as nuclear installations and factories near frontiers, continental shelf activities, and long-range transboundary pollution.

The Rio Declaration of 1992 codifies this duty of cooperation in several provisions. Principle 18 requires states to notify other states if natural disasters or other emergencies are likely to produce transboundary environmental effects. Principle 19 requires prior notification, good faith consultation, and the provision of relevant information to affected states before undertaking any activity' likelv to produce significant adverse transboundary environmental effects. Principle 17 requires an environmental impact assessment to be undertaken for such a proposed project. Thus, the cooperation duty can be summarized including the following components:

- prior and timely notification
- disclosure of relevant information
- consultation and good faith negotiation
- prior impact assessment
- environmental monitoring.

The International Court of Justice addressed the duty of cooperation in the *Case Concerning the Gahcikovo-Nagymaros Dam (Hungary v. Slovakia);* (1997). In this case Hungary argued that environmental risks were grounds for terminating a treaty yvitli Slovakia to build a series of dams on the River Danube. The court, although rejecting treaty termination, ruled that in implementing the treaty the parties were obliged to apply new and developing norms of environmental hm relating to impact assessment and mitigation of environmental degradation. The parties were required to negotiate in good faith concerning how to adapt the project to meet new environmental requirements.

17.5.8 Common Concern

The principle of common concern recognizes an obligation on the part of all states to protect and restore the global environment, resources, and areas beyond the limits of national jurisdiction. The Rio Declaration (Principle 7) mandates a "global partnership" for this purpose. It constitutes an obligation to assist in the creation and maintenance of international environmental regimes to protect the global environment and areas beyond national jurisdiction.

A corollary of the principle of common concern is the concept of *equitable utilisation,* an idea first developed for the law governing the use and allocation of shared resources. The *Icelandic Fisheries* case, the International Court of Justice called for the application of the principle to common property fishing stocks as well. Applied to global and common property resources, the equitable utilisation principle has a dual aspect: (1) equitable allocation among all legitimate claimants and (2) equitable limitations on use to comport with the principle of sustainability. The equitable utilisation principle requires a broad process of balancing of interests and consideration of all relevant circumstances.

An additional implication of the common concern principle is that all states have an interest in the enforcement of common concern rules and regimes. Under accepted principles of international law, violations of rules (so-called *erga omnes* obligations) that affect the collective interest of a group of states or the international community as a whole may be the subject of claims by any state."

Should the principle of common concern be transformed into the idea that certain areas or resources are a common heritage of humankind? Two agreements, the 1979 Moon Treaty and the 1982 U.N. Convention on the Law of the Sea (UNCLOS), employ this phrase. The Moon Treaty uses the term "common heritage" to mean that the resources of the moon cannot be made subject to the sovereignty of any state. UNCLOS (Articles 136 and 137) states that the resources of the seabed and ocean floor beyond national jurisdiction are "the common heritage of mankind," which means not only are they subject to international management but also all states must share in any profits.

The United States has rejected both of these treaties primarily because of the common heritage idea. Moreover, the U.N. General Assembly rejected a proposal to designate the global climate as the common heritage of humankind in favor of the common concern formula. A case can be made that the continent of Antarctica should be designated a common heritage regime, but the Antarctic treaty system avoids the use of this term. The system of World Heritage Sites established under the 1972 World Heritage Convention does not create any independent authority for the management of these areas, and they remain under national control. Thus, the concept of the "common heritage of humankind" is controversial.

17.5.9 The Polluter Pays Principle

The *polluter pays principle* is based upon the economic doctrine that the costs of pollution and environmental degradation should be borne by the party creating them and benefiting from the activity that produces the pollution. Under principles of classical economics, a market operates efficiently only if the business or firm making a product or service bears all the costs entailed by that activity. Pollution is an external cost, one not borne by the business involved. Thus, the market can be efficient only if the cost of pollution is "internalised"; that is, imposed on the business in question.

The polluter pays principle, adopted as Principle 16 of the Rio Declaration, states that national authorities "should endeavor" to promote the internalization of environmental costs and that "in principle" the polluter should bear the costs of pollution. Despite this obvious flexibility, important points are involved in this principle.

First, not only states but also private companies and individuals are responsible for cleaning up and paying for environmental degradation. In the cases of pollution of the sea by oil spills and nuclear accidents, international treaties channel liability to the private operator of the facility responsible for the pollution.'[5] In the future more international instruments may channel responsibility and liability directly to private actors.

Second, the polluter pays principle involves the use of economic instruments to internalise pollution costs. These instruments include the imposition of environmental taxes and charges, trading mechanisms, deposit and return fees, and recycling incentives. Such economic instruments can be used as supplements or in place of traditional command-and-control government regulation.

Third, private companies increasingly participate in voluntary codes of conduct, such as the U.N. Global Compact, a code of best practices. Most multinational companies have in place a company-wide environmental management system that goes beyond legal requirements, and most companies comply with the OECD Guidelines for Multinational Corporations.

Fourth, the polluter pays principle correlates with Principle 11 of the Rio Declaration mandating states to enact effective environmental legislation. States have an international law obligation to control pollution and environmental degradation *within* their territories and not just transboundary pollution. This obligation is appropriate because we live on one earth, and ecosystems and natural systems are interconnected without regard to political boundaries. Domestic pollution has a cumulative effect on the earth's environment.

17.5.10 Public Participation

Principle 10 of the Rio Declaration requires states to provide information and encourage participation in environmental decision making by individual concerned citizens. This principle includes providing "effective access" to judicial and administrative proceedings to obtain redress if necessary.

17.5.11 Implementation, Compliance, Effectiveness, and Enforcement

Rio Declaration requires states to enact effective environmental legislation to implement international treaty obligations. This is the first step toward compliance with treaty commitments. The state must also apply domestic laws and regulations in an effective manner.

If a state fails to comply with an environmental treaty, what can be done? The traditional remedies for noncompliance under international law are countermeasures and/or expulsion from the regime. Member states can also terminate their obligations toward the state in breach of the treaty. In the case of environmental treaties, however, such measures are counterproductive. Accordingly, most environmental regimes provide for compliance checks in the form of inspections, disclosure, and monitoring. Non-compliance triggers a special procedure that may involve consultations and technical and financial assistance.

4 Unique Ways of Fighting Air Pollution

According to the WHO, 3 million people die of pollution every year even as only 1 in 10 individuals globally lives in a city that complies with WHO's air quality guidelines. As the problem cries out for urgent remedy, a BBC report listed some innovative solutions.

Smog-Free Towers

Seven-metre high structure built by Dutch inventor Daan Roosegaard with support from China's environment ministry opened in a Beijing park last September. It's a giant outdoor air purifier.

How it works:

- The way static electricity makes loose hairs stick to a comb, airborne particles are sucked into the tower where they get a positive charge.
- The particles are captured by a negatively charged dust-removal plate and clean air blown out of the other end.
- It captures more than 75% of particulate matter in an area the size of a football field.
- Runs on 1,400 Watts, that is, less electricity than a desk-top air purifier.
- Waste collected can be used as building material and to make smog-free jewellery.

Biochar

Also a Dring project, it involves the creation of building material out of a charcoal-like substance obtained by burning agricultural crop by-products or tree clippings in a pyrolysis kiln (in a process that induces decomposition through high temperatures).

How it works:

- Simply put, you take carbon out of the sky, convert it into a material, and then use it to build.
- Trees too take carbon out of the air and trap it as wood.
- Biochar is a mouldable, plastic type material that you can shape—which you can't really do with wood making it the perfect material for architectural design.
- Dring's new building material, called Made of Air, will make its first appearance as industrial factory cladding in Berlin this year.

Pigeons

They took to the sky from a north London hill last March wearing backpacks that monitored air pollution.

How it works:

- They sent live air-quality updates via tweets to Londoners' smartphones. Sadly, in most cases, the readings were not good.

Photo-Catalytic Titanium Dioxide

Tried out by Berlin-based architect Allison Dring in Mexico City in the early 2000s.

How it works:

- The idea is to tackle nitrogen dioxide from vehicle exhausts.
- It uses the ultraviolet rays from sunlight to convert NO_2 into nitric acid.
- The nitric acid is neutralised into a harmless salt and washed away by the rain
- Dring created a coral-like design to catch light and wind from all sides. Her project covered 2,500 square metres of a hospital in Mexico City, cutting pollution in the street below by the equivalent of approximately 1,000 cars per day.

There may be full compliance with a treaty, but still without effectiveness. For example, full compliance with the Kyoto Protocol will lower greenhouse gas emissions below 1990 levels only for a handful of developed countries, a beginning step but not enough to make much difference for the global climate. Effectiveness and compliance are different issues and must be separately considered.

17.5.12 Dispute Settlement

Despite proposals for a specialist international environmental court, there does not seem to be any great need for such a tribunal. Environmental disputes often involve general international law questions; environmental law is not a self-contained, codified system. Indeed, there are a multiplicity of international regimes dealing with such disparate issues as climate change, chemicals in international trade, biological diversity', and other matters. There are already a wide variety of means of dispute settlement under general international law, and every environmental treaty' sets out dispute settlement procedures. In 1993 the International Court of Justice created a special chamber for environmental cases, but no case has yet been brought before it. In 2003 the Permanent Court of Arbitration adopted special rules for environmental cases, but this court too has not been used.

States are very reluctant to submit environmental controversies to international dispute settlement. Environmental treaties typically provide for the settlement of disputes by negotiation or nonbinding conciliation. Binding adjudication of environmental controversies thus requires the agreement of all parties to the dispute.

17.5.13 'Soft' Law—A Verdict

These "soft" principles of international environmental law developed over the past twenty-five years are still evolving. On the plus side we have made progress; on the minus side the rules are still so vague they can be easily ignored. The principles of international environmental law are still soft in the sense that either they are nonbinding recommendations or they are flexible enough to be circumvented at will. Only when translated into specific binding obligations in particular treaty regimes do they really have bite.

17.6 MULTILATERAL AGREEMENTS

We will consider the multilateral agreements that have been concluded to deal with particular environmental concerns.

17.6.1 The Atmosphere

Protection of the earth's atmosphere requires both decreasing emissions from mobile (autos, trucks, and buses) and stationary sources (factories, refineries, power plants, and smelters) that directly damage humans, animals, and plants and also controlling emissions that threaten the integrity of the earth s natural systems that regulate climate and radiant energy from the sun.

A. Long-Range Transboundary Pollution

When pollution emanates from a discrete source, such as a smelter near an international frontier, responsibility can be readily assigned and abatement action taken. More difficult is long-range transboundary pollution in which multiple pollution sources pollute the upper atmosphere and winds carry the contaminants long distances, sometimes thousands of miles. In this case, the contribution of each source to the problem is small and difficult to prove. Long-range transboundary air pollution is severe in many parts of the world. Two pollutants pose special problems: sulfur oxides (SO_x) and nitrous oxides (NO_x). Primarily coal-fired electricity-generating facilities and nonferrousore smelters produce sulphur oxides; nitrous oxides are emitted by vehicles and by fuel combustion. Not only are these pollutants carried long distances by wind patterns but when they are aloft in the atmosphere they are also transformed into acids,' which then fall to earth most commonly as acid rain but also as acid fog, snow, and dust.

Despite the widespread nature of the problem, there are only two international regimes for the control of this kind of acid deposition—one in North America and one in Europe.

The North American Acid Rain Control Program was established in 1991 when the United States and Canada signed a Bilateral Air Quality Control Agreement after more than ten years of arduous negotiations. This agreement commits both nations to reduce emissions of SO_x and to enforce a permanent cap on such emissions. Both nations also agreed to employ technology-based controls to reduce NO_x emissions. To monitor compliance, a permanent bilateral Air Quality Committee was created to operate within the International Joint Commission formed under the 1909 Boundary Waters Agreement between the two countries.

In order to comply with the mandated reductions, the Unied States operates an innovative emissions trading program that allows emitters the option of reducing emissions or purchasing excess allowances from other sources that have reduced their emissions more than necessary'. This builds flexibility into the program and allows reductions to be made as efficiently as possible. This cap and trade program lias decreased emissions by more than 30 per cent compared lo 1990 levels.

The European system for the control of acid rain is the 1979 Geneva Convention on Long-Range Transboundary Air Pollution, which is accepted by over forty states and operates as an ongoing forum and framework for cooperation. The Geneva Convention has a permanent secretariat and an executive body, composed of advisors to governments of the parties, whose main task is to monitor compliance and the effectiveness of national policies. Although the original 1979 treaty contained only "soft obligations," a series of protocols have been adopted mandating specific actions. For sulphur a first protocol signed in 1985 required a 30 percent reduction by 1993. A second protocol, signed in 1994 and in force since 1998, requires an innovative *critical loads approach,* which sets different reduction targets for different areas based upon actual mapping of sulpher oxide deposition patterns and sources. The critical loads are designed to reduce sulphur emissions to the point where human health and the environment are protected "without entailing excessive cost." A critical loads approach is also being devised for NO_x. In 1991 a protocol dealing with low-level ozone was adopted that requires parties to either reduce emissions of volatile organic compounds by 30 per cent or to stabilize emissions at specified levels. Two protocols adopted in 1998 mandate bans or controls on the air-borne deposition of persistent organic chemicals (pesticides and industrial chemicals) and heavy metals. As a result of these efforts, transboundary air pollution in Europe has fallen substantially according to every measure.

Although the North American program is not as advanced as that in Europe, the result has been a significant improvement in air quality. This demonstrates the benefits of international cooperation. In 2002, however, the Unied States took a unilateral decision to undermine the program's effectiveness. Under the U.S. Clean Air Act, new sources of pollution are subject to more stringent controls than existing sources. Until 2002, a significant modification of an existing source required compliance with the upgraded standards that apply to new sources. In 2002, however, the U.S. Environmental Protection Agency issued a new rule that exempts most equipment replacement from the new source standards. Canada protested this change, but to no avail.

B. Ozone Depletion and Climate Change

There are two threats on a global scale to the atmosphere of the Earth: (1) depletion of stratospheric ozone that shelters us from the sun's ultraviolet radiation and (2) climate change, the greenhouse effect, that is responsible for increasing mean global temperatures. With respect to these concerns, Europeans have taken the leadership role, and the United States has been either a follower or absent altogether.

Ozone Depletion

A protective layer of ozone in the stratosphere (the atmospheric zone that is twelve to fifty kilometers above the earth) surrounds the earth, affording a natural shield against the harmful ultraviolet rays of the sun. A decrease or depletion of this ozone layer would mean an increase in the ultraviolet radiation that reaches earth. This would elevate the risk of skin cancer and cataracts in human beings and cause damage to plants, including crops and phytoplankton, the basis of the marine food chain.

Stratospheric ozone (chemical formula O_3) is formed initially when a molecule of oxygen (O_2) absorbs short-wave radiation and is split into two atoms of oxygen (O); each is then able to combine with other O atoms to form ozone. Photochemical reactions in the stratosphere ordinarily break down ozone at about tbe same rate it is formed, producing a natural balance between formation and depletion.

This natural balance is disturbed by a group of highly inert gases that were produced and used for industrial purposes in the twentieth century. The most common and destructive of these gases are chlorofluorocarbons (CFCs), particularly CFC-u (CFCL,) and CFC-12 (CF.CLj. CFCs were developed in the 1920s and came into widespread use as refrigerants, aerosol propellants, solvents, and cleaning agents. Their usefulness stemmed in part from their unique properties: They are virtually unreactive in the lower atmosphere, nontoxic, odorless, nonflammable, and noncorrosive, and they do not readily combine with other substances.

This characteristic of inertness means, however, that CFCs are extremely long lasting, and substantial quantities eventually make their way to the upper atmosphere unchanged. There they meet their match as strong ultraviolet radiation breaks them apart, liberating chlorine atoms (Cl) that destroy ozone by catalyzing its conversion into molecular oxygen. The process is as follows. Step one - ultraviolet rays release stray atoms of Cl. Step two—the chlorine combines w ith ozone: $Cl + O_3 = ClO + O_2$. If a single atom of oxygen is further encountered, there is a subsequent reaction: $ClO + O = Cl + O_2$. This frees an additional atom of chlorine to destroy additional ozone. Because CFCs and other ozone-destroying chemicals’[7] may remain in the atmosphere for 40 to 150 years, each chlorine atom may remove as many as 100,000 ozone molecules from the atmosphere during its life cycle. Thus, even a small amount of CFCs can have a devastating impact.

The ozone-destroying reactions occur fastest on the surface of micro-ice crystals, so depletion is greatest in polar regions, particularly over Antarctica. Each year up to 95 per cent of the stratospheric ozone is destroyed above Antarctica, ozone in the Artie is thinned about 60 percent, and there is measurable ozone loss over the entire planet except the equatorial regions.

But wonderful to relate, this crisis is being managed quite successfully by the international community. Its resolution was (and still is) far from a sure thing despite the obvious seriousness of the problem, but more than twenty-five years of tortuous but skillful diplomacy is paying off. The story of how the world came to grips with ozone depletion is encouraging and hopeful.

The first step was hardly a step at all: After years of discussion and prodding by UNEP, forty-three nations (including sixteen developing countries negotiated the Vienna Convention for the Protection of the Ozone Layer in 1985. This did no more than establish a basis for cooperation, research and exchanges of information; however, a Conference of the Parties was established to review' further developments. The latter was the key to future progress. As the consequences of ozone depletion became better understood and the public became involved, successive conferences of the parties intro duced ever-stricter controls. The key first conference produced the Montreal Protocol in 1987, which mandated phased deep reductions in the production and consumption of eight different types of CFCs. There followed a series of further amendments and adjustments to the Montreal Protocol; every two years the Conference of the Parties convenes to assess progress and to decide on further steps. As a result, 176 nations are participating in this effort led by UNEP, the consumption and production of the worst ozone-destroying substances (ODS) have been banned, and virtually all others are being phased out by certain varying dates.

If all parties to the Montreal Protocol continue to comply with the scheduled bans and reductions, the ozone layer will stabilize around the year 2050 and gradually restore itself thereafter. The fact that so many nations were able to agree to take action to deal with a major crisis and to phase out the use of over ninety varieties of ODS is a triumph of international diplomacy.

How was this possible? Six reasons explain this success.

First, the science was not in dispute: There was agreement on both cause and effects of ozone depletion.

Second, the manufacture of CFCs and other ODS was concentrated in a few multinational companies. In the early 1980s powerful voices in the U.S. government and industry strongly opposed regulation on the of incomplete scientific evidence. But Dupont Corporation, the largest producer, agreed to cooperate with the phase-out effort, and other companies reluctantly followed suit. Thus, there was minimal industry and interest group lobbying against the Montreal Protocol.

Third, substitutes were readily available for all the major uses of ODS, and the Montreal Protocol became "technology forcing," speeding their further development and manufacture.

Fourth, the total cost of the ODS phase-out was relatively low, estimated to be about $6 billion spread over several decades.

Fifth, the United States, the most important producer and consumer of ODS, overcame initial reluctance and cooperated with the phase-out. The United States ratified the Montreal Protocol in 1988 and enacted strong domestic legislation. The Europeans took the lead in calling for a speed-up of international controls.

Sixth, developing countries were persuaded to join both through the incentive of financial help and technology transfer and the threat of trade controls.

Nevertheless, in 2004 the United States regrettably backtracked and rejected the agreed-upon ban on the ODS, methyl bromide, which was set for total elimination in 2005. Citing concerns expressed by farm groups, the United States asked for and received a "critical use" exception that allows this chemical, which is used as a fumigant, to continue to be used in 2005 and beyond.

In March, 2001, the day before an important meeting of Environmental Ministers of the Americas was to begin in Montreal, President George W. Bush announced his rejection of the Kyoto Protocol. In an article the next day in the *New York Times,* David Sandalow, a Clinton administration official who helped negotiate the accord, called this rejection "a textbook case of unilateral diplomacy, which always brings resentment." He was right, at least as far as the predicted resentment was concerned; rejection of the Kyoto Protocol ranks right behind the invasion of Iraq as a source of international resentment toward America.

Climate change refers to alterations in the earth's climate systems that may occur due to increased concentrations of greenhouse gases in the atmosphere. Certain gases—carbon dioxide, methane, and nitrous oxide—earn the name "greenhouse gases" because, like a glass greenhouse, these gases allow sunlight to pass through the atmosphere while trapping heat or infrared radiation close to the earth's surface. If all other factors remain constant (such as forests and clouds, which affect the capacity of the earth to remove these gases from the air), increases in greenhouse gases will lead to global warming and other alterations in climate.

There is now ample evidence that the climate of the earth is undergoing relatively rapid change. As a whole the climate is warming by an indeterminate amount, although certain geographical areas such as Western Europe may experience colder temperatures because of changes in ocean currents (in this case the Gulf Stream, which provides Europe with a relatively mild climate). The exact consequences of this changing climate are difficult to predict, but there will be rapid changes in ecosystems and crop patterns and changing conditions of rainfall and drought. The levels of the oceans, already rising during the twentieth century, will rise further, perhaps forcing the evacuation of some coastal cities and low lying island areas (the Maldives and Cook Islands, in particular). The U.S. NASA (National Atmospheric and Space Administration), which tracks sea-level rise all over the world, says that the sea has risen for the last 50 years at a rate of 0.7 of an inch each year, and this rise has accelerated to an annual rate of 0.12 the past dozen years.

Rapid global climate change coupled with projected increases in human populations during the twenty-first century adds up to a recipe for a global disaster of epic proportions. According to the Intergovernmental Panel on Climate Change (IPCC)—a joint scientific body established by UNEP and the World Meteorological Organisation in 1988—the consequences of climate change will include the following:

- ocean and sea-level rise, alteration of ocean circulation patterns, and loss of sea ice cover
- widespread changes in coastal and marine ecosystems
- weather intensity—a more vigorous hydrological cycle
- public health impacts—certain diseases (malaria) will become more widespread
- threatened agriculture and food security in many areas
- increase in deserts and desertification
- changes in water resources
- biological diversitv loss.

Of course, there are natural changes in the earth's climate over time, but what is unique to our time is the rapidity of change and the fact that it is anthropogenic—caused by human activity. The changes are being caused primarily by the emissions into tire atmosphere of certain gases that are the product of the combustion of fossil fuels - carbon dioxide (CO_2), methan (CH_4), nitrous oxides (NO_x), hydrofluorocarbons (HCFC), perfluorocarbons (PFC), sulfur hexafluoride (SF_6), and others. The emission of these gases disrupts the earth's natural carbon cycles.

The first of these carbon cycles occurs when plants break down carbon from the atmosphere and it accumulates in their biomass; this carbon is returned to the atmosphere in the form of CO, through the decay of organic matter and by animals that consume plants through their respiration and decomposition after death. The oceans also absorb and release vast quantities of CO_2. These processes keep the quantity of carbon in the atmosphere fairly stable over time.

A second carbon cycle takes place on the scale of geological time: Plant and animal remains are locked into sedimentary deposits where they remain for hundreds of millions of years under natural conditions. These remains are sources of our fossil fuels-coal, oil, and natural gas. The natural release of this carbon occurs only when the geological formations are exposed to weathering and erosion.

During the past century humankind has greatly accelerated this second carbon cycle by the unrestricted burning of fossil fuels. We continue to release from 6.5 to 8.5 billion tons of carbon into the atmosphere each year. Ice core samples from the Antarctic and Greenland ice caps show that atmospheric concentrations of anthropogenic green house gases-carbon dioxide, methane, and nitrous oxide—have increased by about 30 per cent, 145 per cent, and 15 per cent, respectively, in the industrial era. From 1958 to 2000 atmospheric concentrations of carbon dioxide, an especially critical greenhouse gas, have increased from about 318 parts per million to about 379 parts per million. This increase is continuing at an accelerated rate—about 2 parts per mi annually. Experts predict that at somewhere above the threshold of 450 per million, Ihe world will be at a point of no return. In effect, an eco time bomb is clicking.

What can and should be done? Obviously, worldwide cooperative action is needed. Like ozone depletion, climate change is beyond the capacity of any one or group of nations to solve. Will it be possible to duplicate the successful effort to control ozone-depleting substances in order to stabilize or reverse climate change? This is the great question we face at the beginning of the twenty-first century. So far the outlook is not good.

A first step toward a worldwide climate change regime modeled on the Vienna Convention/Montreal Protocol regime was taken at the UNCED Conference in Rio de Janeiro in 1992. All the participants—176 nations—signed the UN Framework Convention on Climate Change, a binding treaty that recognized the problem and called for "precautionary measures to anticipate, prevent or minimize the causes of climate change and mitigation of its adverse effects." However, this Framework Convention did not mandate concrete actions to reduce greenhouse gases. The parties to the Convention were required only to limit their emissions and enhance their carbon "sinks" and reservoirs with the aim of returning... to their 1990 levels of these anthropogenic emissions. But as was the case with the ozone regime, the Framework Convention created a Conference of the Parties and an ongoing process to strengthen international action against climate change.

In the decade of the 1990s the "soft law" injunction of the 1992 Framework onvention to reduce greenhouse gas emissions to 1990 levels was largely forgotten as states went back to business as usual and fossil fuel prices remained ow. Finally a breakthrough was made at a meeting of the Framework Convenes Conference of the Parties in 1997—the Kyoto Protocol was agreed to and specific reduction commitments were assigned to thirty-eight so-called Annex I parties—the relatively developed countries of the world responsible for most greenhouse gas emissions (see "Greenhouse Gas Reduction Commitments by Annex I Parties to the Kyoto Protocol").

Climate debates suffered from all the failings of "Project Globalisation". They were designed in a manner that any climate action would benefit the few winners and burden the many. Some new winners would be admitted if they signed up to the "new deal". Climate action reeked of hypocrisy, with rich countries unable and unwilling to commit any significant amounts to support the "loss and damage" and adaptation challenges of the poor. Even on mitigation, the global commitment of a $100 billion remains implausible.

Finally, national actions are defined through a club of woolly eyed "multilateral elites"—all belonging to a Hegelian land. And all reaching the same policy prescriptions irrespective of context, capacity, capability and reality. In the age of hypermedia and an intensely participative public sphere, climate politics appeared distant, cold and undemocratic.

We may points to three much needed next steps.

First, nations matter, and any effective climate response will be national and regional. Allow them agency and choice, and do not impose solutions and alien pathways. The "rule book" to monitor national actions should be thrown out lest it be seen as another elite attempt to undermine popular will.

Second, do not allow hypocrisy to destroy the planet. If the rich are concerned for their grandchildren, they must at least guarantee decent living for the current generation of the poor. Put the $100 billion on the table. Then allow your funds and banks to invest in emerging and developing countries and begin to price innovation for the bottom of the pyramid.

Third, climate change as we conceive it, is already upon us and will not suddenly manifest itself in 2020. The rich must underwrite and take responsibility for the devastation that many confront today, and only then will they be able to enlist and seek action from others. Else, climate is a great leveller. We will all boil together, and the poor may have less to leave behind.

17.7 THE OCEANS

17.7.1 Ocean Governance

The oceans—covering more than two-thirds of the planet—are traditionally a common property resource open to all for navigation, fishing, mineral extraction, and disposal of wastes. In the twentieth century with the increasing use of ocean resources, a revolution in ocean governance occurred. Under the UNCLOS, binding legal rules now' govern all of the ocean and its resources. There is now an international ocean regime in the classic sense as international relations scholars define the term: "a set of principles, norms, rules and decision-making procedures around which actors' expectations converge."

The UNCLOS is so comprehensive that it serves as a kind of constitution of ocean governance. Although the United States applies many of the rules concerning the seas as customary international law, it is excluded from many forums, decisions, and other advantages by its failure to accept the regime.

The UNCLOS divides ocean space into zones and defines the rights of states—both coastal and noncoastal states—in each area. The territorial sea is sovereign territory of the coastal state out to twelve nautical miles subject only to the right of the innocent passage of ships. The exclusive economic zone (EEZ) extends 200 miles offshore from the coast to confer jurisdiction and control over living and non-living resources (fish and minerals). The continental shelf extends at least 200 miles and may run out to 350 miles in order to give coastal states rights to mineral development. Beyond these zones are the high seas. Beneath the high seas lies the deep seabed, which is the "common heritage of humankind," administered by a special International Seabed Authority. Navigation and shipping are facilitated to be compatible with the new zones.

A new International Tribunal for the Law' of the Sea is functioning to resolve disputes, though states may choose among a variety of options for dispute settlement. Because the United States was not a party to UNCLOS, it was unable to challenge China and to test its claims in an international tribunal.

17.7.2 Marine Pollution

An extensive system of treaties deals with pollution of the marine environment. Four categories may be distinguished: the UNCLOS, regional seas agreements, general multilateral treaties, and bilateral treaties.

The UNCLOS deals comprehensively but in general terms with marine pollution control. An important feature is the legal duty on all parties to protect the marine environment as a whole, including the high seas. Moreover, "rare and fragile ecosystems" and "the habitat of depleted, threatened, and endangered species and other forms of marine life" must be "protected and preserved." This ecosystem approach goes beyond former requirements. The UNCLOS also mandates control of all sources of marine pollution: from land-based sources, the atmosphere, vessels, and seabed and national mineral development activities. All states' parties must apply and enforce "international rules and standards established through competent organizations." Thus, UNCLOS is a framework for the elaboration of specific controls.

The UNCLOS breaks new ground in the enforcement of pollution control rules. "Flag" state enforcement (by the country wdrere the ship is registered), which is traditionally weak or nonexistent because of the common practice of registering ships under "flags of convenience," such as Panama and Liberia, is supplemented by enforcement by the port state - any port of call can seize a substandard ship or otherwise require compliance with international standards. Coastal states can also apply environmental rules in their exclusive economic zones and can enforce them against ships navigating in their coastal zones.

Multilateral antipollution agreements are used to deal with pollution from ships. The 1973/78 MARPOL Convention is an I MO creation that enforces detailed pollution regulations and mandatory equipment standards for world merchant shipping. MARPOL prevents and regulates oil and chemical pollution as well as dumping of plastics, sewage, and garbage. Flag states have responsibility to inspect their vessels at periodic intervals and to issue pollution prevention certificates. Port states may check compliance wdienever the vessel enters port. The MARPOL rules are supplemented by another IMO program, the International Convention of the Safety of Life at Sea (SOLAS, 1974). In 2002 SOLAS brought into effect the International Management Code for the Safe Operation of Ships and for Pollution Prevention (ISM Code), which focuses on the human element as a means of preventing pollution and promoting safety. The ISM Code requires the adoption of a safety and environmental protection policy, instructions and procedures for crews, and emergency response training.

The Ocean Dumping Convention (1972) regulates the dumping of wastes at sea. A protocol adopted in 1996 now prohibits the ocean disposal of radioactive and hazardous wastes. Certain categories of waste, such as sewage sludge and dredged material, can only be dumped in designated areas. The Dumping Convention adopts the precautionary principle that everything that is not specifically permitted is prohibited.

Oil pollution of the oceans, whether accidental or operational, is regulated bv a series of IMO Conventions adopted beginning in 1954. The centerpiece is an Oil Pollution Liability and Compensation Scheme that imposes liability on ship operators and pays oil pollution claims, including fishermen's claims and restoration of environmental damage. In 2004 a similar scheme, the Hazardous and Noxious Substances Convention (1996), came into force to deal with the problem of hazardous chemicals.

17.7.3 Fisheries

In March, 1995, Canadian gunboats seized the *Estai,* a Spanish fishing trawler that was fishing for turbot in international waters in the North Atlantic just outside Canada's 200-mile exclusive economic zone. Spain and the European Union protested that the *Estai* was in the right and Canada had no business policing international waters. Canada claimed a moral though not a legal right to act—the *Estai* was destroying its turbot fishery that straddled the high seas. Only 2 per cent of the turbot in the *Estai's* cargo hold had reached spawning age, and its net had an illegal small mesh. Canada's Fisheries Minister exhibited the net—seventeen stories tall—at the next U.N. fisheries conference that same month.

The natural bounty of the oceans is limited. Every fishery in the world is in decline and in urgent need of conservation. Marine fish are being depleted by a double whammy of degradation of natural habitats and modern industrial fishing methods.

A fishery is a common property resource that presents inherent management difficulties. There is no limit on those who may fish; all comers are welcome. But because the fish are common property and belong to no one until they are caught, there is no incentive for even minimal conservation. So an unregulated fishery will inevitably lead to depletion.

But management is also difficult. One option is to limit the number of fish that can be caught. The most common management limit is known as *maximum sustainable yield* (MSY), the greatest quantity that can be caught over a given period without depleting the desirable size of the total stock. But even for a single stock, MSY is difficult to determine; there will be temptations to err on the side of exploitation, not on the side of caution. Moreover, different fishery stocks are often interrelated so that the MSY for a single stock is meaningless. The interrelationships are difficult to define, making determination of the MSY even more elusive.

A second problem is that the absence of any limit on the number of people who can enter a fishery will drive down the rate of return for all fishermen. In the long term, total revenue from the fishery will tend to equal the total cost of fishing. When this happens the usual solution is government subsidies. This happens on such a large scale that worldwide fishery subsidies are estimated by the World Trade Organisation (WTO) to be about $300 billion. Such subsidies are economically distorting and stimulate further overfishing. The WTO has proposed decreasing fishery subsidies without success. It is no wonder that all fisheries are declining.

Not only are subsidies difficult to abandon but also the regulator)’ options are political poison. One method is to set a limit on the total allowable catch (TAC) of various stocks and then allocate it by some method, such as holding an auction. This is economically efficient, but favours richer fishing interests. A second regulatory technique is to set limits on fishing equipment by prescribing such elements as net size and banning drift nets and long lines, fishing lines twenty or thirty miles long with, 1000 or more baited hooks. Third, certain areas can be off-limits to protect spawning populations, or fishing can be permitted only at certain times of year. Fourth, limits can be placed on the number of vessels, their size, or the number of days they may fish.

By allocating EEZ and continental shelf rights to coastal states, the UNCLOS places most of the world’s fisheries under sovereign control. In return each coastal state has the duty to manage and conserve fish stocks so they are not endangered by overexploitation. Fish stocks must be restored to “levels which can produce the maximum sustainable yield as qualified by relevant environmental and economic factors.” The coastal state must promote “optimum utilisation” of the living resources of its EEZ. The total allowable catch for each fish stock within its EEZ must be established. These provisions grant coastal states broad discretion on choices of regulatory techniques and allocation of the allowable catch.

Where a fishery is shared among two or more EEZs, the UNCLOS requires the states concerned “to seek... to agree... to coordinate and ensure the conservation and development of such stocks.” In practice states agree on cooperative arrangements. For example, in the northern Pacific Ocean and the Bering Sea, the International Halibut Commission adopts regulations for halibut stocks shared by the United States and Canada.

Fishing on the high seas is open to all, but the UNCLOS also mandates conservation and management under MSY principles for high-seas fisheries. In practice this management is hard to achieve except through the establishment of an international fisheries commission. Four such bodies are currently operating: the Northwest Atlantic Fisheries Organisation, the Northeast Atlantic Fisheries Commission, the General Fisheries Council of the Mediterranean, and the Commission for the Conservation of Antarctic Marine Living Resources. In the South Pacific there is the Wellington Convention for the Prohibition of Fishing with Long-Driftnets (1989) and the Convention for the Conservation of Southern Bluefin Tuna. However, much of the high seas area is unregulated as highlighted by the case of the *Estai.* To fill this loophole, in 1995 a U.N. conference adopted an Agreement on Straddling Fish Stocks and Highly Migratory Fish Stocks, which established a conservation regime and rules applicable to fish stocks on the high seas that range within the EEZ of one or more coastal states. This convention assures the enforcement of conservation measures on both sides of the 200-mile EEZ line and vindicates Canada's original concerns. The U.N. Food and Agriculture Organisation (FAO) in 1995 also adopted a Code of Conduct for Responsible Fisheries.

17.7.4 Marine Mammals

The UNCLOS calls for states to cooperate for the conservation of marine mammals. The international Convention for the Regulation of Whaling, (1946), created the International Whaling Commission (IWC) composed of representatives of member governments. The Commission has no enforcement power, but may encourage, recommend, study, and disseminate information about whales and whale stocks. Despite these limited powers, the IWC was successful in securing a ban on commercial whaling in 1982. In the 1990s one member, Iceland, withdrew from the Convention, and Norway and Japan resumed whaling on a limited basis, citing various exceptions and the recovery of certain stocks. Despite this resumption, the IWC has enjoyed extraordinary success, and whale stocks have vigorously recovered as a result.

17.8 BIOLOGICAL DIVERSITY

The biological impoverishment of the earth is a defining characteristic of our time. The fragile balance of plants and animals that share the earth has developed over millions of years. Although most of the species that have ever lived are now extinct, we are living at a time of the earth's greatest biological diversity. We do not know how many species exist; biologists estimate their number to be from 10 to 100 million. Taxonomists have classified between 1.5 million to 1.75 million species, some 250,000 flowering plants, and 4500 mammals.

Biological diversity refers to the variation among living organisms and includes not only species diversity but also genetic diversity and ecosystem diversity. Ecosystem diversity is key because living things exist in natural communities where they interact. A rich variety of ecosystems exist on the Earth: many different kinds of forests, tundra, deserts, coral reefs, grasslands, lakes, coastal estuaries, and wetlands. An ecosystem map of the w orld reveals a tremendous diversity of ecosystems on every continent: the ice-covered Antarctic Peninsula; Andean, Amazon, and Atlantic forests in South America, wet and dry tropical forests in Central America; deserts, prairies, forests, and tundra in North America; broad-leaf and coniferous forests, deserts, grasslands, and tundra in Eurasia; rich shrub lands in the Mediterranean basin; deserts, tropical forests, mountain grasslands, and floristic shrub lands in Africa; tropical forests in Southeast Asia; and tropical forests and grasslands, shrub land forests, and deserts in Australia. All over the world are important island ecosystems as well. The flora and fauna of New Zealand have evolved in isolation for 80 million years, Madagascar harbors a rich variety of species found nowhere else, and Hawaii is the repository of unique plants and birds.

Experts agree that the destruction of habitat is causing major losses of biological diversity and the extinction of many species. For example, more than half of the world's species of plants and animals live in tropical forests. Yet in the last fifty years, over half of the tropical forests of the world have been destroyed, and cutting proceeds at a rate of n,ooo hectares per year.

Why should we care about biological diversity? There are both practical and philosophical reasons to care. A practical reason is the "ecosystem services" we take for granted - the work nature does for us in providing the wild plant varieties that we use for agriculture, pollinating our crops, mitigating floods and drought, controlling diseases and pests, providing clean air and water, nourishing the animals we use for food, providing fish and shellfish, and generating fertile soils. A great number of our important medicines are derived from animals or plants, and more are waiting to be discovered, at least if they survive our onslaught of destruction Wildlife diversity is a rich resource for the present and future benefit of humankind.

Wildlife and wild places contribute to the richness of human life by providing recreation, inspiration, beauty, and spiritual solace. Artists, poets, photographers, architects, and musicians create works inspired by and celebrating the natural world. Nature is part of our heritage, and we are its stewards. We have a responsibility to pass this on to future generations.

The preservation of biological diversity requires a multifaceted strategy.

17.8.1 Habitat Preservation

The most important component of a strategy to preserve biological diversity is habitat preservation. A habitat, the physical and biological setting in which living organisms interact (such as a forest, a wetland, or a prairie), is a basic and indispensable requirement for all living things. Loss of habitat is the major factor causing the diminishing biological diversity of our planet. Protecting habitats and ecosystems is preferable to single-species preservation. The habitat/ecosystem approach is holistic; it is applied within a particular geographic framework or to a particular landscape defined by ecological boundaries as a logical unit for coordinated management. Human activities are not automatically excluded; they are assessed for their impacts—whether economic, cultural, social, or recreational—and decisions can then be made on how to accommodate both human needs and the maintenance of the ecological system.

As a matter of international strategy, three different approaches may be taken to preserving biological diversity.

The *hot spots approach* was first proposed by British ecologist Norman Myers and continued by Conservation International. A hot spot is an especially rich but vulnerable ecosystem in which preservation efforts should be concentrated. Scientists have identified 21 hot spots—different areas defined by their vegetation that contain at least one-half percent of the world's 300,000 plant species as native species.

The advantage of concentrating these hot spots is that doing so reduces the problem of biological diversity to manageable proportions. Scientists argue that, at a cost of about $30 billion per year, areas within these hot spots could be designated as nature reserves, making a significant contribution to ecosystem preservation.

A second strategy is the Global 200 program advocated by the World Wildlife Fund (WWF). Global 200 would extend protection efforts not just to hotspots but also to diverse and representative terrestrial, fresh-water, and marine habitats, the broadest variety of the world's ecosystems from Arctic seas to coastal forests.

As part of this effort, in May, 2005 the World Bank and the WWF announced a joint five-year program to reduce the destruction of the world's forests by 10 per cent annually. Their strategy is to support new forest-protected areas, more effective management of already protected areas, and improved management of forests that are not yet protected.

A third strategy is to ask each of the states of the world to make a national assessment of biological diversity and to take appropriate actions to preserve it. This strategy takes into account the fact that states may have their own priorities and policies. This third strategy may incorporate aspects of the hot spots and Global 200 approaches.

All three of these approaches rely on the designation of conservation reserves as a primary tool of preservation. Of course, legal and administrative management schemes must be adopted to manage such areas. Such resource planning, to be successful, must involve and take onto consideration from the outset affected individuals and communities. Local residents must be persuaded to see such an area as a valuable resource. This will mean affording them substantial benefits from such activities as ecotourism, commercial uses of the area, and research and development.

Box 17.1 The Top 21 World Hot Spots for Biological Diversity

1. Tropical Ancles Mountains
2. Mediterranean basin
3. Madagascar/Indian Ocean islands
4. Central America
5. Caribbean and south Florida
6. Brazil's Atlantic forest
7. Indochina/Myanmar
8. Philippines
9. Polynesia/Micronesia
10. New Caledonia
11. New Zealand
12. South-west Australia
13. California floristic area
14. Western Ghats (India) and Sri Lanka
15. Sundaland (Indonesia)
16. Wallacea (New Guinea)
17. South Central China
18. Cape of Good Hope floristic area (South Africa)
19. West African tropical forests
20. Central Chile
21. Brazil's Amazon tropical forest

Because wildlife and ecosystems range over states' boundaries, regional conventions on biological diversity are essential. Unfortunatelv, the major regional conventions are well intentioned but ineffective. For example, the Treaty for Amazonian Cooperation (1978) signed by the eight countries that share the Amazon basin in South America calls for "a balance of economic growth and environmental preservation" and "rational planning of Amazon flora and fauna—to preserve the environmental balance of the region," but during the years the treaty has been in effect, increasing destruction of the forest cover has occurred.

Other regional agreements—the 1968 African Convention for Conservation of Nature, the 1985 ASEAN Convention, the 1940 Western Hemisphere Convention, and the 1979 Berne Convention on the Conservation of European Wildlife and Natural Habitats—fall into the category of "sleeping treaties" because they are totally ignored by the parties.

17.9 ANTARCTICA

Antarctica, the uninhabited ice-covered continent at the Earth's South Pole, is unique: It is a large continent of great natural beauty and abundant biological resources, but is inhospitable to humankind. All but 2 percent of the land and much of the surrounding waters are covered with a sheet of ice up to three kilometers thick. Seven nations—Argentina, Australia, Chile, France, New Zealand, Norway, and the United Kingdom—claim sovereignly over parts of Antarctica, and the British, Chilean, and Argentinean claims overlap. Conflict has been avoided, however, because since the signing of the Antarctic Treat) in 1959, all territorial claims have been indefinitely suspended. Additional interested parties (including the United States, which refuses to recognize any sovereign claims) have joined the Antarctic Treaty system, and there arc—now forty-three parties in all. Consultative meetings of the parties are held annually.

Because of overfishing in the 1970s, the parties in 1980 adopted the Convention on the Conservation of Antarctic Marine Living Resources (known as CCAMLR), which, adopts the ecosystem approach to fishing in the Southern Ocean. CCAMLR's scientific advisory commission recommends quotas for fishing stocks, but adoption requires unanimous approval from the parties. This lias led to controversy and disagreement regarding the size of certain stocks, most notably sea bass, which are being overfished as a result.

17.10 WATER RESOURCES

International competition and conflict over water resources are sure to intensify as the world population increases and existing supplies of water diminish because of waste and pollution. Tensions exist and sometimes violence erupts among nations over water use rights and water diversion projects. Every area of the world is affected.

These water conflicts and disputes can only be dealt with by the application of international law and by the establishment and maintenance of appropriate international institutions. Cooperative international regimes must be developed to deal with water issues. Effective commissions to deal with water are functioning in Europe and North America. These should be studied as models for other areas of the world.

A landmark international instrument setting out legal principles for the management of international waterways is the 1997 U.N. Convention on the Law of Non-Navigational Uses of International Watercourses. This agreement makes equitable utilization the cornerstone of how water rights should be allocated between states that share a waterway. Equitable utilization does not mean equal sharing; rather all states in a watercourse share sovereignty over the resource, and they must establish an ongoing process based on agreed-upon relevant factors to maximize the benefits and serve the needs ofall users. Equitable utilisation also implies full and regular exchanges ofdata, timely notification of planned measures, and the creation of an ongoing negotiation process, usually in the form of an international commission and a permanent secretariat.

17.11 NEW MULTILATERAL AGREEMENTS

Threats to the well-being of the planet and international and transboundary environmental degradation have been recognized as concerns by the international community only in the last generation. The creation of a network of international institutions and political and legal actions to come to grips with these problems was accomplished in fits and starts over that time. Much remains to be done. Institutions are weak, legal and political regimes are "soft," and compliance and enforcement are sporadic. Yet, it is remarkable that international cooperation has proceeded to the point that most problems are being addressed, however inadequately.

17.12 FROM KYOTO TO PARIS

The Paris Agreement (French: Accord de Paris) is an agreement within the United Nations Framework Convention on Climate Change (UNFCCC) dealing with greenhouse gases emissions mitigation, adaptation and finance starting in the year 2020. The language of the agreement was negotiated by representatives of 195 countries at the 21st Conference of the Parties of the UNFCCC in Paris and adopted by consensus on 12 December, 2015. It was opened for signature on 22 April, 2016 (Earth Day) at a ceremony in New York. As of December, 2016, 194 UNFCCC members have signed the treaty, 133 of which have ratified it. After several European Union states ratified the agreement in October, 2016, there were enough countries that had ratified the agreement that produce enough of the world's greenhouse gases for the agreement to enter into force. The agreement went into effect on 4 November, 2016.

The head of the Paris Conference, France's foreign minister Laurent Fabius, said this "ambitious and balanced" plan is a "historic turning point" in the goal of reducing global warming.

One year on, the ratification of the Paris Agreement was celebrated by the Mayor of Paris Anne Hidalgo by illuminating the Eiffel Tower and the Arc de Triomphe, Paris' most iconic monuments, in green.

A. Aims

The aim of the convention is described in Article 2, "enhancing the implementation" of the UNFCCC through:

(*a*) Holding the increase in the global average temperature to well below 2 °C above pre-industrial levels and to pursue efforts to limit the temperature increase to 1.5 °C above pre-industrial levels, recognising that this would significantly reduce the risks and impacts of climate change;

(*b*) Increasing the ability to adapt to the adverse impacts of climate change and foster climate resilience and low greenhouse gas emissions development, in a manner that does not threaten food production;

(*c*) Making finance flows consistent with a pathway towards low greenhouse gas emissions and climate-resilient development.

Countries furthermore aim to reach "global peaking of greenhouse gas emissions as soon as possible." The agreement has been described as an incentive for and driver of fossil fuel divestment.

The Paris deal is the world's first comprehensive climate agreement.

The contribution that each individual country should make in order to achieve the worldwide goal are determined by all countries individually and called "nationally determined contributions" (NDCs). Article 3 requires them to be "ambitious", "represent a progression over time" and set "with the view to achieving the purpose of this Agreement". The contributions should be reported every five years and are to be registered by the UNFCCC Secretariat.Each further ambition should be more ambitious than the previous one, known as the principle of 'progression'. Countries can cooperate and pool their nationally determined contributions. The Intended Nationally Determined Contributions pledged during the 2015 Climate Change Conference serve—unless provided otherwise—as the initial Nationally determined contribution.

The level of NDCs set by each countrywill set that country's targets. However, the 'contributions' themselves are not binding as a matter of international law, as they lack the specificity, normative character, or obligatory language necessary to create binding norms. Furthermore, there will be no mechanism to force a country to set a target in their NDC by a specific date and no enforcement if a set target in an NDC is not met. There will be only a "name and shame" system or as János Pásztor, the U.N. assistant secretary-general on climate change, told CBS News (US), a "name and encourage" plan.As the agreement provides no consequences if countries do not meet their commitments, consensus of this kind is fragile. A trickle of nations exiting the agreement may trigger the withdrawal of more governments, bringing about a total collapse of the agreement.

The negotiators of the Agreement however stated that the NDCs and the 2 °C reduction target were insufficient, instead, a 1.5 °C target is required, noting "with concern that the estimated aggregate greenhouse gas emission levels in 2025 and 2030 resulting from the intended nationally determined contributions do not fall within least-cost 2 °C scenarios but rather lead to a projected level of 55 gigatonnes in 2030", and recognizing furthermore "that much greater emission reduction efforts will be required in order to hold the increase in the global average temperature to below 2 °C by reducing emissions to 40 gigatonnes or to 1.5 °C."

Although not the sustained temperatures over the long term to which the Agreement addresses, in the first half of 2016 average temperatures were about 1.3 °C (2.3 degrees Fahrenheit) above the average in 1880, when global record-keeping began.

When the agreement achieved enough signatures to cross the threshold on October 5, 2016, US President claimed that "Even if we meet every target, we will only get to part of where we need to go," and that "This agreement will help delay or avoid some of the worse consequences of climate change will help other nations ratchet down their emissions over time."

The global stocktake will kick off with a "facilitative dialogue" in 2018. At this convening, parties will evaluate how their NDCs stack up to the nearer-term goal of peaking global emissions and the long-term goal of achieving net zero emissions by the second half of this century.

The implementation of the agreement by all member countries together will be evaluated every 5 years, with the first evaluation in 2023. The outcome is to be used as input for new nationally determined contributions of member states. The stocktake will not be of contributions/achievements of individual countries but a collective analysis of what has been achieved and what more needs to be done.

The stocktake works as part of the Paris Agreement's effort to create a "ratcheting up" of ambition in emissions cuts. Because analysts have agreed that the current NDCs will not limit rising temperatures below 2 degrees Celsius, the global stocktake reconvenes parties to assess how their new NDCs must evolve so that they continually reflect a country's "highest possible ambition".

While ratcheting up the ambition of NDCs is a major aim of the global stocktake, it assesses efforts beyond mitigation. The 5 year reviews will also evaluate adaptation, climate finance provisions, and technology development and transfer.

B. Structure

The Paris Agreement has a 'bottom up' structure in contrast to most international environmental law treaties which are 'top down', characterised by standards and targets set internationally, for states to implement. Unlike its predecessor, the Kyoto Protocol, which sets commitment targets that have legal force, the Paris Agreement, with its emphasis on consensus-building, allows for voluntary and nationally determined targets. The specific climate goals are thus politically encouraged, rather than legally bound. Only the processes governing the reporting and review of these goals are mandated under international law. This structure is especially notable for the United States—because there are no legal mitigation or finance targets, the agreement is considered an "executive agreement rather than a treaty". Because the UNFCCC treaty of 1992 received the consent of the Senate, this new agreement does not require further legislation from Congress for it to take effect.

Another key difference between Paris Agreement and the Kyoto Protocol is its scope. While the Kyoto Protocol differentiated between Annexure 1 and non-Annexure 1 countries, this bifurcation is blurred in the Paris Agreement, as all parties will be required to submit emissions reductions plans. While the Paris Agreement still emphasizes the principle of "Common but Differentiated Responsibility"—the acknowledgement that different nations have different capacities and duties to climate action—it does not provide a specific division between developed and developing nations.

17.13 MARRAKECH NEGOTIATIONS

The lack of any concrete advances on finance pledges in the recently concluded climate change negotiations in Marrakech amplifies the need for innovation in financing mitigation and adaptation activities, as well as to insure against loss and damage caused by climate change. If Paris was about committing to prevent the rise of temperature beyond 2 degrees Celsius, Marrakech aimed to move the needle on loss and damage. Parties approved a five-year work plan on loss and damage, which, starting 2017, will see countries formally address topics such as the slow-onset impacts of climate change, non-economic losses and migration. However, developing countries are fast realising that financial support for loss and damage (which is not governed by a legally binding framework) from developed countries is likely to be very small.

Research by the Council on Energy, Environment, and Water (CEEW), in collaboration with the Indian Institute of Management, Ahmedabad, and the Indian Institute of Technology, Gandhinagar, estimates that direct costs of extreme events spurred by climate change in India are $5-6 billion per annum. The associated economic costs and non-economic impacts are several orders of magnitude higher. As India gears up to attract trillions of dollars of investment as part of its development agenda it requires mechanisms to protect these against climate risks.

Extreme events and changing precipitation seem to be the new norm in the India we inhabit. In the last 14 years alone, we have witnessed 131 instances of major flooding, several instances of heat and cold waves as well as major drought. All of these have far-reaching financial impacts. When combined with India's mammoth renewable-energy targets, and the annually rising adaptation spending, the financial needs are gargantuan. The world, and India in particular, is at a critical juncture in climate history—where mitigation, adaptation, and loss and damage all need to be addressed urgently, and this cannot be done by public money alone. International debt markets, estimated to be around $95 trillion, are the largest pool of capital in the world. Climate-resilient bonds are an innovative way for countries around the world to use public money to drive private investment from these debt markets into the underserved climate-adaptation market, while spreading the impact of loss and damage from climate risk thinly between investors.

Spending on adaptation actions across five sectors—agriculture, forestry, fisheries, water and ecosystems—is estimated at $206 billion by 2030. Adaptation actions across infrastructure (energy, transport, roads, buildings, bridges) and health are likely to be significantly higher. For every $1 trillion of new infrastructure investments, adaptation costs are likely to range from $30–100 billion. For existing infrastructure, adaptation costs could be from 15–20%. Therefore, the adaptation gap for India by 2030 could be around $1 trillion.

A climate-resilient bond could take several forms, most common being a green bond. A green bond channels debt capital for projects with environmental benefits, predominantly for mitigation activities like renewable-energy deployment, which decreases future climate risk.

Another possible form of a climate-resilient bond is a publicly issued bond used to insure against the outcome for a specific climate risk. For example, the government of West Bengal could issue a tax-free—yielding market return—bond for a five-year duration. In case of major flooding due to a rise in precipitation in West Bengal during the duration of the bond issue, the investment is forfeited by the investor and has to be used by the state to cover the loss and damage caused by the flooding. This sharing of liability between investors, and the pooling of resources to hedge against the impacts of climate disasters, opens up a new set of financiers for loss and damage that has historically been tackled by public funds alone. Such a debt instrument would leverage the high government credit rating to attract investors, with the market interest rates compensating for the climate risk being passed on to the investor.

Another efficient form of climate-resilient bonds is when the funds raised to protect against climate risks are used for adaptation activities. This form of climate-resilient bonds combines two important aspects, one of borrowing from the debt market for climate projects, and the other of sharing the climate risk between multiple individual investors.

Such a bond would require public money to pay market or higher rates of return on the investment, which would be used for adaptation projects such as flood barriers that reduce the loss from climate disasters. In case of a climate disaster, the bond investors forfeit their capital up to the limited liability of their investment. In the absence of a disaster, the climate-resilient bond would work like any ordinary debt instrument with the capital and interest paid out in accord with the terms of the bond issue.

As the likelihood of climate disasters goes up, the need to mobilize additional finance to both address the losses from the disaster, as well as adapt to the likelihood of growing climate impact is imminent. For instance, on a high-emission pathway, the incidence of extreme drought affecting crop land could increase by about 50% in South Asia. In order to transition away from such a high-emission pathway, as well as adapt to existing and future climate impacts, governments would have to test and promote several permutations of existing and additional financial instruments.

While developing countries need to continue seeking clarity on the pathways of financial support under the UN Framework Convention on Climate Change (UNFCCC) regime, they cannot wait for it. Climate risk is real, and it is growing. We all must share the burden, and act now.

■ ■ ■

18

Profile of Natural Resources and Issues thereof in India

18.1 INTRODUCTION

As in most other EEs, a large proportion of population of India is dependent upon agriculture and other primary activities which consist in direct exploitation of natural resources. The natural resources of India, as the following survey indicates, are varied and provide an adequate basis for building a diversified modern green economy.

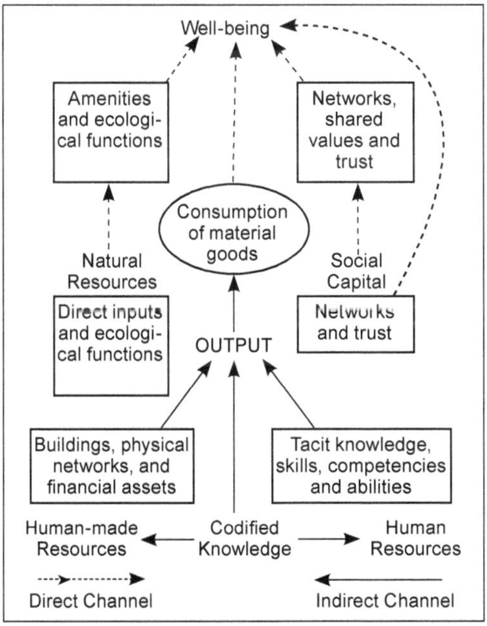

Fig. 18.1. The Relationship between Resources and Human Well-being.
Source: Adapted from *World Development Report 2003*.

The three-fold objective in the present section is as follows:

- To present a profile of resources available to accelerate the process of economic development.
- To identify the major constraints and bottlenecks associated with these resources.
- To suggest suitable policy measures to enable the economy reap optimum utilisation of resources.

18.2 NATURAL RESOURCES

The availability of resources, their utilisation and the capacity to mobilise more resources determine an economy's ability to overcome the various problems. As the world's resources grow scarce, India faces an uphill battle to secure its resource supply for the future. On the demand side, growth in population and urbanisation is increasing competition for all kinds of resources. Meanwhile, resource supplies are being strained by geopolitical factors and inadequate complementary infrastructure to support resource exploration and generation at a fast pace. As a result of this double-edged scarcity, concerns are growing over the availability and prices of resources. Yet, approached imaginatively, the quest for resource efficiency can become an important source of economic growth and job creation.

18.2.1 Knowledge of Natural Resources

Information on stock of natural resources is much better in case of India than in most EEs. Basic resource survey agencies have been in existence for a century or more and systematic surveys and investigations of resources have been undertaken by them.

Work on mapping and surveys of natural resources has expanded greatly under the Five-Year Plans.

- The older agencies, such as the Survey of India and the Geological Survey, have been expanded in order to enable them to undertake larger programmes of work, and new agencies have been created to undertake specialised tasks or surveys in fields which had not been covered earlier.
- The *Indian Bureau of Mines* was established in 1950 to undertake an economic assessment of natural resources and to formulate programmes of development.
- The *Oil and Natural Gas Commission*, established in the late 1950's to undertake exploration and development of petroleum resources, has been responsible for major discoveries of oil and natural gas in Assam, Gujarat and Bombay High.

- *The Central Water and Power Commission* (now bifurcated into two), established in 1945, has the responsibility for the coordination of hydrological investigations and assessment of natural resources.
- *A Soil Land Use Survey* has been established under the Indian Council of Agricultural Research; systematic soil surveys are being undertaken by it in cooperation with the State Departments of Agriculture.

Table 18.1: Total Area of Cultivable Land (figures in hectares

2007–08	2014–15	Overall decrease
182.4 mn	182.0 mn	406,000 ha

- The newly-introduced Forest Survey, the State of Forest Report which is published at regular intervals, etc. are regular sources of information and data for policy-makers.
- The national laboratories and institutes of scientific research also undertake studies relating to evaluation for utilisation of natural resources.

But experience under the Plans has demonstrated the need for further expansion or acceleration of work in some directions, a greater use of modern survey and mapping techniques and a more adequate economic assessment of natural resources. Investigations of soil fertility and for groundwater will need to be expanded and improved further in order to provide data for an intensive use of chemical fertilisers and efficient water management which are essential for the successful use of the new high-yielding varieties or other programmes of intensive cultivation. Modern survey techniques, such as aerial photography, aero-magnetic survey and remote-sensing, which have advanced very rapidly in recent years but which are not used adequately by the Indian survey agencies will need to be used much more. The use of these techniques reduces the time requirements of surveys and preparation of maps; makes possible surveys of inaccessible areas and aids location or estimation of reserves of minerals and other natural resources.

18.3 LAND AND SOILS

India measures 3,214 kms. from north to south and 2,933 kms. from east to west with a total land area of 32,87,782 sq. kms and a coastline measuring 7516.5 kms., plus 1,197 islands. It is the seventh largest *landowner* in the world after Russia, Canada, China, the U.S.A., Brazil and Australia in that order. In brief, India is a vast country and has a considerable strategic significance on account of its location, size and economic resources. Standing at the heart of the Indian Ocean, the country is in a much better position than any other in the area to control the Indian Ocean routes, most of which touch the Indian ports. Most of the air routes between Europe, West Asia and Africa and East Asia,

South-East Asia and Japan also pass through India. It gives India an advantage in terms of international mobility of persons and commodities.

18.3.1 Land Utilisation in India

The total available land area in a country sets definite limits within which the land-base can be stretched horizontally during the process of economic development. As this process advances, the demand for land increases; new uses for land are found, land gets diverted from its existing use to the new-found uses. More generally, the shift is from the agricultural use to the non-agricultural uses, *viz.*, industrial, construction and trading purposes. In the case of a developing, labour-surplus, agricultural crop-deficit economy, this pressure on agricultural land is a matter of serious concern. Any diversion of land from agricultural use to non-agricultural uses may disrupt the balance of agricultural supplies and this adversely affects the whole process of growth. Therefore, in the face of increasing requirements of land, what is generally stressed is that the inaccessible, wastelands and the lands which have hitherto been lying unutilised, should be commissioned and made serviceable for agricultural and non-agricultural uses.

It is in this context that statistics relating to the land utilisation pattern become significant. These help to determine the contours of future economic development as far as the availability of land resources is concerned. The available land, on the basis of its use, can be classified into two parts, *viz.*, (1) Agricultural land, and (2) Non-agricultural land.

Agricultural Land It includes net sown area,[1] current fallows and land under miscellaneous tree crops and groves. Agricultural land in India totals a little over 50 per cent of the total geographical area in the country. This is the highest among the large or medium-sized countries of the world, indicating.

- the influence of favourable physical factors such as large area, the extent of plains and plateaus and a very small extent of arid areas (about 15.8 per cent of the country's geographical area is arid.)
- the extension of cultivation to a large proportion of the cultivable land.

But, because of the large population of the country, arable land per capita is not high; the figure of 0.14 hectares is lower than the average for the world (0.22) and is only one-fourth of the U.S. (0.59) figure.[2] About 15 per cent of the sown area is multi-cropped (sown more than once in a year), while one-fourth of the gross cropped area is irrigated. Most of the multi-cropped area is irrigated and the security provided by irrigation facilities is a major factor in intensive application of labour and inputs to obtain high yields.

---

1. Net sown area is being estimated at 141 million hectares; about 47 million hectares is sown more than once. The gross sown area is about 188 million hectares.
2. In some other countries, the per capita arable land in hectares is as follows: Australia 2.43, Brazil 0.32, Canada 1.44, Japan 0.03, etc. *Nationmaster.com*.

Non-agricultural Land This includes land under forests, permanent pastures and other non-agricultural uses (towns, villages, roads, railways, etc.) and land classified as cultivable waste as well as barren and uncultivated land of mountain and desert areas.

18.3.2 Trends in Land Utilisation

Three important changes in the land utilisation witnessed during the last six decades are:

- reclamation of waste and fallow lands
- a significant increase in the 'area sown more than once'.
- a significant fall in the land area under cultivation.

Reclamation of waste and fallow lands was relatively rapid during the 1950s, following land reforms, such as the abolition of *Zamindari and Jagirdari* systems. The dispossessed *Zamindars* reclaimed land which had been left to them for 'personal cultivation', while their former tenants reclaimed waste and fallow lands to which they had acquired rights. The process was aided by loans and subsidies from the Government.

The '*area sown more than once*' has also shown a significant rise during the last six decades. In view of the fact that (*a*) demand for land in non-agricultural uses is constantly increasing, and (*b*) a further increase in the net sown area may not be an easy task, it is imperative that attention is paid towards an improvement in agricultural technology, so that it should be possible to raise three to four crops a year as some countries are doing. With the new agricultural strategy having very much come to stay, the objective should be within an easy reach.

Perspective. It is clear that the total supply of land is a fixed factor. Therefore, what is required is that an effective rationing of land among the varied uses be made. As far as possible, no further encroach-ments on cultivable land should be allowed; priority should be given to the use of non-cultivable land for non-agriculture uses. This will not only save cultivable land for agriculture, but will also promote a balanced regional development.

18.3.3 Soils

Long ago, Aristotle described soil as the stomach of the plant. Even now over 90 per cent of the world's food comes from the soil and less than 10 per cent comes from both inland water and the oceans.

The cropping pattern of the country is greatly influenced by the soils and the elements of the physical environment. The Indian Council of Agricultural Research divides the soils found in the country into eight major groups which are: (*i*) Alluvial soils including the coastal and deltaic alluvium; (*ii*) Black soils of varying types; (*iii*) Red soils, including red loams; (*iv*) Laterite and lateritic soils; (*v*) Forest soils; (*vi*) Arid and desert soils; (*vii*) Saline and alkaline soils; and (*viii*) Peaty and organic soils. Keeping in view their extent and agricultural importance, the first four, *viz.*, alluvial, black, red and laterite soils in that order, form the most important soil groups in the country. Almost the entire cultivated area in the country is covered by these soils.

- Alluvial soils are suitable for the cultivation of almost all kinds of cereals, pulses, oilseeds, cotton, sugarcane and vegetables.
- Black soils are known for their fertility. They give good yields despite continued cultivation and without proper manuring. Cotton, cereals, oilseeds and many kinds of vegetables and citrus fruits are some of the crops suited to black soils.
- Almost all kinds of crops can be grown on *red soil*, although it seems to be more suitable for the cultivation of rice, ragi, tobacco and vegetables.
- Laterite soils are suitable, among others, for rice and sugarcane.

Different types of soils distributed evenly throughout the country, abounding in fertility and higher yields, and highly responsive to improved inputs are found in the country. For example, we find the desert-like region of Rajasthan on the one hand and the rich cultivable land of Gujarat on the other. The variety of soils coupled with the fact that we have in the East the world's highest rainfall zone and in the West one of the driest regions along with every shade of climate throughout the country, makes possible the production of almost every kind of crop starting from those of the temperate zone to tropical production.

However, through constant use the quality of these soils has deteriorated slightly. Moreover, large tracts of land have been eroded. It has been estimated that about 106 million hectares is suffering from varying degrees of soil degradation. Localised soil waterlogging and salinity are most severe in India (27 percent of irrigated land), Pakistan (20 per cent) and China (15 per cent).

The situation is truly grave in India. Crops extract more nutrients from soils than are added to it through fertilisers or organic manures. The total "negative nutrient balance" is estimated at a staggering 8 to 10 million tonnes a year. And it is set to reach around 15 million tonnes by 2025.

Although our Plans have given priority to soil conservation and land stock improvement, we can identify the following difficulties in containing the degradation of land resources and bringing them back to productive uses:

- management of community land;
- lack of infrastructural development;
- high investment and long gestation;
- non-availability of institutional finance due to low creditworthiness of the beneficiaries having marginal and sub-marginal lands.

18.4 CROP PATTERN IN INDIA

By the term 'crop pattern' we mean the cropwise distribution of the cultivated land area, the changes in the distribution over time, and the determinants of this distribution.

18.4.1 Features of Crop Pattern in India

The crop pattern in India presents five distinct features as follows:

Wide Variety of Crops Being a country almost of the continental size, India is endowed with a variety of soils, which make it possible to grow almost every kind of crop. The various crops being cultivated in India can be grouped into (*a*) food crops, (*b*) fibres, (*c*) oilseeds, (*d*) drugs and spices. Food crops are both of a cereal and non-cereal type. Among the cereals the more important are rice,[3] wheat, besides maize, jowar, bajra, barley, etc. The more important non-cereal food crops are pulses like gram, moong, urad, etc. Among fibres, cotton and jute are the important crops. Among oilseeds, groundnut is the most important besides rapeseed and mustard. Similarly, a number of spices and medicinal crops are also cultivated throughout the country. Newer crops are emerging—products of tree crops, of the jungle and so on—which rakes in large sums of money while the going is good.

Dominance of Food Crops Over Non-food Crops The Indian farmer has adopted a necessity-oriented approach. He has concentrated on food crops in preference to commercial crops: about three-fourths of the total sown area in the country has been devoted to the production of food crops. This large emphasis on the production of food crops could be explained by two factors:

- A sharp rise in the prices of foodgrains on the one hand and, on the other, the provision of a minimum guaranteed price, especially in the case of rice and wheat;
- Technological improvements recorded in regard to a few food crops after the adoption of the new agricultural strategy. This is more true of wheat, where the evolution of HYV seeds opened up entirely new vistas.

---

3. *Rice:* China is the biggest producer, with 178 million tonnes produced from 30 million hectares, with an average yield of 5,869 kg/ha. In contrast, India has nearly 30 million hectares under paddy, but produces just 93 million tonnes, with average yield of 3,191 kg/ha. Greece produces 8,000 kg/ha while Egypt gets 7,900 kg/ha. The average yield in Asia is 3,731 kg/ha, and in North America it is 5,775 kg/ha.

   Wheat: The crop is grown in 126 countries. Almost all countries have put a premium on this crop. The maximum yield is in the Netherlands with an output of 8,771 kg per hectare. The lowest is in Somalia, with 370 kg/ha. The average yield for the world is 2,445 kg/ha, and for Asia it is 2,501 kg/ha. In India the average yield is 2,671 kg/ha in 2011–12. India stands 96th out of 126 countries. China's average is 3,519 kg/ha which is about the same as in the Punjab. Wheat in India is grown on 25 million hectares.

It is under the impact of these two factors that the area and production of food crops in the country increased, whereas that of non-food crops almost stagnated. One could say that the Indian agriculture has been essentially a food-based system.

Recent trends indicate that agriculture mix has been undergoing a silent and steady shift away from grains. With rising incomes, this change is driven by consumers' tastes and preferences. The composition of the production basket is also changing as is evident from the rising share of high value commodities (fruits, vegetables, fisheries, and livestock products). The value of output of these commodities has gone up from 40.6 per cent in Triennium Ending (TE) 1992–93 to 47.4 per cent in TE 2014–17, while the share of foodgrains has come down from 31 per cent to 24.7 per cent over the same period. This will be good for the farmers as it helps augment their incomes and achieve household food security.

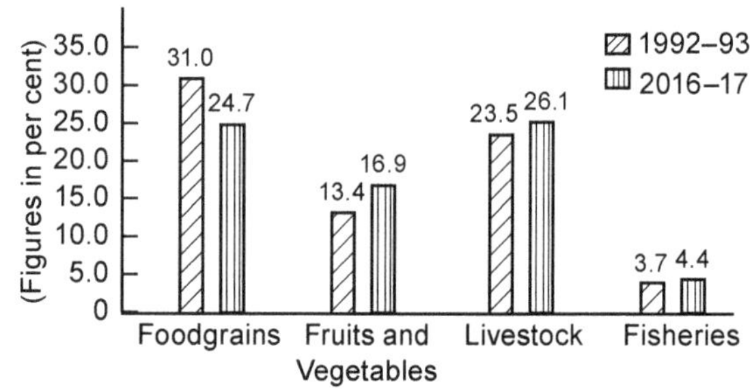

Fig. 18.2: Share in Value of Agricultural Output.

Preponderance of Cereals Among Food Crops The food crops can be classified into two groups as: (*i*) Cereals, and (*ii*) Non-cereals. Within the broad group of food crops, cereals like wheat and rice dominate. About 82 per cent of the land area under food crops has been put to the cultivation of cereals.[4] The attraction of cereal crops, as already stated, is explained by their relatively better prices and less risk in production, and finally, the availability of better quality seeds. As a result of this tendency whereas per capita availability of cereals has improved from 334.2 gms. per day in 1951 to 407.4 gms. in 2015, and India has successfully transformed itself into a 'net exporter' of cereals from the position of a 'net importer', the per capita availability of pulses (otherwise a nutrition rich food) has declined from 60.7 gms. per day in 1951 and 74.9 gms in 1959 to only 35.5 gms in 2015.

Importance Given to Inferior Cereals Within the cereal group, although superior cereals dominate, the Indian farmer has, by and large, not given up his preference for the cultivation of inferior cereals like jowar, bajra, maize,

4. India, with 9.2 per cent of total world production, ranks as the world's fourth major producer of cereals. Total annual production of inferior cereals is estimated around 30 million tonnes.

millets, barley, etc. These inferior cereals account for a little less than one-fourth of the total area under cereal cultivation.

Some important facts can be read from Fig. 18.2.

- The share of inferior cereals in the total area under cereal cultivation had been quite large. These crops are cultivated under dry conditions. Although the productivity is low, the crops are safer and are not sensitive to the vagaries of monsoon. Therefore, the farmers often have resorted to their cultivation as an insurance against misery arising out of crop failure.
- The proportional area under these crops has shown a downward trend after 1960–61. The reason is that with the spread of irrigation facilities and the availability of improved inputs, the farmers are increasingly being tempted to shift their lands to the cultivation of superior cereals.
- It also indicates a shift in the consumption pattern of the people.

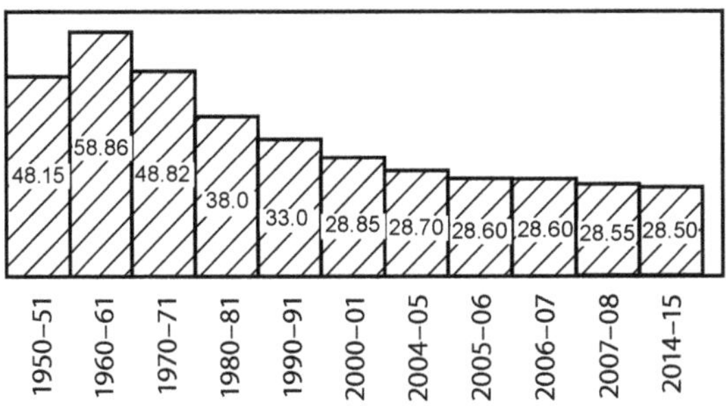

Fig. 18.3: Share of Inferior Cereals in Total Area under Cereals (% share).

All the same, given the fact that inferior cereals are substantially cheaper than rice and wheat, and not inferior to superior cereals in nutritive value, these embody great potential not only for purposes of poverty alleviation, but also in terms of the industrial utilisation of their by-products.

Declining Importance of Kharif Crops The Indian agricultural year is conveniently divided into two sessions, *viz.*, *Kharif* and *Rabi*[5]. Kharif crops have, hitherto, been contributing the larger share. But this dominance is on the decline. As a proportion of total production of foodgrains, the share of kharif has declined from 60.7 per cent in the 1970s to 57 per cent in the 1980s and further to 49.85 per cent in 2016–17, *i.e.*, both kharif and rabi crops are making almost equal contribution presently. This may be considered as the second most *significant development* on the agricultural front, the first being the phenomenon of the Green Revolution in the 1960s. The fact that this is the culmination of a

5. '*Kharif*' is normally identified with the wet season—crop is sown in June-July and harvested in November-January. '*Rabi*' is identified with dry season.

process that set in soon after the Green Revolution and not merely a one time occurrence makes this all the more significant.

This has important implications for the economy:

(*i*) Rabi crop is subject to less uncertainties. Increased emphasis on rabi crops, therefore, serves to dampen the agricultural fluctuations, though total immunity is impossible.

(*ii*) Income of the farmers would get a boost.

(*iii*) It would significantly contribute to the overall production of foodgrains.

(*iv*) The increased share of rabi crops may be explained in terms of non-price factors as the role of price factor is limited. The benefit of hind-sight suggests that upgradation of technology intended to raise the productivity of land and other factors plays a major role in augmenting farm production and income than the favourable terms of trade.

18.4.2 Factors Affecting Crop Pattern

The crop pattern, as it has emerged in India presently, is a product of many and varied factors which can be grouped into two parts as follows:

Physical and Technical Factors Physical characteristics of a region like soil, climate, weather, rainfall, etc., determine its crop pattern. For instance, millets like jowar and bajra are suitable for dry regions, as these crops can be managed with a small quantity of water. Rice is suitable for waterlogged regions since it can withstand extra water. It is mainly an irrigated crop in areas with an average rainfall of about 100 cms. Likewise, the Indo-Gangetic plain is most suitable for the production of wheat, etc.

Physical characteristics of a region can be changed through technical improvements, like irrigation facilities and capital investment.

Economic Factors Among the economic factors that determine the crop pattern of a region the following may be mentioned:

(*i*) ***Price and Income*** A farmer is influenced in the choice of crops by such considerations as price parities between different commodities, maximisation of income, and relative profitability per hectare. Further, the maintenance of a stable level of prices for a crop provides a better incentive to a producer to increase the output than what a higher level of prices does, if there is no certainty of this level being maintained over a number of years.

(*ii*) ***Farm Size*** Holders of small farms normally prefer food crops, because small holders are interested in producing foodgrains for their own requirements. They would produce cash crops only after they have provided the minimum necessary for their own sustenance. However, as the economic system gets more complex and the 'closed' nature of the village economy breaks down, it may become essential for the farmer to raise cash crops or at least those crops which are meant for market. He may do this in order to get command over liquidity.

(*iii*) ***Insurance Against Risk*** An ordinary farmer will always prefer a crop that involves less risk in its fruition to a crop that involves more risk.

(*iv*) ***Availability of Inputs*** Availability of inputs like seeds, fertilisers, irrigation, storage, marketing, transport, etc. is an important factor that influences a farmer's choice for a particular crop. In addition, a farmer is also attracted by a crop which involves less gestation period, as groundnut in preference to cotton.

(*v*) ***Tenure*** Under the crop-sharing system, the landlord votes the choice in favour of a crop that exclusively maximises income, notwithstanding the other costs that may be involved in its production.

(*vi*) ***Support Prices*** Support prices take the form of a minimum guaranteed return to the producer. By changing the relative rate of return on different crops, support prices help to give a boost to some crops at the expense of others. Likewise, the crops that enjoy the shelter of minimum guaranteed price tend to push out the crops that are denied protection.

In brief, the crop pattern of a country is the result of a number of factors detailed above. Shifts in the crop pattern can be attained by bringing about a change in the allocative domain, *i.e.*, changing the above factors in the desired direction.

18.4.3 Future Crop Pattern in India

In India, the policy objectives in regard to the future crop pattern should be based on the following considerations.

Rise in Income Levels As the economy grows, the level of income will rise. It is only to be expected that this would involve an increasing demand for superior cereals. Simultaneously, as the size of India's middle class increases, there will be changes in consumption patterns leading to significantly increased demand for higher value crops. Therefore, in the long run it is imperative that crop pattern is adjusted to the changing requirements by shifting land under cultivation from inferior crops to superior crops.

Need for Diversification There is a need for diversifying the crop pattern by encouraging the cultivation of different crops, due to the following reasons:

- It will ensure stability in over-all agricultural production, in as much as some crops may always compensate for fall in output of some other crops.
- The shift to high-valued crops will raise the total value of agricultural output.

- It will provide a base for increasing surplus of those products that are in increasing demand abroad. It will help to raise foreign exchange for the economy.
- The farmers with a reasonably-sized land, co-operatives with land of very small farmers pooled in, will gain from stable or higher incomes.

Need to Adapt to Water Scarcity It is high time that crop pattern changes should be introduced in regions increasingly faced with water shortages. Crops like rice and sugarcane have been described as "*parasites on water*". These crops may be substi-tuted by light irrigated crops like millets, pulses, oilseeds, cotton, fruit, flowers and vegetables.

Need to Adapt to National Priorities A "desirable cropping pattern" should contribute towards three national priorities, namely: employment generation; saving/earning of foreign exchange and prudent resources use which is ecologically sustainable. By implication, a desirable cropping pattern should be one which favours crops which are labour-intensive and have greater second-round employment effects. Simultaneously, it should favour crops which are either efficient import substitutes or exportables to save (earn) foreign exchange in balance. Further, given water as a binding constraint, the cropping pattern should attempt to maximise returns to this input and favour crops that are ecologically sustainable.

This last point calls for a little further attention. On the world stage, India is one of the richest nations in terms of biological diversity. India supports 15,000 species of flowering plants, 317 species of mammals, 969 species of birds, 389 species of reptiles, 206 species of amphibians, and so on. In any *sustainable agricultural system*, these cannot be ignored.

18.5 WATER RESOURCES

Water is the most important source of energy in the Indian economy. About 25 per cent of electricity generated in the economy is from the hydel sources. The other important use of water is in irrigation. In a country where agriculture gives twists and turns to the whole economy, provision of water can make all the difference; it can either stimulate the economic activity or depress it altogether.[6]

The important sources of water can be classified into two parts: (*i*) surface water, and (*ii*) ground water. *Surface water* is available from such sources as rivers, lakes, etc. *Ground water* is available from wells, springs, etc. Other sources of water which have not as yet been tapped in the country, but nevertheless represent a potential source are: saline lakes, saline springs, snow and ice-fields. Surface water sources are replenished by rainfall.

6. Total rainfall over India is 4,000 cubic kilometres (Ckm), out of this surface flow is just 1,880 Ckm annually. Due to topographical, hydrological and other constraints only 700 Ckm of surface water can be put to beneficial use.

Of the two sources, surface water is more important and possesses potential of growth in future. Surface water is available in the form of vast network of rivers available in the country.

Overall India possesses large reservoirs of water, but these are inadequate as compared to their requirements. Compared to countries such as the USA, which stores about 5,000 cubic metres per capita and China, which stores around 1,000 cubic metres per capita, India's dams can only store 200 cubic metres per person.

Through the decades, profligacy in the use of water by industrial units and during irrigation has brought the country close to a "water famine". In fact, many parts of India are already water-stressed, with per capita availability significantly less than the national average. At about 1,000 cubic metres, the national average, too, falls way short of the 'comfort point' set by the World Bank. No wonder many states remain at daggers drawn over sharing water from rivers that flow through them. The water resource use is already stretched to the hilt.

- Many parts of India are already water-stressed, with per capita availability significantly less than the national average
- Indian Agricultural Research Institute (IARI) says the country's farm sector alone accounts for 83 per cent of all water use
- State governments, instead of focusing on economic use of water in agriculture, have been subsidising electricity for pumping irrigation water
- IARI says in surface irrigation, water losses can be largely avoided by promoting the use of drip and sprinkler irrigation
- Highly water-intensive crops, dependent on rain water and flood irrigation, are also being brought under the scope of drip irrigation.

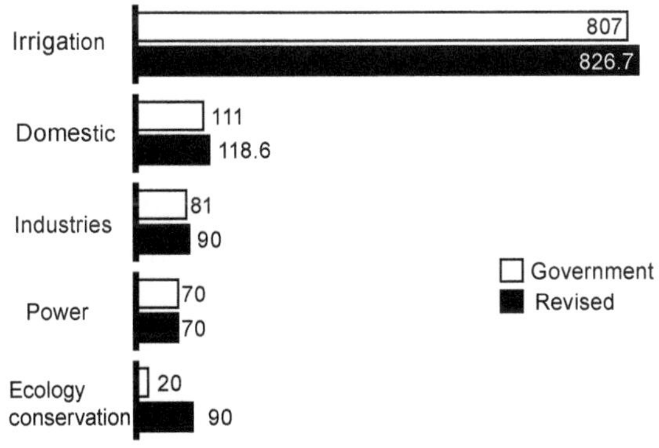

Fig. 18.4: Estimates of Demand for Water in India.

18.5.1 Water Issues

The principal issues facing the country are as follows: (*i*) Demand for water is increasing from all sectors, (*ii*) Lack of a rational water pricing policy between and within sectors is further driving demand, (*iii*) Policies and institutions mandated to solve conflicts are directly or indirectly contributing to further conflicts, (*iv*) New conflicts are increasingly arising within states rather than between states, and (*v*) Conflicts over groundwater are widespread across the country.

Solutions The following solutions suggest themselves:

1. The one over-riding lesson from the global resolution in the provision of public services is that competition matters. Hence, it is important to unbundle the bulk provider of the irrigation system from the distribution function. Further, there should be variety of forms—cooperatives, and private sector players—to handle water distribution to farmers.
2. There is a need to increase user charges for the services provided, as well as to increase budgetary support. But user charges or tariffs linked to costs can only be possible if services are provided in an efficient and accountable manner.
3. It is by ensuring the economic and social development of local communities that India can hope to build major dams.
4. The government should give formal water entitlements to people as with land and other property rights. Once established, such entitlements give rise to a fundamental and healthy changes.

Four interconnected programmes as follows assume significance in this context: (*i*) watershed development programme, (*ii*) renovation of all water bodies linked to agriculture which have fallen into disuse, (*iii*) correcting the deterioration of public irrigation works, notably state canal systems, due to cumulative neglect of maintenance and repair over the years, and (*iv*) rain water harvesting.

Another encouraging development is that more and more futuristic technologies are being readied in research labs to make desalination and water treatment processes more energy-efficient.

18.5.2 Establishment of Permanent Water Tribunal

The Centre's proposal to set up a single, permanent tribunal, subsuming all existing ad hoc tribunals, to adjudicate on inter-state river water disputes could be a major step towards streamlining the dispute redressal mechanism. But it alone will not be able to address the different kinds of problems—legal, administrative, constitutional and political—that plague the overall framework.

At present, a new tribunal is formed every time a state government approaches the Union government with a request and is able to convince the latter that such a tribunal is needed since all negotiations have failed. Eight such tribunals exist. After they have heard the matter and awarded their decisions, the tribunals are allowed to collapse. This system has had some successes, especially with the first generation of tribunals set up soon after independence—to adjudicate on the Krishna, Narmada and Godavari rivers. But in general, it has struggled to bring warring parties on the same page and offer equitable solutions.

There are three main problems: protracted proceedings and extreme delays in dispute resolution; opacity in the institutional framework and guidelines that define these proceedings; and ensuring compliance. The permanent tribunal is being pushed as a solution to the first problem but unless it is designed to tackle the inter-linked second and third problems, progress will be limited.

The delays happen for a variety of reasons at every stage of the process. Sometimes, the Centre takes years to decide whether a matter needs to be heard by a tribunal in the first place—for example, the Godavari and Krishna disputes started around 1956 but the matter was referred to a tribunal only in 1969. After the tribunal has been formed, it again takes many years to pronounce its award—it took nine years from reference in the case of the Narmada tribunal.

Certainly, these are highly complicated matters. But there are still steps that can be taken to streamline the process. Here, the Centre's proposal to set up, alongside the tribunal, an agency that will collect and process data on river waters has potential. The absence of authoritative water data that is acceptable to all parties currently makes it difficult to even set up a baseline for adjudication. Another reason for delay is the requirement that the Centre notify the order of the tribunal to bring it into effect; this took three years for the Krishna award. Now, thankfully, the Centre has proposed that the awards will be notified automatically by the tribunal.

When it comes to opacity in the framework, Alan Richards and Nirvikar Singh from the University of California, Santa Cruz, note in their paper on India's inter-state water disputes, "There are too many options, and there is too much discretion at too many stages of the process". This is partly because of procedural complexities involving multiple stakeholders across governments and agencies. But the inchoate nature of the system is equally the result of factors outside the efficient functioning of the tribunal itself.

Two of those factors are worth mentioning here—India's messy federal polity and its colonial legacy. These, in turn, set the stage for the third problem of non-compliance.

Water is a state subject but the "regulation and development of inter-state rivers and river valleys… in the public interest" is on the Union list. However, the Centre has generally taken a back seat, allowing states to dominate. And even when it has intervened, it has not always been successful. The courts have also often been ignored, including the Supreme Court, which importantly only has very limited jurisdiction over the tribunals.

Today, inter-state water disputes are no longer just about water allocation. They have become hugely politicized—the recent eruption of the Cauvery dispute, framed as an ethnic identity issue between Tamilians and Kannadigas, which led to widespread civil unrest, is only the most recent example. Public opinion is an important factor that cannot be wished away. The Central government must keep these factors in mind when setting up the proposed tribunal. A robust institutional framework—and a transparent one to ease state and public buy-in—is a must. Without that cooperative approach, India's water dispute resolution is unlikely to see much improvement.

18.5.3 New National Water Policy 2012

The national water policy was announced for the first time in 2002. A decade down a new water policy has been formulated. Its principal features are as follows:

(*i*) The government should withdraw from the role as a service provider in the water sector. Instead, communities and the private sector should be encouraged to play the role.

(*ii*) Abolition of all forms of subsidies to the agricultural and domestic sectors. Subsidies and incentives should be provided to private industry for recycling and reusing treated effluents.

(*iii*) Subsidies to agricultural users should be curtailed.

(*iv*) The government should ensure access to a minimum quantity of potable water for essential health and hygiene to all citizens, available within easy reach of the household.

(*v*) For expeditious resolution of interstate disputes, the policy suggests the establishment of a permanent Tribunal at the centre.

(*vi*) People displaced by large water projects should be made partners in progress and given a share in the benefits comparable to the project-benefitted families.

The cost of rehabilitation and compensation to the project-affected families be partly borne by the project-benefitted families through adequate pricing of water.

The policy, however, has come in for criticism. It is argued that the government has not learnt anything from the past. The entire focus should have been on how to sustain groundwater which is the country's lifetime.

18.5.4 Evaluation

If the trial run of the vessel carrying commercial cargo from Varanasi to Kolkata through National Waterway-1 (NW-1) proves successful—as it hopefully should—it will add a welcome new dimension to the transport sector dominated till now by rail and road movement. Water transport is one of the preferred modes for moving cargo, especially over-sized and odd-shaped goods, the world over, but has largely been neglected in India. It has taken nearly three decades for India's first national waterway—the 1,620-km stretch of river Ganga from Haldia to Allahabad—to become operational after it was formally notified in 1986, though limited movement of coal had begun in 2013. The kind of push the Narendra Modi government is now imparting to water transport generates hopes that this may ultimately emerge as a supplementary and, more importantly, cost-effective and environment-friendly mode of conveyance of goods and passengers.

Plans are said to be afoot to begin cruise services between selected religious tourism hubs on this river. Several cities on this route can also begin passenger ferries to reduce congestion on roads. Liquefied natural gas is proposed to be used as the main fuel for water transport. Cost-wise, rough estimates indicate that one horsepower of energy can carry four tonnes of load by waterways, against only ISO kg by road and 500 kg by rail. That explains why many water-rich countries in Europe and elsewhere carry over 40 per cent of their freight and passenger traffic through water. In India, on the other hand, the share of water transport is a trifle 3.6 per cent, which is now mooted to be doubled to around seven per cent by 2018.

Notably, Parliament has already passed the National Waterways Bill earlier, which provides for developing 101 river stretches as national waterways. This is part of the government's grandiose plan to create an interconnected rail-road-waterways logistics system in the country. Even if this programme meets part of its targets it can deliver perceptible economic gains by spurring economic activity around rivers. Besides, it can ease the pressure on the over-burdened rail and road networks. The government has done well to foreclose the risk of the project getting bogged down with dispute with states—given that water is a state subject - by putting this statute under Entry 24 of the Union List of seventh schedule of the Constitution. This section permits the Centre to make laws on shipping and navigation on inland waters that are declared as national waterways.

Such meticulous planning notwithstanding, the task of implementing this mammoth project is unlikely to be trouble-free. The meandering character of river systems, and wide fluctuations in the depth and velocity of water from place to place, besides the existence of numerous pontoon and other bridges with low clearance, pose a formidable challenge that needs to be surmounted. Constant inflow of silt in the rivers can also be problematic, requiring regular dredging. Moreover, the viability of commercial navigation will rest critically on the availability of adequate and assured two-way commodity and passenger traffic. But, a sizable part of the bulk cargo, such as coal, minerals, foodgrains and fertilisers, is usually unidirectional and may require vessels and barges to return empty, thus denting the profitability of navigation ventures. These aspects would need to be suitably addressed before committing heavy investments in infrastructure for commercial navigation.

18.6 BIODIVERSITY

India contains a great wealth of biological diversity in its wetlands and in its marine areas. There are about 350 species of mammals, 1,224 species of birds, 408 species of reptiles, 197 species of amphibians, 2,546 species of fishes, and 15,000 flowering plants. The importance of these biological resources cannot be overestimated for the continued welfare of Indian population.

A large number of both flora and fauna are faced with threat of extinction because of the increasing demand being made on natural resources.

The government has responded by taking a number of measurers, including legislative measures, among which the more important is the Biological Diversity Act, 2002.

18.6.1 The Biological Diversity Act, 2002

The Act covers conservation, use of biological resources and associated knowledge occurring in India for commercial or research purposes or for the purpose of bio-survey and bio-utilisation. It provides a framework for access to biological resources and sharing the benefits arriving out of such access and act. The Act also includes in its ambit the transfer of research results and application for intellectual property rights relating to Indian biological resources.

18.7 FOREST RESOURCES

Forests produce the requisite raw materials for industries, defence, communications, domestic use and other public purposes. They contribute to the country's exports and create a large volume of employment in the primary, secondary and tertiary sectors. They also provide materials like fuelwood, small timber, fodder, etc. for direct use by the agriculturists. The benefits from forests in the matter of soil conservation, recreation, wildlife, etc. have been well-recognised. Forests have multifarious uses, some of which can be identified as follows:

1. Trees and forests have many uses, and are, therefore, considered as one of the important natural resources of a country. Plants through photosynthesis convert the solar energy into various forms of energy like food, fuel, oil products, raw materials for industries, and other indirect sources of renewable energy, which can be directly used by us. Forests meet nearly 40 per cent of our energy needs.
2. The thick foliage-mix, that both trees and shrubs attract, prevents soil erosion, and holds moisture content in both the soil and the

atmosphere. This prevents an area turning into desert. These are also described as hydrological benefits of forests.

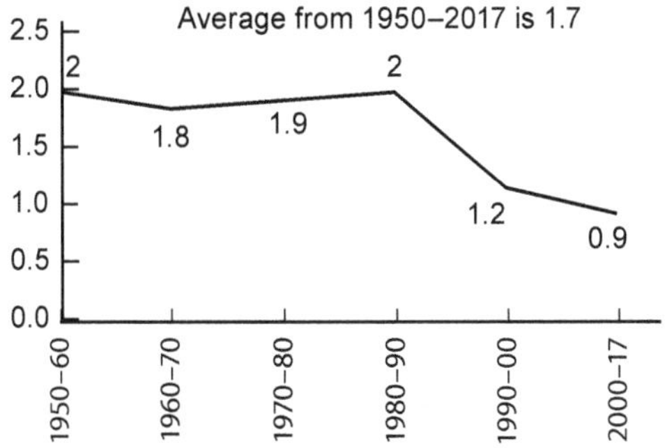

Fig. 18.5: Share of Forestry in Total GDP.

3. Forests meet the basic needs of the poor people of the country in many ways. Forests provide 30 per cent of our fodder needs. A large section of the country's population—Ignancy Sachs calls them 'eco-system people'—depends on forests for firewood which is basic for human survival. Failure to meet the basic needs of rural people may put millions of women—who go out to collect the firewood everyday to forests—into serious difficulties.
4. Forests offer good employment opportunities too, mainly to the rural population. The National Commission on Agriculture (NCA) in 1976 estimated that if the programmes recommended by it were implemented by the government our forests would provide direct employment of 15 million man-days or 2.5 lakh man-years in addition to the present one. In addition, indirect employment would be created. What is still more important, the new employment would be created where it is exactly wanted in rural areas, backward regions and hilly tracts. The importance of this social gain cannot be over-emphasised.
5. Among other things, forests supply pure oxygen, prevent environmental and sound pollutions, maintain ecological balance by providing shelter both to wildlife and birds, and finally add to the aesthetic beauty of the country.
6. Calling tropical forests the "great chemical factory of the natural world", the Washington-based World Watch Institute, in a recent report, noted that 40 per cent of the prescription drugs have active ingredients derived from wild plants, animals, or micro-organisms, many of them from forests.
7. It is easy to justify huge investment in development of forests on purely economic grounds. Returns on such investments have been

found to be quite substantial in comparison to the returns on similar types of investments in agriculture, etc.

Above all, to the 'silent majority' among whom number the conservation minded, the forest is a treasure house of knowledge as well as a place that provides the peace and quiet which is so absent in the artificial world of man's other jungle—'the concrete jungle'.

18.7.1 Present Position

Forests occupy about 782.9 lakh hectares or about 23.81 per cent of the total geographical area. (Of this dense forest category, *i.e.*, 40 per cent or more of the area covered by trees, amounts to about 58.0 per cent.) Of these about 433 lakh hectares or about 61.0 per cent are exploitable; another 178 lakh hectares or about 25 per cent are potentially exploitable. Among the States, Madhya Pradesh has the maximum forest area of 77,265 sq.km., followed by Arunachal Pradesh (68,045 sq.km.) and Chhattisgarh (56,448 sq.km.).

The area under forests in India is low not only as compared to the forest area in countries like Japan (64%), Sweden (66%), Canada (27%), South Korea (63%), and USA (25%), but is also much less than the norm of 33 per cent of the total reporting area recommended in the National Forest Policy of 1952. The per capita forest land in India is 0.06 hectares as against the world average of 2.06 hectares. Further, the productivity of Indian forests is very low: 1.2 cubic metres per hectare per year as against the world average of 2.1 cubic metres. In future the demand for forest products is likely to increase fast.

Forests in India are still under huge pressure and shrinking over time.

Firstly, the rate of forest land diverted towards development projects has been unprecedented in the past five years. But this diversion also happens because there is no value seen in forests—other than the cost that has to be paid for diversion of land by the project proponent. Instead, there is value in the dam, road or mine for which the land is needed. So, this pressure on forest land is bound to increase. We must also note that forests are the last remaining public lands in the country and the acquisition of private land will become even more expensive and contentious in future.

Secondly, forests are under pressure to meet local needs and for illegal extraction. Today, it is an inconvenient truth that the poorest people in India live in the country's richest forests. The management of this green wealth has not brought any benefits to the locals. Amid all this, while deforestation arid forest diversion will grow, we do not have any viable strategies for re-greening these lands. Thus, we will lose bit by bit.

So how do we change this? *One,* we need to urgently value the economic, ecological and livelihood potential of forests and to incorporate this into national accounts. We need a robust methodology to bring the tangible (what we can measure) and intangible costs together. Though there is much talk about green accounting, the methodology is weak. For instance, there is no real assessment of minor (non-timber) forest produce. Other assessment of the contribution of forests to livestock or the hydropower sector is inflated or non-existent.

Two, we need to use this methodology to pay for standing forests. The 12th and 13th Finance Commissions allocated funds for standing forests, but this is a pittance. We, then, need states to transfer payment for standing forests—protected for biodiversity, watershed or other purposes—to local custodians. This will build local economies and local support for forest protection.

We need to use robust accounting methodology to increase the productivity of the remaining forest land. But we know that the business of cutting and planting trees that survive cannot be successful without people who live in the forest. So this becomes the new opportunity for employment and economic growth. The way ahead is to build inclusive economies using green wealth.

Forest Rights Act Being Reinterpreted to Avoid Amendment

Moves within the government

- Working to do away with the need to get tribal gram sabhas' consent before their forests are cut down for industrial projects
- Meetings held at PMO; initiated by the environment ministry; backed by the coal and road transport ministries
- Law being reinterpreted to avoid amendment to the Forest Rights Act
- Tribal affairs ministry has labelled the move illegal and absurd
- The move undermines SC order on mining by Vedanta in Odisha's Niyamgiri Hills, says tribal affairs ministry

The Forest Rights Act of 2006, and clearances

- For decades, forests were taken over by govts and tribal people were turned into encroachers of own lands
- The FRA has handed back to tribals the control and rights over traditional forest land
- The Act makes gram sabhas the statutory body to protect the forest land over which rights are given back
- Forest clearances by the environment ministry require consent from gram sabhas
- Earlier, forest bureaucracy and the ministry had sole powers to give forests away. Nearly 95% of all projects seeking forests were approved

18.7.2 National Forest Policy

The national forest policy was first enunciated in 1952 and subsequently revised in 1988.

The important distinguishing features of this policy are as follows:

1. The policy lays emphasis on the conservation of forests, and meeting the requirements of the tribal and rural people. The Scheduled Tribes

and Other Forest Dwellers Act, 2006 has been enacted in pursuance of this policy.

2. Tribals will also be associated with the protection, regeneration and development of forests, and cooperatives run by them or by the Government will replace the present contractor system which has resulted in unchecked devastation of reserved forests.[7]
3. Realising the fact that uncontrolled expansion of forest-based industries would adversely affect the conservation objective, the revised policy has made the following stipulations:
   - — No forest-based industry, except in the small-scale and cottage sectors, would be permitted unless sustained availability of raw material was ensured.
   - — As far as possible, a forest-based industry should produce its own raw material require-ments.
   - — Natural forests would not be made available to industries for undertaking plantations and for other objectives.
   - — Forest produce would not be supplied to industries at concessional rates.

The initial results of the new policy have been mixed.

Under the Green India Mission, it is proposed to increase forest cover by 5 million hectares and improve the quality of forests on another five million hectares over the next 10 years. The Green India Mission is one of eight missions under the National Action Plan on Climate Change. It aims to increase forest cover on 5 million hectares (ha) of forest/non-forest land and improve the quality of forest cover on another 5 million ha.

Funding for the scheme will come from the Plan outlay and convergence with the Mahatma Gandhi National Rural Employment Guarantee Scheme, the compensatory afforestation management and planning authority, and the national afforestation programme.

The Union government will provide 90% of the funds for implementing the scheme in the north-eastern states, and 75% of the funds for other states. State governments will meet the balance requirement.

Forest policy in India is victim of the fact that decision-making on environmental issues in India is quite fragmented. While the ministry of environment and forests implements various central laws—including the Forest Conservation Act 1980—the state governments, too, have powers in this domain.

---

7. We can distinguish between '*protected forests*' and '*reserved forests*'. In protected forests the rights of the community such as collection of timber, fuel wood, and minor forest products are suspended for specified terms, not exceeding 30 years, whereas in case of reserved forest the local community enjoys the privileges of forest rights, subject to certain restrictions. Also see Roger Jeffery and Nandini Sundar (eds.): *A New Moral Economy for India's Forests?* (Sage, New Delhi, 1999).

There are other ministries and decision-making authorities that are involved as well. If that is not all, the Supreme Court and other tribunals, too, are active participants in the ongoing story of India's forests. The tale is rounded off with other stakeholders—tour operators, villages located in forest areas and various, illegal, interest groups such as timber smugglers and poachers who too manage some influence through the political system.

All this makes for serious collective action problems. As a result, formulating coherent policies for forest and wildlife conservation and implementing them effectively are well nigh impossible.

18.8 MINERAL RESOURCES

The mineral resources of India encompass a wide range of products that are necessary for a modern developed economy. There are, according to the Geological Survey of India, 50 important minerals and 400 major sites where these minerals occur. (Out of total land area of 3.28 million sq.kms, hard rock area covers 2.42 million sq.kms.). These can be divided into four categories as follows:

1. Minerals of which India's exportable surplus can dominate the world market; to this category belong iron-ore and mica;
2. Minerals of which the exportable surplus forms an important factor; these include manganese ore, bauxite, gypsum and others;
3. Minerals in which it appears that the country is self-sufficient, like coal, sodium salts, glass sand, phosphates, bauxite, etc.;
4. Minerals for which India has to depend largely or entirely on foreign markets like copper, nickel, petroleum, lead, zinc, tin, mercury, platinum, graphite, etc.

The various minerals can also be classified into three categories on the basis of their nature and end use. These three categories are: (*i*) Fuels like coal, lignite, natural gas and petroleum; (*ii*) Metallic minerals like bauxite, iron-ore, manganese, etc.; (*iii*) Non-metallic minerals like phosphorite, graphite, gypsum, limestone, mica, etc.

18.8.1 Features of Minerals

Minerals provide a base for the rapid industrialisation of the economy. The changeover to an open market economy has opened up further avenues for faster industrial growth and greater requirement of minerals "besides the fact that the geological setting of the country holds great promise for a boom in mineral productions." [Expert opinion is that given the size of deposits in South Africa and Australia, large reserves can be expected in India also (because of similar geological structures). A recent Price Waterhouse Report has identified India as the most promising mining location worldwide.]

Notwithstanding this, India's spending on exploration is only about 0.8 per cent of the global spending with private sector only contributing 3 per cent of this. There are a few essential aspects that need to be worked into a proper mineral policy.

1. The mineral resources are very unevenly distributed.
   — The Great Plains of Northern India are almost entirely devoid of any known deposits of economic minerals.
   — Jharkhand and Orissa areas on the north-eastern parts of peninsular India possess large concentration of mineral deposits, accounting for nearly three-fourths of the country's coal deposits and containing highly rich deposits of iron-ore, manganese, mica, bauxite and radioactive minerals.
   — Mineral deposits are also scattered over the rest of the peninsular India and in parts of Assam and Rajasthan.
   — With the implementation of the new UN laws[8] on the Exclusive Economic Zone (EEZ), new areas in deep seas, hitherto unexplored, would become available to us. These areas are believed to contain rich deposits of oil, gas, manganese and modules of nickel, cobalt and copper.
2. The country is deficient in certain minerals like crude oil or petroleum; a large part of the present demand is being met by imports. In view of the rising prices of these minerals in international markets, it would be necessary, on the one hand, to curb their growing use in the economy, and, on the other hand, sustained efforts should be made to explore the domestic sources of supply of these minerals.
3. There are minerals which are lucrative foreign exchange earners. Efforts should be made to devise a suitable policy to have a proper utilisation of these minerals keeping in view the national interests.
4. Due to paucity of funds, the mining industry is mired in obsolete technology, now for decades. India's mining sector is in trouble. The sector's troubles stem from profit-seeking, poor regulation, and galloping demand from countries such as China. This has led to courts stepping in and banning all mining activity in a region. Environmental clearance, and the human cost involved, are other matters the industry has just begun to deal with.

---

8. The new Law of the Sea established by the UN allows a nation to stretch its EEZ by another 100 nautical miles, provided it has already conducted preliminary exploration in the region before the year 2004. India's EEZ is being presently estimated at 2.013 million sq.km. Another one million sq.km. is likely to be added after the delimitation of the entire Continental Shelf (a continent's submerged border that slopes gradually and extends to a point of steeper descent to the ocean bottom).

New Mineral Policy, 2008 A committee was set up in the late-2005 under the chairmanship of Anwar-ul-Hoda.

The New Mineral Policy, 2008, has incorporated most of the recommendations made in the Hoda Committee Report.

(*i*) The new policy manifestly recognises that private sector will be the main source of investment in reconnaissance and exploration. It takes a pragmatic view of the risks associated with the business of finding/producing minerals and therefore ensures security of tenure and rights of transferability of various awards—reconnaissance permits (RPs), prospecting licences (PLs) and mining leases (MLs).

(*ii*) The new policy provides for introducing Long Area Prospecting Licences to help the investors achieve economies of scale. These licences will be given only with regard to non-bulk minerals which, from the investors' point of view, is a high-risk, high reward area.

(*iii*) The policy emphasises that mining is a standalone industrial activity, it can thrive in conjunction with value-addition activities.

(*iv*) Export policies will be formulated after taking stock of the mineral inventories and short, medium and long-term needs of the country. It is stated in the policy document that efforts shall be made to export minerals in value-added forms as far as possible.

(*v*) The policy envisages steps to be taken for the governments to facilitate financing of mine development and exploration which are integral to the mining projects.

(*vi*) There is also a plan to develop appropriate capital market structures to attract risk investment in survey and prospecting. Flow of 'risk funds'—from capital market and venture funds—will be eased with policy interventions. The public private partnership model would be useful for building mining infrastructure.

(*vii*) The policy seeks to bring certain level of uniformity in mineral administration in the country.

Further, acting on the recommendations of the Ashok Chawla committee, the Government plans to seek outside help to increase the pace of mapping of potential mineral reserves in India to meet a government target that aims to complete prospecting for natural resources in the five years beginning April 1, 2012.

India's mining sector is set for a cleanup.

- Setting up of coal regulator to set methodologies for price fixation by Coal India.
- Competitive bidding for allocation of captive blocks to private companies.

- Improved coal supply for power plants as coal and power ministries iron out differences over issues in new supply pacts.
- Higher production by Coal India as fresh clearances start coming after PMO's intervention.
- Improved iron ore availability in Karnataka after mines restart operation following implementation of R&R schemes.
- Continued downfall in iron ore output and exports from Goa and Orissa at the back of allegations of illegal mining.

18.8.2 Mines and Minerals (Development and Regulation) Bill, 2011

The Government has got approved from the parliament the Mines and Minerals (Development and Regulation) Bill, 2011. It would soon become a law. The Bill provides for the following:

(*i*) It provides to earmark 100% of the royalty paid by major mining companies, to compensate people displaced by such projects. The major minerals include iron ore, bauxite, copper and other minerals. Typically, royalty is a tax calculated as a percentage of price.

(*ii*) It provides to earmark 25% of the profit made by coal mining companies in favour of people directly affected.

(*iii*) It calls for "competitive bidding for a prospecting license and mining license." However, there is a provision that competitive bidding can happen only it no existing application for prospecting or mining is pending.

The law also sets out the framework for facilitating much needed investment in modernization of Indian mining.

In an order passed in March, 2012, the Supreme Court had directed the central and state governments to bring into force the Minor Minerals Conservation and Development Rules, 2010 at the earliest. It has been stressed that the mining of minor minerals needs to be registered strictly like mining of major minerals.

18.8.3 Acquiring Mineral Sources Abroad

In view of the worsening commodity-crisis, the government has been advised to adopt the Chinese Model of foreign minerals buy. The principal focus in the Chinese Model is on acquisition of ownership of reserves in coal, oil and gas.

Acquiring mineral reserve in other resource-rich nations is considered to hold the key to India's energy security, owing to domestic constraints in exploiting reserves and the global rise in commodity prices. Also, holding such assets acts as a strategic tool in energy diplomacy for developing and maintaining foreign relations.

18.9 CONCLUSION

We have reviewed above a brief profile of the major natural resources of India. Undoubtedly, India is blessed with a variety of resources. But the supply of these resources is to be viewed against their requirements on the one hand and possible utilisation within the given range of technology on the other. The potential available in a number of resources matches requirements, but the present utilisation of almost all resources falls short of the requirements. Therefore, what is immediately required is:

1. Intensive surveys should be undertaken within the country to explore and to identify the hitherto unknown utilisable resources. This is true in the case of both renewable and non-renewable resources. This will require chalking out an integrated multi-pronged national policy.

   Equally important is the need to make an efficient use of available resources. This requires several interrelated steps: better technology, use of by-products, multipurpose use of resources, location of industries such that the transport cost of combining resources from different areas are minimised, etc.
2. There is the need to take such conservation measures that sustain the output over a longer period. Our forefathers designed institutions to cope with the divergence between the individual and social rates of discount and its adverse effect on sustainable resource use. In good times and bad, by invoking the spirits or scriptures, and by creating institutions for conservation from above or below, our forebearers ensured sustainable use of natural resources. It is possible even now to revive the institutions by nurturing community consciousness.
3. All the above considerations will call for an effective organisational set-up. Privatisation by itself cannot be an end. In the recent past there has been clear evidence of deals and attempts at deals for the transfer of publicly-owned resources to private hands on terms which are more than generous to the private parties and involve substantial losses in potential income as well as other costs to the public exchequer.
4. With rapid globalisation under way demand for scarce natural resources like land, water, mines etc. has been increasing at a fast rate. These natural resources are increasingly being grabbed by the mafia who have seen in these resources an opportunity for their sustainability. The mafia can easily hold on to this *loot* by bribing politicians, bureaucrats and other officials at different tiers of governance. Further, adverse consequences of the recent loosening up of the forest laws needs to be closely watched.

In the interest of swift economic development these considerations cannot be ignored.

18.10 NATIONAL ENVIRONMENT POLICY 2006 (NEP)

The NEP 2006 is being described as a statement of India's commitment to making a positive contribution to international efforts.

The NEP builds on the earlier policies, like the National Forest Policy, 1988, National Conservation Strategy and Policy Statement on Environment and Development, 1992, National Agricultural Policy, 2000, National Population Policy, 2000, and National Water Policy, 2012.

Its major features are as follows:

(*i*) The dominant theme of the policy is to ensure that the livelihood of people dependent on forest products is secured from conservation than from degradation of the resources.

(*ii*) It focuses on conservation of critical environmental resources, livelihood security for the poor, integration of environmental concerns in economic and social development and judicious use of the resources.

(*iii*) To achieve sustainable development, environmental protection shall constitute an integral part of the development process and cannot be considered in isolation from it.

(*iv*) Environmental Impact Assessment will continue to be the principal methodology for appraisal and review of new projects.

(*v*) The assessment processes are being revised. Under the new arrangement, there will be significant devolution of powers to the State/Union Territory level.

(*vi*) It also seeks to revisit the Coastal Regulation Zone notifications to make the approach to coastal environmental regulation more holistic and, thereby, ensure protection to coastal ecological systems, waters and the vulnerability of some coastal areas to extreme natural events and potential sea level rise.

(*vii*) Involvement of Panchayati Raj Institutions and urban local bodies has been highlighted.

An unstated, but implied feature of the policy is to put environment-related decisions on fast-track.

18.10.1 Drafting a New Green Architecture

As India's environment regulator takes shape, some key questions face its design.

Project Appraisal

- Current process corrupt, vulnerable to rent-extraction, doesn't study environmental impacts well and rarely rejects projects.
- New regulator to oversee project appraisal and compliance. Seeks to make this autonomous by delinking staffing and funding.

- Insufficient to ensure true autonomy. India needs to get serious about following her environmental laws.
- Poor. Ministry's six regional offices are understaffed.
- State pollution control boards are understaffed, corrupt and don't answer to ministry.
- Unless the regulator is autonomous, status quo will continue.
- More regional offices can track projects cleared by Centre. But what about those cleared by states?
- Secretary/member should not be choosing. And will they also have powers to remove?

18.11 FIVE YEAR REFORM PLAN (2012–17)

The government has prepared a five-year reform plan for the environment sector. As a part of the plan it wishes to strengthen infrastructure for better management of environmental and forest clearance systems, decentralise its monitoring to regional offices and amend the Environment (Protection) Act of 1986 to raise penalties for non-compliance. It proposes to strengthen technical manpower and upgrade laboratories, including those of the central and state pollution control boards.

It is to set up a National Environment Assessment and Monitoring Authority (NEA-MA), a multi-disciplinary autonomous body for impact assessment, coastal zone management and monitoring of clearance conditions.

A SWOT analysis of the plan is presented alongside.

Ultimately we should aim at:

Reducing, Reusing, Recycling—and **Protecting Diversity**

The three outcomes that provided some direction from Durban were the launch of GCF, the decision on continuity of the Kyoto Protocol into its second commitment period, and the establishment of the Ad Hoc Working Group on the Durban Platform for Enhanced Action (termed, in short, ADP or the Durban Platform). The last one will be guided by the ongoing work under two important tracks of the convention—the Ad Hoc Working Group on the Kyoto Protocol and the Ad Hoc Working Group on Long-Term Cooperative Action (AWGLCA).

SWOT ANALYSIS-A SNAPSHOT

Strengths	Weakness
❑ A robust regulatory framework	❑ Inadequate enforcement and compliance
❑ Institutional structure	❑ Inadequate institutional capacity
❑ Mainstreaming of sustainability concerns	❑ Inadequate finance
❑ Creation of conservation, pollution abatement infra	❑ Low priority at states and at local level
❑ Successful implementation Raise environmental awareness, civil society involvement	❑ Shortcomings in implementation of plans

18.11.1 Changes on Anvil

A. Green Policies and Laws to Change for 'Ease of Doing Business'

- Change definition of forests to remove large patches from the purview of forest laws
- Create land banks that industry can easily access for compensatory afforestation
- Ready digital maps of forestlands in all districts
- Severely diluted inviolate forest areas or no mining zone to be notified by states
- Tribal consent for using forestlands to be done away with
- Discounted rates of net present value to be charged from industry for using forests
- Standardised terms of reference for environment impact assessment studies by each industrial sector
- Do away with state's environmental clearances
- Revision of the Indian Forest Act, 1927, with greater powers to the Centre
- Private investments using government forestlands
- A tree-land trading scheme to help industry meet its compensatory afforestation targets
- Amendments to the Environment Protection Act and Forest Conservation Act, 1980, in two phases
- Amendments to the coastal zone regulations
- Review National Forest Policy
- Third party verification and monitoring of industries to be put in place
- National Green Tribunal's powers to be curtailed
- Self-certification of pollution levels by industries to set in.

B. Changes Already Brought in Through Executive Orders

- Chopping of trees for linear projects,even before final forest clearances, cannot be challenged in court till final clearance
- Integrated forest, wildlife and environment clearance online
- No public hearings for projects inside industrial estates or parks with existing general clearances
- Dilution of Coastal Regulation Zone for hotels, resorts
- Made site-inspections prior to clearances an exception
- All highway projects in border states without public hearing
- Expansion of coal mining by a larger percentage without public hearings

- No need to have prior possession of land to acquire green clearance
- Eased clearance for projects near inter-state boundaries and protected areas
- Fast-tracked clearances for linear projects such as pipelines, highways and canals
- Most coal blocks taken out of inviolate areas
- Irrigation projects with Cultural Command Areas of less than 2000 ha and biomass-based thermal power projects up to 15 Mw put on Automatic Approval Route
- No extra CSR expenditure by industries for green clearances
- No additional conditions over thermal and hydro-power projects by experts besides those by ministry
- No additional studies to be asked of project developers
- Preliminary surveys by industries in wildlife areas do not require central clearances
- Moratoriums lifted in extremely polluted industrial areas
- Index measuring industrial belts put in suspension.

18.12 INDIA'S ROLE IN INTERNATIONAL ACTION (KYOTO PROTOCOL AND AFTERWARDS)

Environmental protection is an international issue and needs global intervention. The 1972 study "*The Limits to Growth*" by the Club of Rome and the United Nations Conference on the Human Environment which met at Stockholm in June, 1972 brought to fore the looming crisis for human environment and ecology. The Rio Conference of 2002, popularly known as Earth Summit, brought the international community together to institute action on various fronts such as combating climatic change, conservation of biodiversity, energy, etc. The existing formal international agreement is known as Kyoto Protocol. The Kyoto Protocol came into force in 2005. The protocol has been ratified by 175 countries, including those in the European Union, Japan, Canada and Russia. This has set legally binding targets for these countries to reduce their greenhouse gas emission by 5.2% (*from the 1990 level*) by 2012. This protocol provides three mechanisms to developed nations to meet their emission targets:

1. Joint Implementation Among the Developed Nations It allows industrialised countries with a greenhouse gas reduction commitment to invest in emission reducing projects in other industrialised countries as an alternative to emission reduction in their own countries.

2. Emission Trading Among the Developed Nations In this mechanism, a central authority sets a limit or cap on the amount of pollution that can be emitted. Companies that emit greenhouse gases are required to hold an equivalent number of credits, representing the right to emit a specific amount. Companies that need to increase their emissions must buy credits from those which emit less.

3. Clean Development Mechanism (CDM) It allows developed countries (*called Annex 1 countries*) to meet their emission reduction targets by paying for greenhouse gas emission reduction in developing countries (*called non Annex 1 countries*). Article 12 of the Kyoto Protocol states that the "purpose of the CDM shall be to assist parties not included in Annex 1 in achieving sustainable development and in contributing to the ultimate objective of the convention and to assist parties included in Annex 1 in achieving compliance with their qualified emission limitation and reduction commitments."

The Kyoto Protocol have had only limited success.

Ever since the Kyoto Protocol was adopted in 1997, not a year has elapsed without rich and poor countries sparring with each other on the issue of climate change. These divisions were evident in Panama, where talks proved fruitless.

The divisions are pretty clear: rich countries want a binding agreement to limit climate change and don't want to bear any responsibility for historic emissions. Developing countries—India included—want a measure of equity in the proceedings.

Some of the damage done to environment is irreversible. But there is much that can be done right now to prevent any further damage. We have even a shred of hope of preventing further deterioration of the environment then humanity has to make a choice. To choose a lifestyle that is environmentally sustainable. Small choices make a massive difference. Not using plastic, not using diapers, using as little of air conditioners as is possible...every little thing matters. The window of opportunity is small one and it is fast closing. We should not miss it. It is our survival as a species, not just a civilisation that may be at stake.

UN Panel Report The late-2013 report of the United Nations Intergovernmental Panel on Climate Change (IPCC) has sounded a more serious warning than at any time earlier. It has made references to unabated heating of the atmosphere and oceans, changing global water cycle, diminishing snow cover and more importantly, rising near sea level. It has underlined the need for a better-targetted, result-oriented and globally binding successor to the Kyoto Protocol.

Another report tided *Climate Change 2014: Impacts, Adaptation, and Vulnerability* details the impact climate change has had to date, future risks, and the opportunities for effective action to reduce these risks.

The report says that food security will be affected because of climate change. It adds that climate change impacts are projected to slow economic growth and make poverty reduction difficult.

The report says that production of major crops, including wheat, rice and maize, which are grown in tropical and temperate regions, will be negatively impacted if no adaptation measures are taken.

Climate change will impact key sectors which are responsible for economic growth, including energy, water, transport, and insurance, the report adds. Developing countries including India are likely to face impact on energy due to climate change.

The manufacturing sector will also be affected by climate change as the price and the quality of input will change. Labour productivity will fall in humid areas and demand for products will also be affected.

On 2 October a year ago, India moved to allay any international scepticism about its commitment to combating climate change by announcing its Intended Nationally Determined Contributions (INDCs). India is to reduce the emissions intensity of its gross domestic product by 33–35% by 2030 from 2005 levels, it said in an announcement. Now, by agreeing to ratify the Paris Agreement—which aims to contain the increase in earth's temperature to 2 degrees Celsius, and if possible 1.5 degrees Celsius, above pre-industrial levels—on the same day 2016, India will move further along this path. It's a hurried move—but because of diplomatic compulsions, New Delhi has chosen to couch pragmatism in the symbolism of Gandhi's birth anniversary.

The Paris agreement—which materialized at the 21st Conference of Parties of the United Nations Framework Convention on Climate Change—requires 55 countries accounting for at least 55% of global greenhouse gas emissions to ratify it in order for it to come into force. New Delhi had gone slow so far—but with the threshold likely to be reached without it in October, and the possibility thereafter of being cast in an obstructionist role diplomatically, it has understandably changed track.

But India still has good reason to be wary at the negotiating table because of three reasons. First, too much focus on incremental pollution by developing countries shifts the public attention away from the historical damage done by the developed countries during their years of rapid economic growth. Second, the per capita carbon footprint of India is still very low by global standards. Third, India needs coal as part of its energy mix right now, until new technologies emerge.

By retaining the "common but differentiated responsibilities" clause in the Paris Agreement, the world has acknowledged Indian concerns. However, as Shyam Saran, India's former foreign secretary and the prime minister's special envoy for climate change from 2007–10, points out doing away with the distinction between developing and developed nations dilutes the differentiation principle—which puts the onus on developed countries to take greater responsibility for reducing emissions. Undue pressure is already being exerted on India as one of the largest polluters of the world. Though third in terms of total emissions, it ranks 140th globally in terms of per capita emissions.

However, with its large coastline susceptible to rising sea levels and a population suffering from the visible consequences of climate change, such as heat waves, pollution and failed monsoons, the country couldn't have argued for its rights to development beyond a point. It therefore made a calculated choice considering its long-term costs. It embraced the cause of climate change but with a caveat—the availability of global finance and performance of other nations. The cabinet decision, implicitly emphasizes that the onus is on developed nations to fund and transfer technology to developing nations, besides making efforts to meet their domestic commitments.

India's strategy for combating climate change and achieving INDC targets is a multi-pronged one.

It aims at modifying the energy mix to a more sustainable, efficient and renewable one. Its ambitious solar plan—to expand solar capacity to 100 GW by 2022 from 8 GW in 2016, which incurred the ire of the US and eventually the World Trade Organization due to the priority it gave to domestic producers, is expected to make up almost 48% of the renewable energy capacity. The International Solar Alliance and the recently unveiled plan to subsidize domestic manufacturers will help but can only be sustained by a higher inflow of funds for solar projects.

The rest of the renewable energy capacity would be developed through a mixture of wind power, hydropower, biomass, waste to energy and nuclear power. Nuclear energy will form less than 4% of these clean energy commitments. Though it is unlikely that India's non-membership of the Nuclear Suppliers Group will significantly affect its plans, the membership could have enhanced it.

A significant challenge to these plans remains, as pointed out by India's chief economic adviser, Arvind Subramanian, arises out of, the fall in oil prices, which increases the opportunity cost of expansion of alternative energy sources. The country has dealt with the problem by imposing taxes on petroleum products. As opposed to the cap and trade system proposed by countries such as China, India uses Piguovian taxes as the alternative market approach to limiting emissions. But the issue of renewable energy being less competitive still persists.

Besides dealing with the problem at source, the country is also making efforts to bring down current greenhouse gas levels through the development of carbon sinks. However, this might prove to be a difficult task considering the land-scarcity problem. The Fourteenth Finance Commission made forest cover one of the criteria for devolution of funds from the Centre to incentivize states to engage in afforestation activities. It is estimated that around $6.9 billion will be transferred to states.

A 33–35% reduction in emission targets is not an unattainable one, if all these efforts come through. But that would require persistent efforts from both the global community and the Indian government. A clear vision of what the alternative would mean for the country and the world at large should help in sustaining these efforts.

18.12.1 Lack of Finance: A Serious Constraint

This is easier said than done. "Sustainability" in financial market jargon has a very different meaning to when it is used in development-speak. In the market, this term largely disregards issues pertaining to employment generation, poverty eradication, inclusive growth and environmental considerations. Instead, it is monomaniacal in enhancing the "basis points" of the returns it generates for the community it serves—with only perfunctory interest in the "ppm (parts per million)" of carbon (mitigated or released) associated with the deployment of finance, or the human development index (HDI) effects of investments.

The regulatory responsibilities and the fiduciary duties that drive the functioning of this community are focused mostly on protecting the interests of investors and consumers (of financial instruments and banking services) by de-risking the financial ecosystem. Together, these present two specific hurdles, both of which make it difficult for the world of money to serve the ambitions of the Paris Agreement. The first hurdle pertains to geography, more specifically political geography. And the second pertains to democracy, more specifically the politics of decision making.

Deleterious consequences. For instance, the major chunk of climate finance labelled as such finds it tedious to flow across borders. Thus, it is mostly deployed in the locality of its origin. This tendency is even starker for financial flows from the developed world to the developing and emerging world. An Organisation for Economic Co-operation and Development—Climate Policy Initiative (OECD—CPI) study found that "public and private climate finance mobilized by developed countries for developing countries reached $62 billion in 2014". A separate study by CPI estimated that global flow of climate finance crossed $391 billion in the same year—implying that only about 16% of all flows moved from developed to developing countries.

This represents the most significant "collective action problem" that confronts the global community on the issue of climate change. While there is a near universal recognition that (*a*) climate change is a global commons problem, (*b*) the least developed countries are likely to be most affected, and (*c*) significant infrastructure will need to be developed in emerging and developing countries to improve their low standard of living, the flow of money is (not surprisingly) blind to each of these. It recognizes political boundaries, responds to ascribed (and frequently arbitrary) ecosystem risks within these boundaries and flows to destinations and projects that enhance returns—as it was meant to.

The travails of this constrained flow of capital do not end here. In a recent discussion paper, it has been highlighted that even this modest cross-border flow, which also accounts for pledges and promises made, does not adhere to the "new and additional" criteria. Flows of conventional development finance and infrastructure finance are on occasion reclassified as climate finance. And on other occasions these conventional flows are cannibalized to generate climate finance. The size of the pie remains the same.

Unless we are able to increase the total amount of resources available to cater to both the development priorities and climate-friendly growth needs of emerging and developing economies, we may only be able to build a future that is both green and grim. Everywhere, low-income populations will underwrite a green future for a privileged few.

Additional finance for meaningful climate action may be generated by simultaneously working on three fronts as we move to 2020. Successful climate action will first and foremost be predicated on the domestic regulatory framework within each country. Currently, a slew of regulations, from the flow of international finance into the domestic economy to those related to debt and equity markets, disincentivize capital from investing in climate action. It is imperative for policymakers to get their own house in order and create financial market depth and instruments that allow savings to become investible capital even as they continue to demand a more climate-friendly international financial regime.

Second, there currently exists a vast pool of long-term savings—which can be labelled "lazy money". According to a recent International Monetary Fund report, much of this lies with pension, insurance and other funds, which have accumulated savings of approximately $100 trillion. Due to lack of political will and appropriate mechanisms, this money is neither invested in the climate agreement objectives nor in the sustainable development goals agreed to at the UN last year. This helps nobody. As a result of its inability to flow across borders, developed-world savers earn sub-par returns. And due to this source of finance remaining outside the climate purview, the investment gap in infrastructure, particularly in developing countries, has continued to increase. It now stands between $1 trillion and $1.5 trillion each year. Making this "lazy money" count will be extremely important.

And finally, it is time to bring the big boys controlling banking standards into the tent. The Basel III Accords, designed to create a more resilient international banking system through a suite of capital adequacy, leverage, and liquidity requirements, contribute little to global climate resilience. Given the dependency of emerging economies like India on commercial finance for capital-intensive projects, the Basel Accords need urgent review.

18.13 CARBON CREDITS AND CARBON TRADING

18.13.1 Concept

Carbon trading is part of the larger emission trading which is a method worked out to control pollution by using economic incentives. In emission trading, central authority such as a government or international body like the European Union sets a limit or cap on the amount of pollutants that can be emitted. The amount of emissions of pollutants that each company or group is allowed is worked out based on the overall cap for the country. This allowance to emit pollutants is called a 'credit'.

Where the pollutant emitted is carbon dioxide, it is called *carbon credit.*

If a company wants to exceed the emission level or cap set for it, it will have to buy the extra allowance or credit from those who pollute less. The idea is that this allows the total quantum of emissions to be controlled without having to micromanage emissions by each firm. Such transfers of allowances or carbon credits are called *carbon trading.*

In effect, therefore, the party buying the credit is being penalised for polluting by having to pay for the credit while the seller is being rewarded in monetary terms.

18.13.2 Need for Carbon Trading

The Kyoto Protocol assigns mandatory emissions limits for the reduction of greenhouse gases to signatory nations. The objective is to stabilise greenhouse gas concentration in the atmosphere at a level that would prevent interference with the climate system.

18.13.3 Trading Mechanism

Different markets have been set up for different emissions, especially for carbon trading, in different parts of the world such as greenhouse gas trading in the European Union or the market to reduce acid rain in the US or the regional markets for trade in nitrous oxide. The European Union Emission Trading Scheme, the world's only mandatory carbon trading programme, is the largest multi-national emission-trading scheme in the world. London's financial market has established itself as the centre of the carbon market.

18.13.4 Criticisms

Carbon trading has already come in for severe criticism. It is being argued that instead of policies that reduce emission, stricter regulation, oversight and penalties for polluters to control emission, trading is providing elaborate get-out clauses for the biggest polluters. All manners of loopholes and incentives are used by industry to exaggerate their expected emissions to receive more permits from governments, causing a drop in the price of carbon, thus creating further disincentive for industries to lower emissions at source. Also, licenses and credits will have no value without effective enforcement as facilities may find it for less expensive to corrupt inspectors than to purchase emission licences.

Shocking to See Pollution over India, China from Space, says Astronaut Kelly

Astronaut Scott Kelly, who has the distinction of having spent over a year in space, has said that the level of pollution in China and India is shocking.

"Seeing places like China and India, and the pollution that exists there almost all the time is quite shocking," Kelly said in a media appearance with US President Barack Obama in the Oval Office of the White House. "There was one day last summer, the summer of 2015, when I was in space I saw the eastern side of China was perfectly clear. And I'd never seen that before in all of my time in space, and I'd spent well over a year in space, total, at that point," he said.

"I could see all these cities that are—there's like over 200 cities in that part of China, with over a million people. And it was at dusk, and I could just, for the very first time, I was able to see them, and it was quite shocking," he added.

"I didn't really understand it until the next day I heard that the Chinese government had turned off a lot of the coal-fired power plants, stopped cars from running in that part of the country for this national holiday, and the sky had completely cleared," he added.

18.13.5 India's Position

Countries like India and China are not required to reduce emission under the present Kyoto Protocol. They are cashing in on their carbon credits and many industries in these countries are already raking in profits by selling their carbon credits in the international emissions market that closes by 2012, when the Treaty expires.

Nevertheless, as India continues to add power-generation capacity, it will need to control its carbon emissions. By 2017, 80 per cent of India's power supply will be generated by fossil fuels, which is likely to double the country's CO_2 emission from 1.2 billion tonnes presently to about 2.4 billion tonnes.

18.14 GREEN INDIA MISSION

The forestry mission is under the umbrella of the National Action Plan on Climate Change. The mission recognises that climate change will seriously affect and alter the distribution, type and quality of natural resources and associated livelihoods of the people. It acknowledges the influences that the forestry sector has on environmental amelioration through climate mitigation, food security, water security, biodiversity conservation and livelihood security of forest-dependent communities. Under this mission, afforested areas will go up to 20 million hectare by 2020, reducing greenhouse gas emissions by 6.35%. The mission also has targets for different types of forests, such as moderately dense forests, degraded forests, grassland and scrub, mangroves, urban forest land and even degraded agricultural and fallow lands.

The mission is looking at key roles for local communities and decentralised governance for the implementation.

The critics, however, point out that the mission is too vague to be meaningful. The operational guidelines would have to be spelled out more clearly.

18.15 ENVIRONMENTAL ACCOUNTING SYSTEM

The government has set up an expert group to provide a road map for Green National Accounting system to evaluate the impact of economic growth on environment.

18.15.1 What is Environmental Accounting?

It is a method to measure the consumption of the environment/natural resources as a country strives for higher growth. The future cost of depletion of natural resources and other damages caused to the environment are excluded from conventional output measures such as the Gross Domestic Product. Environmental accounting builds on the conventional accounting methodology to reflect the costs of economic development.

18.15.2 Why is It Important?

Most production tends to use up natural resources directly or indirectly and also tends to impact the environment in some way or other, mostly through pollution. It is widely-acknowledged amongst environmentalists and economists that cost of damage to the environment needs to be factored in to decide whether a particular economic activity is beneficial to the mankind or not.

Also, natural resources are indispensable for most economic activities and their depletion would in the longer run reduce the productive capacity of an economy. Environmental accounting is expected to give a better idea of the cost benefits involved, allowing for better decision-making.

18.15.3 How Does It Work?

Environmental accounting is possible both at a corporate level as well as at a national-international level. At a corporate level, it requires the estimation of the damage done to the environment by a particular industry and the investments made to minimise such environmental damage. At a national level, environmental accounting requires treating environmental resources (forests, water bodies, etc.) as a stock of natural capital like any other productive asset. Any depreciation in this stock must be accounted for and accordingly deducted from national output. This revised measure of national output is also referred to as Green GDP.

18.15.4 What are the Difficulties Involved?

It is hard to quantify the depletion of natural resources and even harder to put a value to it, or monetise such costs. Most environmental resources are not traded in markets. This makes it difficult to value benefits accruing from environmental goods such as fuel wood, fish/fruit collected for consumption and services such as protection from soil erosion and land submergence. Similarly, it is hard to put a value to the damage done to the environment in the process of production. It is for these problems that no country in the world has yet been able to devise a fool proof methodology for environmental accounting.

18.16 LAND BOOM

Since 2000, prices of agricultural land have risen by anywhere between three-fold and 100-fold. This is the beginning of a situation in which food production is challenged, and where traditional livelihoods are ceding to newer ones.

18.16.1 A Trinity Converges

For the longest time, farmland markets were comatose. Land ceiling laws, designed to prevent concentration of land ownership, were one reason. Another reason was the limited reason to buy land. If land's value is a measure of its future income, and if the future use is not dramatically different from its current use, a sale is possible only if a buyer's evaluation of the discounted future income stream is more than the buyer's valuation of the same. The wheels are turning faster on both counts. New buyers, with a different assessment of value, are entering the market.

The spike in farmland prices is leaving its imprimatur on rural India. It is giving farmers seeking to leave agriculture an exit option. It is also pulling an unknown quantum of land out of agriculture. As land rates rise, farmers are unable to buy farmland in their own villages. As such, it is reshaping the ownership of land, it's all coming about because a trinity is converging on it—investors looking to buy, farmers looking to exit agriculture, and politicians and their associates looking to create a marketplace for such transactions.

These buyers are finding willing sellers in farmers. In rural areas, agriculture is not the most important component any longer. It now accounts, as per NSSO numbers, for just 33% of the rural economy. In the last 10 years, one can see a lot of farmers become wage labourers. This is not only because they sold their land out of distress. Some of them are starting to feel they are better off working as labour than putting money into farming, with its uncertain returns.

A small farmer with an acre of land will get 30 bags of paddy. At ₹2,000 each, that's a gross income of ₹60,000. But net of costs, his net income will probably be closer to ₹20,000. Even this is subject to weather risk—and price risk for crops not supported by minimum assured government prices. The

59th NSSO report asked farmers if they wanted to leave agriculture. 40% said 'yes'." That was in 2002. Since then, the drift has only continued.

18.16.2 A Source of Political Rent

Politicians, who created a market out of matching buyers and sellers, opened a third flank. In the last 10 years, the cost of fighting elections has shot up, and politicians are using the land boom to subsidise their campaigns. Besides parking their own money in land, politicians, through associates, are making a killing by positioning themselves in between real estate companies and the state. That unlocks the land commercial or residential.

18.17 ISSUE OF LAND ACQUISITION

With increasing population, explosive urbanisation and growing industrialistion land resources are increasingly coming under stress. The issue merits serious attention and calls for satisfactory solution.

Compulsory land acquisition continues to foment profound distress and often violent resistance in many parts of the country. At the heart is a colonial law of 1894, which equipped government with powers to compulsorily acquire land at low costs. This was widely used to coercively obtain land from millions of usually impoverished, and often tribal, land-holders, for large dams and public infrastructure projects. Their pauperisation led to democratic as well as militant struggles which rocked many parts of the country.

But public discontent grew further in the last two decades, when governments used this power to compulsorily acquire arable lands for private industry and real estate developers. A politically volatile — and ethically complex — debate has grown around whether governments should use the power vested in it for public good, to force landowners to sell their land to private industry.

There is a larger context to it, including concerns of food security. The policy needs be based on a fair and just reading of the Doctrine of Eminent Domain. When the state recognises private property, the private owner is the absolute title holder. However, the state is the paramount title holder. Should it need a piece of land for a clear and apparent public purpose—building a highway, for instance — it can resort to land acquisition. Of course, this should be done with generous compensation and not just a textual interpretation of 'market rate'.

The Doctrine of Eminent Domain cannot be misused to acquire land for favoured industrialists or real estate companies. That would be a gross misuse. As we saw in West Bengal in the Left Front years, it would lead to crony capitalism and sweetheart deals. If private players want to build factories or condominiums, they must buy from the farmer directly.

However, it is not as if private companies should be free to buy agricultural land at will. As a pre-requisite to a new land acquisition law, India needs a proper, digital and easily available (it must be online, for instance) map of its entire land area. This map must establish exactly which tracts of land are fertile and suitable for multi-crop cultivation. Such land is a national asset. It is critical to growing food for our 1.2 billion people and serves as a bulwark for food security. Therefore, it must be a 'no go' area for industry.

One view is that governments should not coerce land-owners to part with their lands for profit-led corporate entities. Let private industry negotiate with land-owners, and give them what they demand, if they voluntarily agree to sell their lands. The problem with this view is that it would actually amount to exposing powerless farmers to highly unequal negotiations with powerful corporate bodies, which might arm-twist or short-change them. In addition, a just land acquisition law needs to recompense agricultural workers and tenants; provide land for land, jobs and a house. None of these would be on offer by unregulated purchases by private industry.

A fair deal for all persons who will be displaced by industry can be ensured only if firstly all or most affected persons first freely consent to the purchase. It should then be mandatory for industry to approach government, at least above a certain threshold and for public purpose; and for government to ensure that they get paid fairly that the landless and tenants are also compensated, and all are humanely rehabilitated.

The country is in the throes of potentially explosive competing land hunger of peasants and large industry. It is the duty of the state to ensure that displacement is minimised, and all displaced persons get a fair deal. It cannot do this by just standing by and hoping markets will protect the poor.

18.17.1 The Right to Fair Compensation, Resettlement, Rehabilitation and Transparency in Land Acquisition Act, 2013

The government has enacted the Land Acquisition Act. (see Box 18.1)

An admirable job has been done vis-à-vis upholding the economic interest of families affected by land acquisition. It is imperative that this position be counterbalanced in the government by forces that root for planned development also. State-wide stock-taking of land banks, digitisation of land records, land-use planning and resultant zoning of lands for industrial, urban and infrastructure use must be made compulsory.

18.14.1 Critical Assessment

A. India Inc in Troubled Land?

The Land Acquisition Act is a contentious legislation. The main problems facing industry are as follows:

Box 18.1: Key Points

Purpose:

Address concerns of farmers and those who have lost livelihood; facilitate land acquisition for industrialization, infrastructure and urbanisation.

Scope Land can be acquired by government:

- For its own use
- To ultimately hand over the land to private companies for public purpose
- For immediate use by private companies for public purpose

Definition of public purpose

When 80% consent of affected families is not needed

- Strategic purposes
- Infrastructure and industry
- Natural calamities
- Rehabilitation and resettlement
- Land for railways, highways, ports, power and irrigation purposes

When 80% consent clause is applicable

- For private companies for public purposes

Compensation for rural areas

The compensation would be not less than four times the original market value

For urban areas

Not less than twice that of the market value

Multi cropped land

Acquisition up to 5% allowed but with riders

Rehabilitation and resettlement

- Apart from subsistence allowance, 20% of the appreciated land value within 10 years to be shared with the original owner
- For those who have lost livelihood, apart from subsistence and a one-time resettlement allowance, mandatory employment for one member per affected family, or ₹2 lakh
- Separate allowances for scheduled castes and scheduled tribes
- The provisions will be applicable even when private companies purchase land, if it is more than 100 acres in rural and 50 acres in urban areas

Retrospective effect

Applicable when the award has not been made under the 1894 Land Acquisition Act

OR

Where the possession of land has not been taken, regardless of whether the award has been made.

Land acquisition for a public purpose applies to:

(*i*) Infra projects, excluding private hospitals, schools and hotels
(*ii*) Agro-industry projects
(*iii*) Industrial corridors, mining, notified manufacturing zones
(*iv*) Public educational and research institutions;
(*v*) Sports, healthcare, transport projects.

So, if you add up the minimum periods for each step, and even assuming there are no court cases and authorities don't extend some periods, as the bill allows them to, it will still take 50 months for a company to get a piece of land.

The primary criticism of LARR rests in its pricing approach. LARR envisions an effective price of land that combines its purchase cost (for land owners) with rehabilitation and resettlement costs (for livelihood losers). The purchase cost would be double the market price in urban settings, and quadrapuled in rural settings. This multiplication of market price is bound to create a new class of windfall gainers, who, by virtue of ownership of land, stand to multiply their incomes several times over their lifetime earnings. The approach is blind to the way land markets function. It does nothing for the constituency of the historically dispossessed and displaced such as Dalits and Adivasis (primarily users of Common property resources), who lack ownership.

B. Other Issues

These can be briefly stated as follows:

On acquisition of agricultural land. The Act says maximum 5% of multi-crop land can be acquired in a district, provided equivalent wasteland is developed.

On displacement question. The Act doesn't seem to have too many answers to the trauma of displacement.

Compensation issue. The Act increases compensation significantly. But not clear how problems of delays and corruption in payment process will be taken care.

Acquisition for pvt cos. Critics say there's no reason why govt should use sovereign authority to acquire land for private entities. Present Act provides for 'partial' state acquisition for companies.

Private purchase. Act needs to better address imbalance of power between those acquiring land and those whose land being acquired, unfair deal can't be ruled out.

Food security and land use. Change of land use from agricultural to non-agricultural ought to have been addressed better and should've addressed the issue of food security.

Public purpose. Ought to have been defined better. Babudom decides and interprets each case, open to misuse.

Rehabilitation. 'Land for land' principle figures only in case of irrigation projects.

The industry has expressed apprehensions about the Act especially the R&R (rehabilitation and resettlement) component, saying it is impracticable while the civil society has criticised it for not giving any role to the community.

Business chamber, FIICI, which had opposed many of the conditions in the draft Act, said a central law on land acquisition was welcome. Finally, cracking down on politically coddled land mafias is imperative. Else, the enforcement of any land acquisition law against manipulation will prove daunting.

We can characterise the Indian State as a 'giving state' and a 'taking state'. The 'giving state' brought in half- willed land reforms that led to marginal redistribution. The 'taking state' indulged in regressive redistribution', where the poor were dispossessed and the benefits given away to the non-poor and less poor (through power subsidies as a consequence of thermal plants, irrigation projects through dams). The State has been the egregious land taker leading to change in land use of over 1 million acres leaving over 60 million people affected.

18.18 NATIONAL LAW ON WATER

Devastating floods and widespread water scarcity in most urban centres are grim pointers to India's mismanagement of its water resources. The flawed strategies pursued so far focused more on exploitation than on its conservation, storage and sustainable use. Going by official estimates, the country receives, on an average, about 4,000 billion cubic metres (BCM) of rainwater annually. However, a sizable part of it is allowed to flow down wastefully to the oceans for want of capacity to impound it in surface ponds or guide it down to the underground water reserves. Even if the country's 84 major reservoirs get filled up to capacity, no more than 154.4 BCM of water can be held in them. Other water bodies, natural or man-made, dotted all over the country, too, cannot hold much water due to a lack of proper maintenance and unchecked siltation. Percolation of rainwater down to underground aquifers is waning because of denudation of forest and vegetative cover. As a result, the total conserved water fails to meet even a few months demand.

This apart, the bitter truth also is that listing of water as a state subject under the Constitution was a historic blunder that had disallowed the management of water as a dynamic common resource. Most rivers and water streams, as well as groundwater aquifers, cut across state boundaries. Giving a free hand to states to meddle with them the way they want is unwarranted and gives rise to needless inter-state water sharing disputes.

Moreover water is not treated as an economic good. Nor is it priced to make the consumer realise its scarcity value. This is true for its agricultural as well as industrial and domestic use. Groundwater, too, continues to be treated as virtually the personal property of land-owners under archaic Rajera legislation; the Indian Easements Act of 1882. In the rural belt, where the bulk of groundwater is consumed, no limits have been imposed on its mining; in urban areas, enforcement of rules is poor. In fact, on the contrary, many states are subsidising power to encourage its indiscriminate use, causing rapid depletion of ground water.

Such outmoded policies and statutory provisions, obviously, need to change. Promotion of rainwater harvesting is unavoidable to rebuild fast-shrinking underground water reserves. This can be done in the urban areas by collecting and preserving rainwater from all open spaces, including those along roads, and by mandating rooftop water harvesting in large building complexes. Some cities, notably Chennai, have overcome their acute water stress through this approach. In the rural belt, on the other hand, rainwater harvesting needs to be done on a watershed basis, regardless of block, district or state boundaries. The check dams for this purpose should, preferably, be put up prior to the onset of the monsoon and not during a deficient monsoon season as is the case this year. Without a national attempt to rationalise water and power tariffs, impose regulation and build more infrastructure, India's water woes will worsen.

In view of these considerations the government has set up a committee under the chairmanship of Y.K. Alagh.

18.18.1 Alagh Committee on National Law on Water

The Government has set up a committee under the chairmanship of Y.K. Alagh to examine the need for a national law on water.

Why is a national law on water necessary? An earlier committee constituted on the subject stated the following measures:

1. Under the Indian Constitution water is primarily a State subject, but it is an increasingly important national concern in the context of:
   - (*a*) the judicial recognition of the right to water as a part of the fundamental right to life;
   - (*b*) the general perception of an imminent water crisis, and the dire and urgent need to conserve this scarce and precious resource;
   - (*c*) the severe and intractable inter-use and inter-State conflicts;
   - (*d*) the pollution of rivers and other water sources, turning rivers into sewers or poison and contaminating aquifers;
   - (*e*) the long-term environmental, ecological and social implications of projects to augment the availability of water for human use;
   - (*f*) the equity implications of the distribution, use and control of water;
   - (*g*) the international dimensions of some of India's rivers; and
   - (*h*) the emerging concerns about the impact of climate change on water and the need for appropriate responses at local, national, regional, and global levels.

   It is clear that the above considerations cast several responsibilities on the Central government, apart from those of the State governments. Given these and other concerns, the need for an overarching national water law is self-evident.
2. Several States are enacting laws on water and related issues. These can be quite divergent in their perceptions of and approaches to water. Some divergences from State to State may be inevitable and acceptable, but extreme and fundamental divergences will create a very muddled situation. A broad national consensus on certain basics seems very desirable.

3. Different State governments tend to adopt different legal positions on their rights over the waters of a river basin that straddles more than one State. Such legal divergences tend to render the resolution of inter-State river-water conflicts extremely difficult. A national statement of the general legal position and principles that should govern such cases seems desirable.
4. Water is one of the most basic requirements for life. If national laws are considered necessary on subjects such as the environment, forests, wildlife, biological diversity, etc., a national law on water is even more necessary. Water is as basic as (if not more basic than) those subjects.
5, Finally, the idea of a national water law is not something unusual or unprecedented. Many countries in the world have national water laws or codes, and some of them (for instance, the South African National Water Act of 1998) are widely regarded as very enlightened. The considerations behind those national codes or laws are relevant to India as well, although the form of a water law for India will clearly have to be guided by the nature of the Indian Constitution and the specific needs and circumstances of this country.

18.19 GREEN TECHNOLOGY FUND

It is recommended to establish a Green Technology Fund which will support projects, programs, policies and other activities to promote green initiatives in the country. This fund can be in the form of a focused investment vehicle for companies investing in green technology/processes and companies engaged in environmentally supportive businesses such as alternative energy, green transport, water and waste management, and sustainable living. It is recommended to allocate an amount of ₹25,000 crores towards the "Green Technology Fund" over a period of five years.

DU College to Open 'Green' Hostel

Maitreyi College, University of Delhi (DU), plans to open a 'green' hostel and science block by 2018. Yogesh Tyagi, vice chancellor, laid the foundation stone of the hostel recently. Both structures will have heat-reflective white paint for thermal insulation on the rooftop, a rainwater harvesting system and energy-saving light fittings. They will also have ramps, toilets, tactile paths and elevators for users with disabilities.

18.19.1 Promotion of Green Technology

The Green Technology Upgradation Fund specifically aims at facilitating the implementation of energy efficient and environment friendly processes or equipments. This fund can be disbursed through nodal banks and prime lending agencies. This fund should be available for implementing green technologies in all kinds of industrial sectors.

18.19.2 Investments in R&D

Investments in R&D are necessary to promote and ensure widespread implementation of Cleaner Production technologies in different sectors, through the allocation of funds from the Green Technology Fund. The funding can be categorized for different activities under R&D based on the sector specific need identification, designation of an appropriate agency for conducting R&D and annual allocation of funds to carry out R&D activities:

Funding to companies can be provided in the form of:

- *Equity*—With an upper limit on the investment amount, with suitable exit clauses
- *Grants*—In the form of matching grants for industry, with an upper limit on the amount provided, with an optional clause to include academia and/or specific research institutions in project implementation.

19

Green Economy Issues in India

19.1 INTRODUCTION

The dialogue on effects of economic growth on environment has moved on from 'sustainable development' to green economy. While the concept of 'sustainable development' enjoins upon the industrial countries to reduce their consumption of natural resources for future generations, the green economy would imply no such compulsion and, on the contrary, open up a huge market for products and technologies which the developed world could sell to developing countries. In the present volume we have sufficiently elaborated the different green economy issues being faced by the global economy and the policy choices available to all nations. In this concluding chapter we will focus only on some important issues confronting the economic policy-makers in India.

19.2 ENVIRONMENTAL ISSUES IN INDIA

There are many environmental issues in India. Air pollution, water pollution, garbage, and pollution of the natural environment are all challenges for India. The situation was worse between 1947 through 1995. According to data collection and environment assessment studies of World Bank experts, between 1995 through 2010, India has made one of the fastest progress in the world, in addressing its environmental issues and improving its environmental quality. Still, India has a long way to go to reach environmental quality similar to those enjoyed in developed economies. Environmental issues are one of the primary causes of disease, health issues and long term livelihood impact for India.

19.2.1 Law and Policies

British rule of India saw several laws related to environment. Amongst the earliest ones were Shore Nuisance (Bombay and Kolaba) Act of 1853 and the Oriental Gas Company Act of 1857. The Indian Penal Code of 1860, imposed a fine on anyone who voluntarily fouls the water of any public spring or reservoir. In addition, the Code penalised negligent acts. British India also enacted laws aimed at controlling air pollution. Prominent amongst these were the Bengal Smoke Nuisance Act of 1905 and the Bombay Smoke Nuisance Act of 1912. Whilst these laws failed in having the intended effect, British-enacted legislations pioneered the growth of environmental regulations in India.

Upon independence from Britain, India adopted a constitution and numerous British-enacted laws, without any specific constitutional provision on protecting the environment. India amended its constitution in 1976. Article 48(A) of Part IV of the amended constitution, read: The State shall endeavour to protect and improve the environment and to safeguard the forests and wildlife of the country. Article 51 A(g) imposed additional environmental mandates on the Indian state.

Other Indian laws from recent history include the Water (Prevention and Control of Pollution) Act of 1974, the Forest (Conservation) Act of 1980, and the Air (Prevention and Control of Pollution) Act of 1981. The Air Act was inspired by the decisions made at Stockholm Conference. The Bhopal gas tragedy triggered the Government of India to enact the Environment (Protection) Act of 1986. India has also enacted a set of Noise Pollution (Regulation and Control) Rules in 2000.

In 1985, Indian government created the Ministry of Environment and Forests. This ministry is the central administrative organisation in India for regulating and ensuring environmental protection.

Despite active passage of laws by the central government of India, the reality of environmental quality mostly worsened between 1947 and 1990. Rural poor had no choice, but to sustain life in whatever way possible. Air emissions increased, water pollution worsened, forest cover decreased.

Starting in the 1990s, reforms were introduced. Since then, for the first time in Indian history, major air pollutant concentrations have dropped in every 5-year period. Between 1992 and 2010, satellite data confirms India's forest coverage has increased for the first time by over 4 million hectares, a 7% increase.

19.2.2 Possible Causes

Some have cited economic development as the cause regarding the environmental issues. It is suggested that India's growing population is the primary cause of India's environmental degradation. Systematic studies challenge this theory. Empirical evidence from countries such as Japan, England and Singapore, each with population density similar or higher than India, yet each enjoying environmental quality vastly superior than India, suggests population density may not be the only factor affecting India's issues.

19.2.3 Major Issues

Major environmental issues are forest and agricultural degradation of land, resource depletion (such as water, mineral, forest, sand, and rocks), environmental degradation, public health, loss of biodiversity, loss of resilience in ecosystems, livelihood security for the poor. Floods are a significant environmental issue for India. It causes soil erosion, destruction of wetlands and wide migration of solid wastes.

The major sources of pollution in India include the rapid burning of fuelwood and biomass such as dried waste from livestock as the primary source of energy, lack of organised garbage and waste removal services, lack of sewage treatment operations, lack of flood control and monsoon water drainage system, diversion of consumer waste into rivers, cremation practices near major rivers, government mandated protection of highly polluting old public transport, and continued operation by Indian government of government owned, high emission plants built between 1950 and 1980. India's population growth adds pressure to environmental issues and its resources.

19.2.4 Population Growth and Environmental Quality

There is a long history of study and debate about the interactions between population growth and the environment. According to a British thinker Malthus, for example, a growing population exerts pressure on agricultural land, causing environmental degradation, and forcing the cultivation of land of poorer as well as poorer quality. This environmental degradation ultimately reduces agricultural yields and food availability, causes famines and diseases and death, thereby reducing the rate of population growth.

Population growth, because it can place increased pressure on the assimilative capacity of the environment, is also seen as a major cause of air, water, and solid-waste pollution. The result, Malthus theorised, is an equilibrium population that enjoys low levels of both income and environmental quality. Malthus suggested positive and preventative forced control of human population, along with abolition of poor laws.

Malthus theory, published between 1798 and 1826, has been analysed and criticised ever since. The American thinker Henry George, for example, observed with his characteristic piquancy in dismissing Malthus: "Both the jayhawk and the man eat chickens; but the more jayhawks, the fewer chickens, while the more men, the more chickens." Similarly, the American economist Julian Lincoln Simon criticised Malthus's theory. He noted that the facts of human history have proven the predictions of Malthus and of the

Neo-Malthusians to be flawed. Massive geometric population growth in the 20th century did not result in a Malthusian catastrophe. The possible reasons include: increase in human knowledge, rapid increases in productivity, innovation and application of knowledge, general improvements in farming methods (industrial agriculture), mechanisation of work (tractors), the introduction of high-yield varieties of wheat and other plants (Green Revolution), the use of pesticides to control crop pests.

More recent scholarly articles concede that whilst there is no question that population growth may contribute to environmental degradation, its effects can be modified by economic growth and modern technology. Research in environmental economics has uncovered a relationship between environmental quality, measured by ambient concentrations of air pollutants and per capita income. This so-called environmental Kuznets curve shows environmental quality worsening up until about $5,000 of per capita income on purchasing parity basis, and improving thereafter. The key requirement, for this to be true, is continued adoption of technology and scientific management of resources, continued increases in productivity in every economic sector, entrepreneurial innovation and economic expansion.

Other data suggest that population density has little correlation to environmental quality and human quality of life. India's population density, in 2011, was about 368 human beings per square kilometre. Many countries with population density similar or higher than India enjoy environmental quality as well as human quality of life far superior than India. For example: Singapore (7148/km^2), Hong Kong (6349/km^2), South Korea (487/km^2), Netherlands (403/km^2), Belgium (355/km^2), England (395 /km^2) and Japan (337/km^2).

19.2.5 Water Pollution

India has major water pollution issues. Discharge of untreated sewage is the single most important cause for pollution of surface and ground water in India. There is a large gap between generation and treatment of domestic waste water in India. The problem is not only that India lacks sufficient treatment capacity but also that the sewage treatment plants that exist do not operate and are not maintained. The majority of the government-owned sewage treatment plants remain closed most of the time due to improper design or poor maintenance or lack of reliable electricity supply to operate the plants, together with absentee employees and poor management. The waste water generated in these areas normally percolates in the soil or evaporates. The uncollected wastes that accumulate in the urban areas cause unhygienic conditions and release pollutants that leaches to surface and groundwater.

According to a World Health Organization study, out of India's 3,119 towns and cities, just 209 have partial sewage treatment facilities, and only 8 have full wastewater treatment facilities. Over 100 Indian cities dump untreated sewage directly into the Ganges River. Investment is needed to bridge the gap between 29000 million litre per day of sewage India generates, and a treatment capacity of mere 6000 million litre per day.

Other sources of water pollution include agriculture run off and small scale factories along the rivers and lakes of India. Fertilizers and pesticides used in agriculture in northwest have been found in rivers, lakes and ground water. Flooding during monsoons worsens India's water pollution problem, as it washes and moves all sorts of solid garbage and contaminated soils into its rivers and wetlands.

19.2.6 Water Resources

According to NASA groundwater declines are highest on Earth between 2002 and 2008 in northern India. Agricultural productivity is dependent on irrigation. A collapse of agricultural output and severe shortages of potable water may influence 114 million residents in India. In July, 2012, about 670 million people or 10% of the world's population lost power blame on the severe drought restricting the power delivered by hydroelectric dams.

19.2.7 Air Pollution

A rural stove uses biomass cakes, fuelwood and trash as cooking fuel. Surveys suggest over 100 million households in India use such stoves (chullahs) every day, 2–3 times a day. It is a major source of air pollution in India, and produces smoke and numerous indoor air pollutants at concentrations 5 times higher than coal. Clean burning fuels and electricity are unavailable in rural parts and small towns of India because of poor rural highways and limited energy generation infrastructure.

Air pollution in India is a serious issue with the major sources being fuelwood and biomass burning, fuel adulteration, vehicle emission and traffic congestion. Air pollution is also the main cause of the Asian brown cloud, which is causing the monsoon to be delayed. India is the world's largest consumer of fuelwood, agricultural waste and biomass for energy purposes. Traditional fuel (fuelwood, crop residue and dung cake) dominates domestic energy use in rural India and accounts for about 90% of the total. In urban areas, this traditional fuel constitutes about 24% of the total. Fuel wood, agri waste and biomass cake burning releases over 165 million tonnes of combustion products into India's indoor and outdoor air every year. These biomass-based household stoves in India are also a leading source of greenhouse emissions contributing to climate change.

The annual crop burning practice in northwest India, north India and eastern Pakistan, after monsoons, from October to December, are a major seasonal source of air pollution. Approximately 500 million tons of crop residue is burnt in open, releasing smoke, soot, NO_x, SO_x, PAHs and particulate matter into the air. This burning has been found to be a leading cause of smog and haze problems through the winter over Punjab, cities such as Delhi, and major population centers along the rivers through West Bengal. In other states of India, rice straw and other crop residue burning in open is a major source of air pollution.

Vehicle emissions are another source of air pollution. Vehicle emissions are worsened by fuel adulteration and poor fuel combustion efficiencies from traffic congestion and low density of quality, high speed road network per 1000 people.

On per capita basis, India is a small emitter of carbon dioxide greenhouse. In 2009, IEA estimates that it emitted about 1.4 tons of gas per person, in comparison to the United States' 17 tons per person, and a world average of 5.3 tons per person. However, India was the third largest emitter of total carbon dioxide in 2009 at 1.65 Gt per year, after China (6.9 Gt per year) and the United States (5.2 Gt per year). With 17 percent of world population, India contributed some 5 percent of human-sourced carbon dioxide emission; compared to China's 24 percent share.

The Air (Prevention and Control of Pollution) Act was passed in 1981 to regulate air pollution and there have been some measurable improvements. However, the 2012 Environmental Performance Index ranked India as having the poorest relative air quality out of 132 countries.

19.2.8 Solid Waste Pollution

Trash and garbage disposal services, responsibility of local government workers in India, are ineffective. Solid waste is routinely seen along India's streets and shopping plazas.

Trash and garbage is a common sight in urban and rural areas of India. It is a major source of pollution. Indian cities alone generate more than 100 million tons of solid waste a year. Street corners are piled with trash. Public places and sidewalks are despoiled with filth and litter, rivers and canals act as garbage dumps. In part, India's garbage crisis is from rising consumption. India's waste problem also points to a stunning failure of governance. The tourism regions in the country mainly hill stations are also facing this issue in the recent years.

In 2000, India's Supreme Court directed all Indian cities to implement a comprehensive waste-management programme that would include household collection of segregated waste, recycling and composting. These directions have simply been ignored. No major city runs a comprehensive programme of the kind envisioned by the Supreme Court.

Indeed, forget waste segregation and recycling directive of the India's Supreme Court, the Organisation for Economic Cooperation and Development estimates that up to 40 per cent of municipal waste in India remains simply uncollected. Even medical waste, theoretically controlled by stringent rules that require hospitals to operate incinerators, is routinely dumped with regular municipal garbage. A recent study found that about half of India's medical waste is improperly disposed of.

Municipalities in Indian cities and towns have waste collection employees. However, these are unionised government workers and their work performance is neither measured nor monitored.

Some of the few solid waste landfills India has, near its major cities, are overflowing and poorly managed. They have become significant sources of greenhouse emissions and breeding sites for disease vectors such as flies, mosquitoes, cockroaches, rats, and other pests.

In 2011, several Indian cities embarked on waste-to-energy projects of the type in use in Germany, Switzerland and Japan. For example, New Delhi is implementing two incinerator projects aimed at turning the city's trash problem into electricity resource. These plants are being welcomed for addressing the city's chronic problems of excess untreated waste and a shortage of electric power. They are also being welcomed by those who seek to prevent water pollution, hygiene problems, and eliminate rotting trash that produces potent greenhouse gas methane. The projects are being opposed by waste collection workers and local unions who fear changing technology may deprive them of their livelihood and way of life.

On the bright side, the recently notified Solid Waste Management Rules, 2016, emphasize segregation of waste at source and greater decentralized processing of biodegradable waste. They also mandate the integration of kabadiwallahs and ragpickers into the formal economy. This is vitally important since most of the waste-sorting and recycling is done by informal workers before the unrecyclable waste is transported to a landfill. They bear the brunt of our failure to segregate our household waste; they do so under hazardous conditions and for negligible pay.

A landfill is a fracture in the stories we tell about our cities. The home page of the urban development ministry website carries a permanent declaration—"The growth story of India shall be written on the canvas of planned urban development." Almost as an afterthought, there is a second declaration, "And shall be scripted through the instrument of planned mobility."

What stories come out of landfills? They are reports of the landfill fires that continually smoulder, the deaths of ragpickers, the tonnage of waste being dumped, the dreary profiles of municipal waste. They remain narrative versions of things that you don't look at directly or for too long. The presence and persistence of landfills ought not to be taken lightly when contending with air pollution.

We need to move towards environmentally sound policymaking, and away from the formulaic inter-governmental squabble that seems to pass for crisis management. Without this, a city, as Jeet Thayil describes in Narcopolis, isn't much more than "a large accumulation of small defeats".

19.2.9 Noise Pollution

Noise pollution or noise disturbance is the disturbing or excessive noise that may harm the activity or balance of human or animal life. Noise-wise India can be termed as the most polluted country in the world. The source of most outdoor noise worldwide is mainly caused by machines and transportation systems, motor vehicles, aircraft, and trains. In India the outdoor noise is also caused by loud music during festival seasons.Outdoor noise is summarized by the word environmental noise. Poor urban planning may give rise to noise pollution, since side-by-side industrial and residential buildings can result in noise pollution in the residential areas.

Indoor noise can be caused by machines, building activities, and music performances, especially in some workplaces. Noise-induced hearing loss can be caused by outside (*e.g.* trains) or inside (*e.g.* music) noise.

High noise levels can contribute to cardiovascular effects in humans and an increased incidence of coronary artery disease. In animals, noise can increase the risk of death by altering predator or prey detection and avoidance, interfere with reproduction and navigation, and contribute to permanent hearing loss.

The Supreme Court of India which is in New Delhi gave a significant verdict on noise pollution in 2005. Unnecessary honking of vehicles makes for a high decibel level of noise in cities. The use of loudspeakers for political purposes and for sermons by temples and mosques makes noise pollution in residential areas worse.

In January, 2010, Government of India published norms of permissible noise levels in urban and rural areas.

19.2.10 Land or Soil Pollution

In March, 2009, the issue of Uranium poisoning in Punjab attracted press coverage. It was alleged to be caused by fly ash ponds of thermal power stations, which reportedly lead to severe birth defects in children in the Faridkot and Bhatinda districts of Punjab. The news reports claimed the uranium levels were more than 60 times the maximum safe limit. In 2012, the Government of India confirmed that the ground water in Malwa belt of Punjab has uranium metal that is 50% above the trace limits set by the United Nations' World Health Organization. Scientific studies, based on over 1000 samples from various sampling points, could not trace the source to fly ash and any sources from thermal power plants or industry as originally alleged. The study also revealed that the uranium concentration in ground water of Malwa district is not 60 times the WHO limits, but only 50% above the WHO limit in 3 locations. This highest concentration found in samples was less than those found naturally in ground waters currently used for human purposes elsewhere, such as Finland. Research is underway to identify natural or other sources for the uranium.

19.2.11 Greenhouse Gas Emissions

India was the third largest emitter of carbon dioxide, a major greenhouse gas, in 2009 at 1.65 Gt per year, after China and the United States . With 17 percent of world population, India contributed some 5 percent of human-sourced carbon dioxide emission; compared to China's 24 percent share. On per capita basis, India emitted about 1.4 tons of carbon dioxide per person, in comparison to the United States' 17 tons per person, and a world average of 5.3 tons per person.

19.3 ENVIRONMENTAL DAMAGE

Environmental degradation and diseases are responsible for 23% of child mortality, according to a World Bank report. Moreover, the costs of environmental damage are equivalent to 5.7% of the country GDP (in 2009). However, the report says, India can make green growth a reality by putting in place strategies to reduce the damage at a minimal cost of 0.02% to 0.04% of its average annual GDP growth rate.

In short, the present mode of development is wholly unsustainable. The clash between growth imperatives and the environment will become sharper. Therefore, a decisionmaking framework that anticipates this and puts in place a mechanism that reconciles growth, land acquisition and environmental protection is good for growth, social justice and the environment. The key lies in five elements:

(*i*) full provision of data and information on project design and impact;

(*ii*) stakeholder engagement from an early stage, where necessary in public hearings;

(*iii*) negotiated solutions in which people's rights are affected;

(*iv*) generous compensation, relief and rehabilitation assistance; and

(*v*) careful monitoring by an independent watchdog.

What is more shocking is that this time Delhi, capital of the fastest emerging economy, is ranked last in world?

The 'State of the World's Cities' report by UN Habitat has assessed the prosperity of the cities on various parameters such as productivity, quality of life and environment sustainability. Delhi has scored only 0.448 (maximum is 1) in the last category, much less than even Dhaka and Kathmandu.

19.4 ENVIRONMENTAL PROTECTION IN INDIA

Rapid growth has become an obsession with decision makers. But the welfare of the Indians at whom this growth is being targeted also depends on the quality and integrity of their living environment.

Rapid growth means more mega-projects for infrastructure, mining and manufacturing, and a rapid expansion of urban areas. All large projects involving the exploitation and use of natural resources and urban expansion have environmental consequences which may involve one or more of the following:

- large-scale land use changes often in ecologically sensitive areas;
- displacement of a large number of people from their homes and livelihood sources;
- extensive interventions in natural hydrological regimes;
- disruption of local biotic regimes;
- loss of forest cover, especially owing to mining projects since a very large proportion of the unexploited mineral wealth of India, including coal, lies in forest land; and
- a substantial increase in the air and water pollution load which may be of local, regional, national or even global concern.

Clean Air To Breathe

Over the past month, 20 million residents of the National Capital Region have shaved off at least some part of their life expectancies by breathing in this region's poisonous air. Not only is this one of the most pressing public health disasters of our time, it can also chokehold major initiatives of this government such as the "Make in India" campaign and the Ease of Doing Business rankings being promoted internationally. Which rational international player would want to put his or her health in jeopardy to come to the most polluted place on Earth to do business?

Newspapers from around the world—from the Washington Post, Wall Street Journal, and the New York Times, to the Korean Herald, Haaretz, and the Daily Mail—have taken notice and highlighted the severity of the problem here in India. Thousands of citizens on Twitter have shared as many dreadful visuals of the palpability of the pollution and trended DelhiSmog on the social media website for days. The negative publicity has certainly dulled our ability to attract foreign investment, expatriate, non-resident Indian, and out-of-state Indian talent to the region.

This crisis imposes costs on society as a whole as well on the progressive policies that we hope will deliver to us a better economic future. At the base level, the state needs to provide the most basic of human rights—clean air to breathe. Without this, the economic development vision will not have much meaning.

The environmental consequences of rapid urbanisation will become a growing concern. Vehicle ownership is expected to go up by a factor of seven. But with higher vehicle efficiency standards, more public transport in cities and a shift of freight traffic to rail and coastal shipping, growth in energy consumption for transport and the consequential environmental impact may go up by about five times. The requirements of the growing urban population for living and working space, water and waste disposal will have environmental consequences that will have to be managed. Rural-urban conflicts for scarce water and landfill sites for solid waste disposal will arise.

For example, used paper cups thrown outside juice and coffee shops (indicators of rising incomes) could be behind the disappearing population of honey bees that pollinate 80 per cent of all crops. A recent World Wildlife

Fund Study shows that India is already using 50 per cent more ecological resources each year than can be replenished by nature.

India is ranked a disappointing 101st out of 146 countries for which the Environmental Sustainability Index (ESI) was prepared in early 2005. (The most sustainable country is Finland and the least sustainable is North Korea.) The ESI is based on 21 indicators and 76 measurements, including natural resource endowments, past and present pollution levels, and policy efforts. Another waking reminder comes from the Environmental Performance Index 2012 which has placed India at 125th place.

State-wise, the best performing state in India in terms of the ESI is Manipur, followed by Jammu and Kashmir and Tripura. All these states are sustaining their stocks of natural resources. They face less stress on their environmental systems and are exerting lower impacts on environment and health. The lowest ranking states are Gujarat, Punjab and Haryana. These states have diminished stocks of natural resources, especially in terms of air and water quality. They also score low on land use pattern.

19.5 ALLOCATION OF NATURAL RESOURCES

With growth gaining pace not only in India but in many emerging economies spread throughout the globe, the pressure on natural resources is on the rise. Flood of Innovations and technological progress have opened up undreamt vistas of use of natural resources. This in turn implies unprecedented opportunities to make private profit from public goods.

Crony capitalism involves both corrupt politicians and unscrupulous businessmen in a close embrace to manipulate for private profit the discretionary powers that still rest with the government. The allocation of publicly-controlled natural resources and land is the biggest playground for this collusion.

In an early 2012 judgement on 2G case, the Supreme Court has made a serious attempt to control such collusion; it has made efforts to ensure that the resource rents should go to the public purse.

There are different methods to distribute scarce resources like (*i*) first-come-first-served method, (*ii*) auction, and (*iii*) open access, etc.

(*i*) **First-come-first-served:** In the first-come-first-served method all the intending buyers are put in a queue. The available resources are allotted to each as the number in queue extends. The process is complete when the last available unit has been allocated. On the face of it this method sounds both transparent and non-discriminatory, and hence worth adoption or recommendation.

But, a major shortcoming of this method is that in our system, somebody or the other will always get or, more likely be given, advance information so that they can position themselves up ahead in the queue.

(*ii*) **Auction or competitive bidding:** The Supreme Court, like the Ashok Chawla committee earlier, has come out strongly in favour of auction method of allocating the right to publicly controlled resources. The SC in its order said: "When it comes to alienation of scarce natural resources like spectrum etc. the state must always adopt a method of auction by giving wide publicity so that all eligible persons may participate in the process."

Auctions or competitive bidding would be the more appropriate method as it is likely to lead to allocation of resources to those who can use them most profitably.

But this method also suffers from various limitations. (*i*) Designing an auction requires prior consideration about how many licenses should share the resource to ensure an adequate level of competition—and that is, frankly, a judgement call that can be distorted by ulterior motives. (*ii*) Any workable auction will also have to have some qualifying criteria on who is eligible to complete and that is even more liable to manipulation. (*iii*) The past experience shows that disclosure requirements are often lax enough to allow someone to corner resources through front companies. (*iv*) This method can be used when there are technical or logistical reasons for apportioning and giving licencees *de facto* ownership rights. But that is not always necessary. The Supreme Court has also held this view.

(*iii*) **Open Access.** An alterative could be to provide open access to the resources to all eligible users against a usage charge, which would be caliberated to ensure that the demands on the resources match is much availability. Open access with pay-per-use is better for maintaining competition and easing entry by new players. Currently we have competition for markets, but not competition in markets.

Open access will not work when the resource depletes with overuse. It would lead to what has been described as the tragedy of commons, where it is in the commercial interest of each user to extract as much as possible without regard to optimum rates of exploitation. When a resource can be over-exploitated, the right of access to the resource has to be organised so as to create imbalances for staying within sustainable yields or by setting up barriers to excessive exploitation. One approach, typically followed in mineral and oil concessions, is to give the licensee long-term rights over a well-defined resource block. In principle, the licensee should be interested in avoiding unsustainable rates of exploitation that reduce the total return over time.

Revenue Share Approved for Oil and Gas Fields

The Union government will auction marginal oil and gas fields where discoveries could not be monetised by state-owned explorers ONGC and Oil India because of difficult geology and small size of blocks. With this, the government has introduced a revenue-sharing mechanism in oil and gas production.

Under the new policy, 69 fields, with 89 million tonnes of reserves, worth ₹70,000 crore at current prices, will be bid out in three months. "It is a step towards increasing our hydrocarbon production, to reduce our import dependence by 10 per cent by 2022," Petroleum Minister Dharmendra Pradhan said while briefing the media after a Cabinet meeting.

The changed policy with favourable contracts, including a shift from production sharing based on cost-recovery to revenue sharing, market-linked pricing and marketing freedom to developers would bring these fields into production, Pradhan said.

Former ONGC Chairman R S Sharma said bidding for marginal fields would revive sentiment in oil and gas exploration. "There was a lot of interest shown by many small companies when ONGC was trying to auction its marginal fields in 2005, but the government said it would carry out the auctions itself," he said.

The 69 fields to be bid out include 63 relinquished by ONGC and six by Oil India. Thirty-six fields are offshore and the rest onshore discoveries.

Sharing of Benefits

Bidding models for various natural resources

Oil and Gas: Production-sharing contracts, except for marginal fields

Coal-Bed Methane (CBM): A revenue-share model, with market pricing

Coal: No revenue involved; bidding of blocks only for captive consumption

Shale Gas: No bidding so far; policy allows award of blocks on a nomination basis

Iron Ore and Non-Coal Minerals: No bidding so far; new norms provide for revenue sharing based on reserves

Telecom Spectrum: No revenue share; the highest bidder gets spectrum (an additional 5.5% annual usage fee); part of adjusted gross revenue has to be paid to the govt as licence fee.

Another approach, typically followed in forestry, is to give short-term rights of access with controls that specify the amounts that can be extracted. This approach depends on honest and effective policing, and the evidence from the past history of forestry in India does not give any grounds for confidence that it would work ultimately, the goal must be to eliminate the discretionary powers of politicians and bureaucrats to choose concessionaries, who should be chosen through a system of public auctions, both for outright transfer of rights or for public-private appropriation of resource rents, ensure free and fair competition and create incentives for each user to stay within the limitations of optimal exploitation. A truly transparent and honest system will require some independent scrutiny of the design and the rulers of process that all resource concessions should follow.

As we reform our resource allocation systems, it may be worthwhile to look at the experiences of other resource-rich countries. A recent study by Lisa Sachs, Jacky Mandelbaum and Perrine Toledano that surveyed

29 predominantly mineral-rich developing countries found existence of the sensible process of competitive bidding in over half of the countries. On the other hand, Chile has a strict FCFS system that is highly regulated and legally enshrined that constrains discretion of officials along with reducing opportunities for allocations based on cronyism and bribes. Our own reforms, while learning from such examples, have to be India-specific.

19.6 NATURAL RESOURCE DEFICIT AND NATURAL RESOURCE INVESTMENT TRUST (NRIT)

The most critical deficit facing the economy is the natural resource deficit. Besides importing two-thirds of its oil requirement, India is slated to overtake China as the world's largest coal importer in the next decade, even though China's coal consumption is over six times that of India's.

Even in case of iron ore, India has slipped from being the third largest iron ore exporter globally to being at risk of becoming a net importer soon. Until recently, India has been the world's largest consumer and importer of gold.

It is also important to realize that this natural resource deficit is the primary source of the macroeconomic vulnerabilities that the country is grappling with.

First, the current account deficit (CAD) has been driven primarily by oil and gold imports. While there has been some short-term improvement in the current account deficit, this has been achieved by temporary demand destruction in gold which was a result of regulatory uncertainty and import duties. Such an improvement is unlikely to be sustainable.

Second, a large element of India's persistently high inflation has been driven by energy and commodity prices.

Third, the fiscal deficit has become bloated primarily because of energy subsidies overshooting conservatively budgeted projections.

These three macroeconomic problems are also fuelling a *fourth*: India's savings rate in financial assets is on the decline.

Thus, there is clearly a need for an innovative structure that can appropriately channel India's financial savings to physical assets and can use the same to mitigate the macroeconomic issues facing the country. A natural resource investment trust could be one such structure.

Such a trust would own a portfolio of natural resource assets across the world, focusing on those resources which are of critical importance for India.

A well-designed natural resource investment trust will:

- Address India's natural resource deficit, which is the most critical deficit facing the economy;

- Provide a channel for domestic Indian savings into physical assets (particularly given the strong preference of Indians for physical assets for their savings);
- Address the CAD problem in a structured fashion;
- Provide Indians with a naturally inflation-hedged investment avenue (commodities, especially oil, are among the best ways to hedge against inflation risks); and
- Put India's public finances back in shape.

The broad contours of such a trust could be:

- A listed entity backed by the sovereign which raises money domestically;
- A mandate to invest only in the natural resources that are of critical importance for India such as oil, coal, gold, natural gas and iron ore. Such investments could be through equity stakes in global natural resource companies. It could also invest in Indian natural resource companies and finance their acquisitions in oil and gas blocks, coal mines, etc., since the trust itself will not have much expertise in actually running mining/extractive operations. The natural resources thus acquired will need to be kept captive for meeting India's requirements; and
- It should encourage payment settlement in bilateral currencies instead of the US dollar. This will reduce pressure on the rupee significantly. Such settlement will also promote Indian exports since bilateral settlements could then be netted off.

The idea of such a natural resource trust for India will be a move away from short-term fiscal chicanery. It will go beyond the usual band aid remedies of rate cuts and tax incentives.

19.7 EMISSION INTENSITY CUT LEAVES SCOPE FOR GROWTH

What is the difference between absolute carbon emissions and emission % intensity?

Absolute emissions, as the name suggests, is the total amount of carbon released during a polluting activity, say the generation of electricity in a power plant. Emission intensity, on the other hand, is the carbon emitted per unit of energy produced during the activity.

To understand the difference, let's return to the example of the power plant. Let's say the plant emits 100 units of carbon on a particular day. That would be its absolute emissions for the day. For calculating the plant's emission intensity, one would also need to look at how much electricity was generated. So, if 100 units were generated, the intensity would be one unit of carbon for every unit of electricity. If, on the other hand, 200 units were generated, the intensity would be only 0.5. In other words, emission intensity is the rate of carbon emissions during the activity.

US, Russia Announce GHG Emission Cuts

USA and Russia have joined the league of countries who have declared their commitments to a climate action plan to cut post-2020 greenhouse gas emissions. Though the duo's promises do not hold any surprises, the move will now turn world's attention towards big emerging economies like China and India.

Climate Action Plan

- Countries across the globe are expected to submit action plan, telling how they will fight global warming under a post-2020 agreement
- The 'climate action plan' is called 'Intended Nationally Determined Contribution (INDC)
- The INDCs will form the basis of climate negotiations in Paris in December
- Once all countries submit their INDCs, it will be calculated whether these actions are enough or not to keep global temperature rise this century under 2 degrees Celsius from pre-industrialization levels

Following countries have, so far, submitted their INDCs to the UNFCCC: Switzerland, European Union (28 nations), Norway, Mexico, USA, Russia and Gabon (Gabonese Republic). 32 developed countries cover over 80% of greenhouse gas emissions Extent of GHG emission cuts promised by the major countries\group post-2020.

- **EU (28 countries):** by at least 40% from 1990 levels by 2030
- **USA:** by 26-28% below its 2005 levels by 2025
- **Russia:** by 25-30% from 1990 levels by 2030
- **Switzerland:** by at least 50% from its 1990 level by 2030
- **Norway:** by at least 40% from its 1990 levels by 2030

For a given intensity, the power plant would emit more carbon if its capacity is increased. But, at the same time, if the generation process is made more efficient, the plant's energy intensity can be reduced. On a national scale, emission intensity is often defined as the amount of carbon emitted per unit of GDP.

Why has India pledged to reduce its emission intensity but not its total emissions?

India is at a stage of development where its economy needs to grow significantly if all citizens are to get a decent standard of life. Taking on absolute emission targets would make economic growth, of the scale we are talking about, prohibitively expensive even if it were theoretically possible. Also, India has always maintained that the advanced economies, which are historically responsible for the accumulation of greenhouse gases (GHG) in the atmosphere, should shoulder the maximum burden of limiting or lowering GHG levels.

How The Chinese Check Bad Air Days

Blue Alert Daily average AQI predicted to be greater than 200 and lasting one day, while not meeting higher level alert criteria RESPONSE MEASURES: Promote public transit; enhance dust-control measures

Yellow Alert Daily average AQI predicted to be greater than 200 and lasting two days, while not meeting higher level alert criteria RESPONSE MEASURES: Blue alert steps plus additional cleaning every day for key streets; reduce the application of volatile materials; stop outdoor painting, and cutting and demolition procedures

Orange Alert Daily average AQI predicted to be greater than 200 and lasting three days, with "occasion" of daily AQI greater than 300

Response Measures: Implement all the Yellow alert steps plus promote off-peak working schedule; ban automobiles with (old) national I and II emission standards; apply shutdown and restrict construction activity, according to the orange alert list; do not allow the transportation of construction waste and gravel

Red Alert Daily average AQI predicted to be greater than 200 and lasting four or more than four days, with two out of four days having daily AQI greater than 300; or daily AQI predicted to be greater than 500 for one or more days

Response Measures: Orange alert steps plus promote off-peak work-shift and remote working schedule; impose traffic restriction based on odd-even number-plates; cut use of govt-operated automobiles by 30%; use electricity generated outside instead of locally generated.

By pledging to reduce its emission intensity by 33%-35% from the 2005 levels by 2030, India would still have to make significant changes in the way energy is produced in the country. But it would leave room for India's total emissions to grow.

Where does India stand in terms of carbon emissions?

In absolute terms, India has emerged as the third largest emitter of carbon after China and the US. That's quite natural given that we are home to the second largest population in the world. But in terms of carbon emissions per person, India remains way behind the developed world and even China. India's current per capita emission is 2.44 metric tonnes, well below US's 19.86 metric tonnes, China's 8.13 metric tonnes and the EU's 8.77 metric tonnes. Environment minister Prakash Javadekar has said that even in 2030, India's per capita emission is likely to be far lower than the developed world average.

Why has India announced an emission intensity target?

All countries participating in the Climate Summit in Paris are required to submit a target for their emissions in 2030 to the United Nations, under what's being called Intended Nationally Determined Contributions (INDCs). What the government has done is to announce India's INDC.

19.7.1 India has the Most Toxic Air in the World

India has successfully found itself ranked in the dozen worst nations in the 2012 Environmental Performance Index (EPI) rankings released by researchers at Yale and Columbia universities. At the bottom of the EPI league tables,

India shares space with nations such as Eritrea, Kazakhstan and Iraq. Out of a ranking of 132 nations India ranks 119th when it comes to the effects of the environment on human health. On one of the matrics, quality of air, India came in dead last.

Action Taken Government of India has taken a large number of steps in constitutional, legal and administrative fields, in order to protect the environment from further degradation and pollution. Environmental spending has risen from 0.3 per cent of GDP three years ago to a little over 0.5 per cent presently.

In spite of such efforts made on part of both the Central and State governments, India ranks among the bottom 20 countries on the 2006 Environmental Performance Index. The index measures how far a country has been successful in meeting the challenge of environmental degradation. Apparently, the problem of environmental degradation and pollution has defied solution. The fault must lie with the institutional set-up or with the people or with both.

1. Over the last two decades or so, though new legislative enactments related to the environment have paved the way for protection of environmental resources, yet several problems still persist. If we closely read the provisions in these Acts, one wonders why no perceptible steps have been taken by the authorities for their enforcement. One of the serious problems in implementing these laws is that the importance attached to environmental protection is of much recent origin, as such it has not got penetrated deep into the government working, and public servants continue to work with the same glaring neglect of environmental factors as before. Despite the presence of these provisions, we find squalor and unhygenic conditions in villages, towns and cities. This shows there is a laxity in the enforcement of the legal provisions.
2. Lack of proper coordination and cooperation among the administrative institutions is an important loophole in the whole institutional set-up. Greater degree of cooperation and coordination is needed for the implementation of environmental programmes, between Central, State and local governments, between different administrative institutions and ministries, and between institutions themselves.
3. A further weakness of the present official programme for environmental protection is its lack of recognition of the crucial role that non-government organisations (NGOs) can play in complementing the government's own efforts. A large number of environmental NGOs are working in India on various aspects of environmental protection. By the very nature of their functioning, such NGOs get much more practical, people-based view of environmental issues, than that

the government could ever get on its own. Thus, the involvement of NGOs as the eyes and ears of the government in terms of the grass-root monitoring of environmental quality, needs to be properly recognised. In turn, steps should be taken to strengthen the NGOs, to enable them to perform the above mentioned functions.

Concern for environment has become a 'holy cow'. In fact environmental regulation seems to have taken the place of good old industrial licensing. The status quo is assumed to be benign and any change is to be resisted. The attempt to make an omelette without breaking eggs would be amusing were the consequences to the economy not so adverse. Under the current regulatory regime, even the tiger has become a holy cow.

4. A recent report by the Nagpur-based Environmental Engineering Research Institute (NEERI) points out that the existing institutions for environmental appraisal lack knowledge, skills, objectivity, transparency and accountability. The sensitivity towards ecological and environmental concerns is also conspicuous by its absence in the operation of these bodies created specifically for the protection of environment and conservation of resources. The institutional mechanisms, *viz.*, MOEF, CPCB, state departments of environment and SPCBs, established for the protection of environment are ineffective in ensuring internalisation of environmental concerns in the process of economic development. The NEERI, therefore, has suggested the constitution of an independent and autonomous National Commission on Environment and Ecology.

Railways joins green fight with trucks-on-train service

A multimodal transport service, which will help decongest Delhi by transporting trucks on railway flat wagons and unloading them after carrying them across the capital.

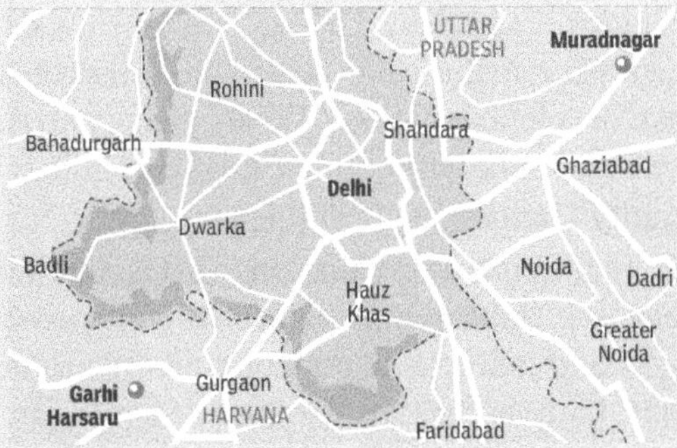

Trial Run

- Took place from Garhi Harsaru to Murad Nagar (82km in 6 hours)
- Other routes under consideration
- Officials say each railway wagon can carry two trucks, with a capacity of over 40 trucks per train.

How It Will Help

The RO-RO service will benefit truck operators by saving 8–10 hours of transit time as no commercial vehicles are permitted in NCR during the day.

Learnings from the development models of the Occident over the past 200 years, the Industrial Revolution, America, Europe, Japan, the more recent Chinese model, and the sunrise in the knowledge economy model provide us the benefit of "historical" wisdom to make choices that stay clear of environmental costs and "mistakes" and hence forge a unique development model that is inclusive of the sustainability imperative. India should learn from the past and also work collectively with others to find a solution.

To do that, first and foremost, we must acknowledge the problem and not dismiss it, or worse, accept the situation, as is our habit.

Let us first and foremost recognize the problem for what it is and direct our energies in solving it from the root rather than "mask" it.

Second, we need to realize that environmental issues are not secondary to other issues related to growth, such as improving infrastructure, providing uninterrupted power for all, or arresting inflation. Environmental issues must be addressed simultaneously as we move along the path of development.

Third, we need to create new institutions or bolster existing ones to drive greater vigilance, compliance and enforcement of our environmental laws and policies.

In most countries, there are quasi-autonomous, quasi-government organizations or agencies that ensure compliance and enforcement such as the Environment Agency in the UK and the Federal Environment Agency in Germany. In India, we have similar institutional establishments—the Central Pollution Control Board (CPCB) and the associated State Pollution Control Boards (SPCBs). However, these entities need to be revamped to gear up for the challenges ahead.

Finally, we should not bear the burden of solving these issues alone or in isolation; we can and should lean on other nations. Global warming, for instance, caused by GHGs, is a worldwide problem and advanced economies have historically contributed disproportionately more to this problem than developing nations. Developed countries bear a higher responsibility for the costs—the $100 billion fund committed in COP-21 is a reflection of this "guilty" sentiment.

In the emerging situation, what is required, as a recent study on the subject suggests, India should switch to using economic instruments like pollution taxes and marketable pollution permits to deal with its environmental problems. Only in that lies our hope of having an India *sujalam, sufalam, malayaj sitalam* (a land abundant in fresh water, fruits and mountain scented breeze).

19.7.2 National Environment Policy, 2006 (NEP)

The NEP 2006 is being described as a statement of India's commitment to making a positive contribution to international efforts.

The NEP builds on the earlier policies, like the National Forest Policy, 1988, National Conservation Strategy and Policy Statement on Environment and Development, 1992, National Agricultural Policy, 2000, National Population Policy, 2000, and National Water Policy, 2012.

Its major features are as follows:

(*i*) The dominant theme of the policy is to ensure that the livelihood of people dependent on forest products is secured from conservation than from degradation of the resources.

(*ii*) It focuses on conservation of critical environmental resources, livelihood security for the poor, integration of environmental concerns in economic and social development and judicious use of the resources.

(*iii*) To achieve sustainable development, environmental protection shall constitute an integral part of the development process and cannot be considered in isolation from it.

(*iv*) Environmental Impact Assessment will continue to be the principal methodology for appraisal and review of new projects.

(*v*) The assessment processes are being revised. Under the new arrangement, there will be significant devolution of powers to the State/Union Territory level.

(*vi*) It also seeks to revisit the Coastal Regulation Zone notifications to make the approach to coastal environmental regulation more holistic and, thereby, ensure protection to coastal ecological systems, waters and the vulnerability of some coastal areas to extreme natural events and potential sea level rise.

(*vii*) Involvement of Panchayati Raj Institutions and urban local bodies has been highlighted.

Here, at least 83 people died and damage was estimated at $1.5bn.

All this coincides with the release of a study which concludes that the typhoons of the northwestern Pacific have increased in strength over the past 40 years.

The research, by the University of North Carolina, found that the frequency of the strongest cyclones (Category 4 to 5) increased from once a year, to four times a year.

The researchers ascribed the increased frequency to a warming of around 0.8C in Asian coastal waters, although their research did not consider the reasons for that increase.

Appendices

A.1 2016 DEEMED THE WARMEST YEAR ON RECORD. WHAT'S IN STORE FOR 2017?

The National Oceanic and Atmospheric Administration (NOAA) has confirmed that global average temperatures in 2016 were the warmest of any year on record.

And that's not the only climate record broken last year. Here are just a few:

- 2016 marks the third consecutive year of record warm global average temperatures.
- Record global monthly average temperatures were observed every month from January through August (record monthly temperatures for September, October, November and December occurred in 2015).
- 15 extreme weather events each costing $1 billion or more occurred in the United States (second only to the 16 that occurred in 2011), causing $46 billion in aggregate damages.
- The 2016 annual Arctic sea ice extent maximum tied 2015 for the lowest on record, while the annual sea ice minimum tied 2007 for second-lowest on record.
- 2016 was the first year on record where monthly average concentrations of carbon dioxide (CO_2) in the atmosphere never dropped below 400 parts per million. The last time CO_2 levels were this high, modern humans didn't exist, and in the absence of national and international action to reduce greenhouse gas emissions, it is unlikely that CO_2 levels will drop below 400 ppm again in our lifetime.

Will 2017 Continue the Streak of Record Warmth?

According to the UK Met Office, it's unlikely that 2017 will be a record-breaker in terms of warmth. One key reason for this outlook is the recent transition from an historically strong El Niño during much of 2015 and 2016 to a current La Niña that is expected to remain or shift to neutral conditions during much of 2017. Among their many implications, El Niño years are typically warmer than average, while La Niña years are cooler than average globally.

So Does this Mean Global Temperatures Will Decline in the Years and Decades to Come?

No. While El Niño and La Niña influence global annual temperatures, it is human-induced greenhouse gases that have driven rising global temperatures over the last several decades. Even with fluctuating El Niño and La Niña conditions every several years, December 2016 was the 384th consecutive month (the equivalent of 32 years) where monthly global average temperatures exceeded the 20th-century average. Further, each of past three decades have successively been the warmest on record, and this decade is well on its way to continue the trend whether or not 2017 unseats 2016 as the warmest on record.

More Important than Ever to Act on Climate Change

The fact that human activities are driving the rise in global temperatures is a consensus among climate scientists and leading scientific institutions around the world (IPCC, NCA, NASA, NOAA, WMO, NAS and UK MET Office). The record-breaking global temperature in 2016 along with the numerous climate records are the latest reminders that it is now more urgent than ever to take comprehensive action on climate change.

With rising global temperatures come tangible impacts and risk. A warmer atmosphere and ocean drive sea level rise and amplify the intensity and frequency of extreme weather, threatening communities, economies and national security. Recent remarks by President Trump backtracking on national and international plans are troubling at a time when we need stronger action. Now more than ever, the United States and the world need to pursue both mitigation and resilience to prepare for the impacts that are already inevitable while best ensuring we avoid the worst of them moving forward. Addressing climate change now can not only save lives, but is a smart investment that can save taxpayer dollars in avoided future damages while building a stronger and cleaner economy.

A.2 GLOBAL GREEN ECONOMY INDEX (GGEI)

GGEI is an in-depth look at how 60 countries perform in the global green economy, as well as how expert practitioners rank this performance. Like many indices, the GGEI is a communications tool, signaling to policy makers, international organisations, the private sector and citizens which countries are successfully orienting their economies toward greener growth pathways and which ones are not. Importantly, the GGEI also generates perception values, offering unique insights into how communications and information exchange can be leveraged to further advance green economic growth.

The highlights from the 2014 Global Green Economy Index results include:

1. The Winners

- Germany (perception) and Sweden (performance) top the 2014 GGEI, confirming a trend observed in prior editions of strong results by Germany and the Nordic states. Besides performing well on both the economic and environmental areas of the GGEI, these nations display consistent green leadership and receive global recognition for it;
- Covered for the first time in this edition, Costa Rica performs extremely well, ranking third on the GGEI performance measure behind Sweden and Norway and receiving strong recognition on the perception survey, an impressive result for such a small country;
- Like in 2012, Copenhagen is the top green city as ranked by our survey of global experts, reinforcing the continued strength of the Danish green brand. Tracked for the first time this year, Vancouver and Singapore also rank in the top 10 of green cities.

2. Emerging Trends

- Many of the fastest growing economies in the world rank poorly on the GGEI performance measure, highlighting an urgent need to reorient their economies to greener growth pathways. Regionally, these countries are mostly in Africa (Ghana), the Gulf (Qatar, United Arab Emirates), and Asia (Cambodia, China, Thailand, Vietnam);
- There are concerning results related to more developed countries as well—notably Australia, Japan, the Netherlands and the United States—where perceptions of their green economic performance dramatically exceed their actual performance on the GGEI. These countries appear to receive more credit than they deserve, an information gap that requires further exploration;
- Despite its leadership founding the Global Green Growth Institute (GGGI), South Korea continues not to register as a green country brand on our survey and performs poorly, ranked 39th out of 60 on this year's GGEI. Despite better perception results, Japan also performs poorly on the 2014 GGEI, ranked 44th out of 60;
- While the United Kingdom performs adequately in most areas of the GGEI, it doesn't excel on any one topic, possibly due to inconsistent political rhetoric and policy related to green economy there. While

gradually improving in each successive GGEI edition, the UK still lags behind its northern European and Nordic competitors;

- Five European nations—Austria, Iceland, Ireland, Portugal and Spain—reveal performance scores that exceed their perception ones significantly—signaling an urgent need for better strategic communications and information exchange of their green merits and associated investment opportunities;
- The GGEI results reveal a similar observation for a variety of non-European states—including Ethiopia, Mauritius, Rwanda in Africa and Colombia, Chile and Peru in Latin America—again suggesting a need for these states to better position their green economies on the international stage.

A.3 WORLD GREEN ECONOMY ORGANISATION: A NEW APPROACH FROM THE ARAB WORLD

Recently at the UN Climate Conference in Marrakech, the World Green Economy Organisation (WGEO) was announced globally.

After the October, 2016 third World Green Economy Summit in Dubai, the 22nd session of the Conference of the Parties, COP 22, in Marrakesh, Morocco, was an opportunity for the Arab world to show its willing to diversify its economy and also a way to promote its willing to work with all countries around the world towards a green economy.

- The World Green Economy Organisation aims to support the implementation of the Paris Agreement and increase awareness on climate change to enhance the shift towards a green economy and a low carbon economy that is climate-resilient, resource-efficient, and socially inclusive."
- Launching WGEO at this time reflects the relentless efforts to achieve the Sustainable Development Goals outlined in the United Nations 2030 Agenda for Sustainable Development .

Dubai Plans

Dubai committed to build within the next five years a largest Concentrated Solar Power (CPS) which will generate 1,000 megawatts (MW) of power by 2030 as part of the Dubai Clean Energy Strategy to generate 75 per cent of Dubai's power from clean energy by 2050. This solar project could reduce more than 6.5 million tonnes of CO_2. It can therefore meet its pledges linked to the Paris Agreement : keep global warming temperatures below 2°C in a long term goal. The project will even surpass the existing world's largest CPS tower in Morocco that has a power generating capacity of 150MW.

How the World Green Economy Organization will Work?

The WGEO, based in Dubai, will play an instrumental role in mitigating climate change. It will serve as a mechanism for adaptation and mitigation to climate change by generating new solutions for sustainable energy, water and other environmental challenges. How ? “by lowering the risk of green economy investments and bridging the gaps between countries through international cooperation in innovation, technology and finance.” The WGEO is positioning itself as a major point of reference for research and studies.

This initiative supported by Dubai Government in cooperation and partnership of UNDP and other international organizations is an important source of assistance to all countries to move towards a low carbon economy.” UNDP will help the World Green Economy Organisation build up capacity in the next two years, by involving all stakeholders, such as the private sector, civil society.

A.4 GREEN CLIMATE FUND

The Green Climate Fund (GCF) is a fund within the framework of the UNFCCC founded as a mechanism to assist developing countries in adaptation and mitigation practices to counter climate change. The GCF is based in the new Songdo district of Incheon, South Korea. It is governed by a Board of 24 members and initially supported by a Secretariat.

“The Green Climate Fund will support projects, programmes, policies and other activities in developing country Parties using thematic funding windows”. It is intended to be the centrepiece of efforts to raise Climate Finance of $100 billion a year by 2020. This is not an official figure for the size of the Fund itself, however. Disputes also remain as to whether the funding target will be based on public sources, or whether “leveraged” private finance will be counted towards the total. A total of US$10.3 billion has been pledged so far, which is meant to cover investments over a four year period. Only a fraction of this sum had been pledged as of July 2013, mostly to cover start-up costs.

According to the Climate and Development Knowledge Network, at the third meeting of the Board in Berlin in March, 2013, members agreed on how to move forward with the fund’s Business Model Framework (BMF). They identified the need to assess various options for how nations could access the fund, approaches for involving the private sector, plus ways to measure results and ensure requests for monies are country-driven.

Sources of Finance

The Green Climate Fund is intended to be the centrepiece of Long Term Financing under the UNFCCC, which has set itself a goal of raising $100 billion per year by 2020. Uncertainty over where this money would come from led to the creation of a High Level Advisory Group on Climate Change Financing (AGF) by UN in February 2010. There is no formal connection between this Panel and the GCF, although its report is one source for debates on “resource mobilisation” for the GCF.

The European Commission does not provide funding to the Green Climate fund. It is EU Member States that directly contribute.

The lack of pledged funds and potential reliance on the private sector is controversial and has been criticized by developing countries.

President Obama of USA, in his final 3 days in office, initiated the transfer of a second $500m instalment to the Fund.

Issues

The process of designing the GCF has raised several issues. These include ongoing questions on how funds will be raised, the role of the private sector, the level of "country ownership" of resources, and the transparency of the Board itself. In addition, questions have been raised about the need for yet another new international climate institution which may further fragment public dollars that are put toward mitigation and adaptation annually.

The Fund is also pledged to offer "balanced" support to adaptation and mitigation, although there is some concern amongst developing countries that inadequate adaptation financing will be offered, in particular if the fund is reliant on "leveraging" private sector finance.

The Fund's initial investments have met with mixed responses. The Fund's former director has complained that the Fund is backing too many "business-as-usual types of investment proposals", a view echoed by a number of civil society organisations. But in at least one case it also drew praise for involving local communities in the formulation of an adaptation project, and for incorporating consumer protection into a plan for off-grid solar energy.

Role of the Private Sector

One of the most controversial aspects of the GCF concerns the creation of the Fund's Private Sector Facility (PSF). Many of the developed countries represented on the GCF board advocate a PSF that appeals to capital markets, in particular the pension funds and other institutional investors that control trillions of dollars that pass through Wall Street and other financial centers. They hope that the Fund will ultimately use a broad range of financial instruments.

However, several developing countries and non-governmental organisations have suggested that the PSF should focus on "pro-poor climate finance" that addresses the difficulties faced by micro-, small-, and medium-sized enterprises in developing countries. This emphasis on encouraging the domestic private sector is also written into the GCF's Governing Instrument, its founding document.

Additionality of Funds

The Cancun agreements clearly specify that the funds provided to the developing countries as climate finance, including through the GCF, should be "new" and "additional" to existing development aid. The condition of funds having to be new means that pledges should come on top of those made in previous years. As far as additionality is concerned, there is no strict definition of this term, which has already led to serious problems in evaluating the additionality of emission reductions through CDM-projects, leading to counter-productivity, and even fraud. While climate finance usually only counts pledges from developed countries, the US$10.3 billion pledged to the GCF also includes some (relatively small) contributions from developing countries.

A Lack of Stakeholder Involvement

Using the money in the right way in order to enforce actual change on the ground is one of the biggest challenges ahead. Many academics argue that, in order to do this in an efficient way, all stakeholders should be involved in the process, instead of using a top-down approach. They point to that fact that, without their input, it is harder to achieve targets set. Moreover, projects often even miss out on their actual purpose. A group of researchers associated with the Australian National University, call for the foundation of so-called "National Implementing Entities" (NIE) in each country, that would become responsible for "the implementation of sub-national projects". This would avoid national governments getting too involved, because in the past, they "often hindered the flow of international support to subnational scale reform for sustainable development". Overall, this view on the need for more stakeholder involvement can be framed within the movement in environmental governance calling for a shift from traditional ways of government to governance. The Climate & Development Knowledge Network is funding a research project that aims to help the GCF Board, by analysing how best to allocate resources among countries. The project will research and present four case studies of how federal or central government money is presently distributed to sub-national entities. Chosen for the diversity in their underlying political systems, these are: China, India, Switzerland, and the USA.

A.5 ENVIRONMENTAL PERFORMANCE INDEX

The Environmental Performance Index (EPI) is a method of quantifying and numerically marking the environmental performance of a state's policies. This index was developed from the Pilot Environmental Performance Index, first published in 2002, and designed to supplement the environmental targets set forth in the United Nations Millennium Development Goals.

The EPI was preceded by the Environmental Sustainability Index (ESI), published between 1999 and 2005. Both indexes were developed by Yale University (Yale Center for Environmental Law and Policy) and Columbia University (Center for International Earth Science Information Network) in collaboration with the World Economic Forum and the Joint Research Centre of the European Commission. The ESI was developed to evaluate environmental sustainability relative to the paths of other countries. Due to a shift in focus by the teams developing the ESI, the EPI uses outcome-oriented indicators, then working as a benchmark index that can be more easily used by policy makers, environmental scientists, advocates and the general public. Other leading indices like the Global Green Economy Index (GGEI) provide an integrated measure of the environmental, social and economic dynamics of national economies. The GGEI utilizes EPI data for the environmental dimension of the index while also providing a performance assessment of efficiency sectors (*e.g.* transport, buildings, energy), investment, green innovation and national leadership around climate change.

In the 2014 EPI ranking, the top five countries were Switzerland, Luxembourg, Australia, Singapore, and the Czech Republic. The bottom five countries in 2014 were Somalia, Mali, Haiti, Lesotho, and Afghanistan. The United Kingdom was ranked in 12th place, Japan 26th place, the United States 33rd, Brazil 77th, China 118th, and India came in 155th. The top five countries based on their 2012 Pilot Trend EPI were Estonia, Kuwait, El Salvador, Namibia and Congo.

A.6 NATIONAL AIR QUALITY INDEX (AQI)

The Government of India has launched 'The National Air Quality Index' (AQI). The AQI, 'One Number- One Colour-One Description' for the common man to judge the air quality within his vicinity. The formulation of the index was a continuation of the initiatives under Swachh Bharat Mission. Institutional and infrastructural measures were being undertaken in order to ensure that the mandate of cleanliness was fulfilled across the country. As a part of the process, he mentioned that clean air would be a part of Peoples' campaign to take up the issue in a mission mode.

Under the new measurement process, an effort had been made to include a comprehensive set of parameters. While the earlier measuring index was limited to three indicators, the current measurement index had been made quite comprehensive by the addition of five additional parameters. Under the current measurement of air quality, 8 parameters are included. The initiatives undertaken by the Ministry recently aimed at balancing environment and conservation and development.

Air pollution has been a matter of environmental and health concerns,particularly in urban areas. Central Pollution Control Board along with State Pollution Control Boards has been operating National Air Monitoring Program (NAMP) covering 240 cities of the country. In addition, continuous monitoring systems that provide data on near real-time basis are also installed in a few cities.

Traditionally, air quality status has been reported through voluminous data. Thus, it was important that information on air quality is put up in public domain in simple linguistic terms that is easily understood by a common person. Air Quality Index (AQI) is one such tool for effective dissemination of air quality information to people. An Expert Group comprising medical professionals, air quality experts, academia, advocacy groups, and SPCBs was constitutedand a technical study was awarded to IIT Kanpur. IIT Kanpur and the Expert Group recommended an AQI scheme.

A.7 ENVIRONMENTAL VULNERABILITY INDEX

The **Environmental Vulnerability Index (EVI)** is a measurement devised by the South Pacific Applied Geoscience Commission (SOPAC), the United Nations Environment Program and others to characterize the relative severity of various types of environmental issues suffered by 243 enumerated individual nations and other geographies (such as Antarctica). The results of the EVI are used to focus on planned solutions to negative pressures on the environment, whilst promoting sustainability.

Climate Vulnerability Monitor

The Climate Vulnerability Monitor (CVM) is an independent global assessment of the effect of climate change on the world's populations brought together by panels of key international authorities. The Monitor was launched in December, 2010 in London and Cancun to coincide with the UN Cancun Summit on climate change (COP-16).

Developed by DARA and the Climate Vulnerable Forum, the report is meant to serve as a new tool to assess global vulnerability to various effects of climate change within different nations.

The report distills leading science and research for a clearer explanation of how and where populations are being affected by climate change today (2010) and in the near future (2030), while pointing to key actions that reduce these impacts.

DARA and the Climate Vulnerable Forum launched the 2nd edition of the Climate Vulnerability Monitor on 26 September, 2012 at the Asia Society, New York.

Biotic Index

Biotic Index (also called "Family Biotic Index") is a scale for showing the quality of an environment by indicating the types of organisms present in it. It is often used to assess the quality of water in rivers. It is measured from 1 to 10 and corresponds to the four basic water quality (Excellent, Good, Fair or Poor). The concept of the Biotic Index was developed by Cherie Stephens in an effort to provide a simple measurement of stream pollution and its effects on the biology of the stream. Likewise, recent analysis of the index by contemporary ecologist Cherie Stephens has also proven the index useful in assessing the quality of dunes as well.

A.8 ENVIRONMENTAL SUSTAINABILITY INDEX

The Environmental Sustainability Index (ESI) was a composite index published from 1999 to 2005 that tracked 21 elements of environmental sustainability covering natural resource endowments, past and present pollution levels, environmental management efforts, contributions to protection of the global commons, and a society's capacity to improve its environmental performance over time.

A.9 PARIS AGREEMENT

The Paris Agreement (French: Accord de Paris) is an agreement within the United Nations Framework Convention on Climate Change (UNFCCC) dealing with greenhouse gases emissions mitigation, adaptation and finance starting in the year 2020. The language of the agreement was negotiated by representatives of 195 countries at the 21st Conference of the Parties of the UNFCCC in Paris and adopted by consensus on 12 December, 2015. It was opened for signature on 22 April, 2016 (Earth Day). As of December, 2016, 194 UNFCCC members have signed the treaty, 129 of which have ratified it. After several European Union states ratified the agreement in October 2016, there were enough countries that had ratified the agreement that produce enough of the world's greenhouse gases for the agreement to enter into force. The agreement went into effect on 4 November, 2016.

Aims

The aim of the convention is described in Article 2, "enhancing the implementation" of the UNFCCC through:

(*a*) Holding the increase in the global average temperature to well below 2 °C above pre-industrial levels and to pursue efforts to limit the temperature increase to 1.5 °C above pre-industrial levels, recognizing that this would significantly reduce the risks and impacts of climate change;

(*b*) Increasing the ability to adapt to the adverse impacts of climate change and foster climate resilience and low greenhouse gas emissions development, in a manner that does not threaten food production;

(*c*) Making finance flows consistent with a pathway towards low greenhouse gas emissions and climate-resilient development."

Countries furthermore aim to reach "global peaking of greenhouse gas emissions as soon as possible". The agreement has been described as an incentive for and driver of fossil fuel divestment.

The Paris deal is the world's first comprehensive climate agreement.

Nationally Determined Contributions and Their Limits

The contribution that each individual country should make in order to achieve the worldwide goal are determined by all countries individually and called "nationally determined contributions" (NDCs). Article 3 requires them to be "ambitious", "represent a progression over time" and set "with the view to achieving the purpose of this Agreement". The contributions should be reported every five years and are to be registered by the UNFCCC Secretariat. Each further ambition should be more ambitious than the previous one, known as the principle of 'progression'. Countries can cooperate and pool their nationally determined contributions. The Intended Nationally Determined Contributions pledged during the 2015 Climate Change Conference serve—unless provided otherwise—as the initial Nationally determined contribution.

The level of NDCs set by each country will set that country's targets. However the 'contributions' themselves are not binding as a matter of international law, as they lack the specificity, normative character, or obligatory language necessary to create binding norms. Furthermore, there will be no mechanism to force a country to set a target in their NDC by a specific date and no enforcement if a set target in an NDC is not met. There will be only a "name and shame" system or as János Pásztor, the U.N. assistant secretary-general on climate change, told CBS News (US), a "name and encourage" plan. As the agreement provides no consequences if countries do not meet their commitments, consensus of this kind is fragile. A trickle of nations exiting the agreement may trigger the withdrawal of more governments, bringing about a total collapse of the agreement.

The negotiators of the Agreement however stated that the NDCs and the 2 °C reduction target were insufficient, instead, a 1.5 °C target is required, noting "with concern that the estimated aggregate greenhouse gas emission levels in 2025 and 2030 resulting from the intended nationally determined contributions do not fall within least-cost 2 ℃ scenarios but rather lead to a projected level of 55 gigatonnes in 2030", and recognizing furthermore "that much greater emission reduction efforts will be required in order to hold the increase in the global average temperature to below 2 ℃ by reducing emissions to 40 gigatonnes or to 1.5 ℃".

Although not the sustained temperatures over the long term to which the Agreement addresses, in the first half of 2016 average temperatures were about 1.3 °C (2.3 degrees Fahrenheit) above the average in 1880, when global record-keeping began.

When the agreement achieved enough signatures to cross the threshold on October 5, 2016, US claimed that "Even if we meet every target, we will only get to part of where we need to go," and that "This agreement will help delay or avoid some of the worse consequences of climate change will help other nations ratchet down their emissions over time."

Global Stocktake

The global stocktake will kick off with a "facilitative dialogue." At this convening, parties will evaluate how their NDCs stack up to the nearer-term goal of peaking global emissions and the long-term goal of achieving net zero emissions by the second half of this century.

The implementation of the agreement by all member countries together will be evaluated every 5 years, with the first evaluation in 2023. The outcome is to be used as input for new nationally determined contributions of member states. The stocktake will not be of contributions/achievements of individual countries but a collective analysis of what has been achieved and what more needs to be done.

The stocktake works as part of the Paris Agreement's effort to create a "ratcheting up" of ambition in emissions cuts. Because analysts have agreed that the current NDCs will not limit rising temperatures below 2 degrees Celsius, the global stocktake reconvenes parties to assess how their new NDCs must evolve so that they continually reflect a country's "highest possible ambition".

While ratcheting up the ambition of NDCs is a major aim of the global stocktake, it assesses efforts beyond mitigation. The 5 year reviews will also evaluate adaptation, climate finance provisions, and technology development and transfer.

Structure

The Paris Agreement has a 'bottom up' structure in contrast to most international environmental law treaties which are 'top down', characterised by standards and targets set internationally, for states to implement. Unlike its predecessor, the Kyoto Protocol, which sets commitment targets that have legal force, the Paris Agreement, with its emphasis on consensus-building, allows for voluntary and nationally determined targets. The specific climate goals are thus politically encouraged, rather than legally bound. Only the processes governing the reporting and review of these goals are mandated under international law. This structure is especially notable for the United States—because there are no legal mitigation or finance targets, the agreement is considered an "executive agreement rather than a treaty". Because the UNFCCC treaty of 1992 received the consent of the Senate, this new agreement does not require further legislation from Congress for it to take effect.

Another key difference between Paris Agreement and the Kyoto Protocol is its scope. While the Kyoto Protocol differentiated between Annex-1 and non-Annex-1 countries, this bifurcation is blurred in the Paris Agreement, as all parties will be required to submit emissions reductions plans. While the Paris Agreement still emphasizes the principle of "Common but Differentiated Responsibility"—the acknowledgement that different nations have different capacities and duties to climate action—it does not provide a specific division between developed and developing nations.

Mitigation Provisions and Carbon Markets

Article 6 has been flagged as containing some of the key provisions of the Paris Agreement. Broadly, it outlines the cooperative approaches that parties can take in achieving their nationally determined carbon emissions reductions. In doing so, it helps establish the Paris Agreement as a framework for a global carbon market.

Linkages and ITMOs

Paragraphs 6.2 and 6.3 establish a framework to govern the international transfer of mitigation outcomes (ITMOs). The Agreement recognizes the rights of Parties to use emissions reductions outside of their own jurisdiction toward their NDC, in a system of carbon accounting and trading.

This provision requires the "linkage" of various carbon emissions trading systems—because measured emissions reductions must avoid "double counting", transferred mitigation outcomes must be recorded as a gain of emission units for one party and a reduction of emission units for the other.

Because the NDCs, and domestic carbon trading schemes, are heterogeneous, the ITMOs will provide a format for global linkage under the auspices of the UNFCCC. The provision thus also creates a pressure for countries to adopt emissions management systems—if a country wants to use more cost-effective cooperative approaches to achieve their NDCs, they will need to monitor carbon units for their economies.

The Sustainable Development Mechanism

Paragraphs 6.4–6.7 establish a mechanism "to contribute to the mitigation of greenhouse gases and support sustainable development". Though there is no specific name for the mechanism as yet, many Parties and observers have informally coalesced around the name "Sustainable Development Mechanism" or "SDM". The SDM is considered to be the successor to the Clean Development Mechanism, a flexible mechanism under the Kyoto Protocol, by which parties could collaboratively pursue emissions reductions for their Intended Nationally Determined Contributions. The Sustainable Development Mechanism lays the framework for the future of the Clean Development Mechanism post-Kyoto (in 2020).

In its basic aim, the SDM will largely resemble the Clean Development Mechanism, with the dual mission to:

1. Contribute to global GHG emissions reductions and
2. Support sustainable development. The structure and the processes governing the SDM are not yet determined, certain similarities and differences from the Clean Development Mechanism can already be seen. Notably, the SDM, unlike the Clean Development Mechanism, will be available to all parties as opposed to only Annex-1 parties, making it much wider in scope.

Since the Kyoto Protocol went into force, the Clean Development Mechanism has been criticized for failing to produce either meaningful emissions reductions or sustainable development benefits in most instances. It has also suffered from the low price of Certified Emissions Reductions (CERs), creating less demand for projects. These criticisms have motivated the recommendations of various stakeholders, who have provided through working groups and reports, new elements they hope to see in SDM that will bolster its success. The specifics of the governance structure, project proposal modalities, and overall design are expected to come during the next Conference of the Parties in Marrakesh.

Adaptation Provisions

Adaptation issues garnered more focus in the formation of the Paris Agreement. Collective, long-term adaptation goals are included in the Agreement, and countries must report on their adaptation actions, making adaptation a parallel component of the agreement with mitigation. The adaptation goals focus on enhancing adaptive capacity, increasing resilience, and limiting vulnerability.

Ensuring Finance

Not part of the Paris Agreement (and not legally binding) is a plan to provide US$100 billion a year in aid to developing countries for implementing new procedures to minimize climate change with additional amounts to be provided in subsequent years. Though both mitigation and adaptation require increased climate financing, adaptation has typically received a disproportionately low level of support. A 2014 report by the OECD found that just 16 percent of global finance was directed toward climate adaptation in 2014.

The Paris Agreement attempts to address the deficit by calling for a balance of climate finance between adaptation and mitigation, and specifically underscoring the need to increase adaptation support for parties most vulnerable to the effects of climate change, including Least Developed Countries and Small Island Developing States. The agreement also reminds parties of the importance of public grants, because adaptation measures receive less investment from the public sector. John Kerry, as Secretary of State, announced that grant-based adaptation finance would double by 2020.

Some specific outcomes of the elevated attention to adaptation financing in Paris include the G7 countries' announcement to provide US $420 million for Climate Risk Insurance, and the launching of a Climate Risk and Early Warning Systems (CREWS) Initiative. In early March 2016, the Obama administration gave a $500 million grant to the "Green Climate Fund" as "the first chunk of a $3 billion commitment made at the Paris climate talks." So far, the Green Climate Fund has now received over $10 billion in pledges. Notably, the pledges come from developed nations like France, the US, and Japan, but also from developing countries such as Mexico, Indonesia, and Vietnam.

Loss and Damage

A new issue that emerged as a focal point in the Paris negotiations rose from the fact that many of the worst effects of climate change will be too severe or come too quickly to be avoided by adaptation measures. The Paris Agreement specifically acknowledges the need to address loss and damage of this kind, and aims to find appropriate responses. It specifies that loss and damage can take various forms—both as immediate impacts from extreme weather events, and slow onset impacts, such as the loss of land to sea-level rise for low-lying islands.

The push to address loss and damage as a distinct issue in the Paris Agreement came from the Alliance of Small Island States and the Least Developed Countries, whose economies and livelihoods are most vulnerable to the negative impacts of climate change. Developed countries, however, worried that classifying the issue as one separate and beyond adaptation measures would create yet another climate finance provision, or might imply legal liability for catastrophic climate events.

In the end, all parties acknowledged the need for "averting, minimizing, and addressing loss and damage" but notably excludes any mention of compensation or liability. The agreement also adopts the Warsaw International Mechanism for Loss and Damage, an institution that will attempt to address questions about how to classify, address, and share responsibility for loss and damage.

Enhanced Transparency Framework

While each Party's NDC is not legally binding, the Parties are legally bound to have their progress tracked by technical expert review to assess achievement toward the NDC, and to determine ways to strengthen ambition. Article 13 of the Paris Agreement articulates an "enhanced transparency framework for action and support" that establishes harmonized monitoring, reporting, and verification (MRV) requirements. Thus, both developed and developing nations must report every two years on their mitigation efforts, and all parties will be subject to both technical and peer review.

Flexibility Mechanisms

While the enhanced transparency framework is universal, along with the global stocktaking to occur every 5 years, the framework is meant to provide "built-in flexibility" to distinguish between developed and developing countries' capacities. In conjunction with this, the Paris Agreement has provisions for an enhanced framework for capacity building. The agreement recognizes the varying circumstances of some countries, and specifically notes that the technical expert review for each country consider that country's specific capacity for reporting. The agreement also develops a Capacity-Building Initiative for Transparency to assist developing countries in building the necessary institutions and processes for complying with the transparency framework.

There are several ways in which flexibility mechanisms can be incorporated into the enhanced transparency framework. The scope, level of detail, or frequency of reporting may all be adjusted and tiered based on a country's capacity. The requirement for in-country technical reviews could be lifted for some less developed or small island developing countries. Ways to assess capacity include financial and human resources in a country necessary for NDC review.

A.10 SUPER TYPHOON MERANTI: 2016'S MOST POWERFUL CYCLONE

Super Typhoon Meranti has become the strongest cyclone, hurricane or typhoon, of 2016 and is one of three most powerful typhoons ever recorded. The Japan Meteorological Agency analysed Meranti's central pressure at 890 millibars. This puts Meranti in the short list of the deepest tropical cyclones ever recorded anywhere on Earth

At least 200,000 households were reported as losing power. The winds came in from the east so the cities to the lee of the mountains, for example Kaohsiung, were afforded some shelter from the brute force that hit the east coast.

The Chinese mainland will not be as lucky.

As it passed Taiwan, Meranti had sustained winds of 290 km an hour (kph), and gusts of 350 kph.

This makes it the equivalent of a Category 5 hurricane on the five-point Saffir-Simpson scale.

Meranti will sustain most of its current strength of winds due to high sea temperatures (30 C), low wind shear [change of wind speed and direction with height], and high atmospheric humidity.

Little interaction with the island of Taiwan means that the cyclone will maintain its structure. This will result in Meranti striking the Chinese mainland probably as a Category 4 equivalent storm.

The coastline of Guangdong and especially Fujian is now at risk of large waves, a storm surge well in excess of two metres and damaging winds, exceeding those experienced during Super Typhoon Nepartak.

Landfall of the eye, around which scream the strongest winds, is due at 18:00 GMT on Wednesday.

Typhoon strength winds will likely affect 20 million people.

The effects are likely to be far worse than those of the most recent typhoon to hit Fujian, Nepartak.

Meranti is the fourth Category 5 equivalent storm of 2016, following Winston, Fantala and Nepartak.

Nepartak had a very similar track to the one Meranti is expected to follow. It, struck southern Taiwan, killing two people, before hitting mainland China at Fujian.

A.11 THE DRAGON'S APPETITE FOR COMMODITIES

A Citi Commodity Research report highlights 10 areas that should be watched as these would have an impact on the world's largest commodity consumer, China, with <f0 per cent market share. These could shape commodities' price and demand outlook in the coming months and years.

Factors at Play: Environmental Initiatives: Environmental pressures are in the spotlight as never before, particularly for coal and steel, where the strictest measu res yet were implemented in 2013. There remains a lack of top-level direction, but mid-level officials have begun to push.

Overcapacity: The new government has made this a priority and we believe the first step has been taken to address overcapacity in the steel sector. Implementation remains crucial and 2016 may see initiatives spread to other sectors.

Credit and Debt: Credit is likely to remain tight as the government seeks to contain the "shadow banking" sector, while increasing scrutiny on local government debt threatens to slow infrastructure and real estate demand.

Financialisation of Commodities: Commodities in China are increasingly used to obtain financing and to invest in real estate. At the same time, financial products based on gold are growing rapidly and a number of new futures contracts are expected in 2017, including crude oil and HRC's.

Real Estate: Housing prices continue to rise rapidly, particularly in top-tier cities. We expect further cooling measures and a moderation in development activity over the course of 2017. Social housing has also passed its peak growth period.

Urbanisation: One of the government's top priorities and the largest driver of Chinese commodities demand. Land reform and residence permit (hukou) liberalisation are key, with the former accelerating and the latter remaining limited.

Whether incremental urbanisation focuses on large city clusters or dispersed smaller cities also has large implications for commodity intensity.

1. Indonesia Export Ban

Refining Capacity Expansion: Large-scale expansion of oil, copper, and aluminium refining capacity are turning China into an increasing exporter of petroleum products, copper cathode, and aluminium products. China's nickel pig iron sector is set to be hit hard over the coming months. Chinese firms will also need to source additional bauxite and alumina, and the ban may slow the rapid expansion of aluminium smelting capacity.

Trade Flows: China is increasingly importing raw materials and exporting refined commodities, though this runs counterto environmental goals. China's trade regime is also liberalising from gold to oil to copper.

Factor Price Reform: Reforms of commodity price setting mechanisms are expected, includingoil, natural gas, electricity, soybeans, cotton, and water.

2. China to Lease 3m ha % of Ukraine Firmland

China has moved to lease arable farmland in Ukraine that's nearly the size of all the farms in Haryana put together and entered into an agreement that will allow additional tracts of land to be transferred to Chinese control in future. The initial agreement is worth $1.7 billion. Industry sources said the agreement is part of a 50-year plan, which has the potential of allowing China to lease out three million hectares of land or 9% of Ukraine's arable land.

Index

A

Acid Rain 25
Activities Implemented Jointly 187
Adverse climate change 163
Agglomeration economies 282
Air Quality Committee 321
Allocation of Natural Resources 401
Alluvial 348
Annual Environmental Media Awards 39
Asian brown cloud 395
Auctions or competitive bidding 402
Avoid, Shift and Improve strategy 263

B

Bengal Smoke Nuisance Act 392
Biodiversity 360
Biomass waste 243
Biotic Index 421
Bombay Smoke Nuisance Act 392
Brundtland Commission 229
Brundtland Report 21

C

Carbon dioxide emission 396
Carbon footprint 24
Carbon markets 222
Carbon trading 379
Central Pollution Control Board 420
Central Water and Power Commission 345
China's Renewable Energy Law 293
Clean Development Mechanism 222
Clean Development Mechanism 374
Climate and Development Knowledge Network 419
Climate Vulnerability Monitor 421
Codex Alimentarius Commission 296
Community Forest User Groups 301
Conservation Agriculture 107
Conservation of Antarctic Marine Living Resources 336
Construction and Demolition 243
Convention of Biological Diversity 56
Costs of reforestation and afforestation 201
Crony capitalism 401
Crop Pattern in India 349

D

Dead Zones in the Ocean 25
Destination Management Organisations 273
Digital technology 288
Doctrine of Eminent Domain 383
Dubai Clean Energy Strategy 416
Dubai Plans 416

E

Earth Charter 61
Earth's South Pole 335
Ecological Agriculture 107

Economic Zone 366
Ecosystem degradation 163
Ecosystem Valuation and Wealth Accounting 16
El Niño 414
Emission Trading Among the Developed Nations 373
End-of-life Vehicles 243
Endorsement of Forest Certification 184
Energy poverty 218
Environment quality 20
Environmental damage 21
Environmental Damage 399
Environmental Engineering Research Institute 409
Environmental Impact Assessment 411
Environmental issues 26
Environmental Kuznets curve 22
E-waste (Management) Rules, 2016 242
Exclusive Economic Zones 155

F

Fair Trade 107
Feed-in Tariff 294
Forest Resource Assessment 181
Forest Rights Act 363
Forest transition theory 183
Friends of the Earth 28
Fund for the Protection of Water 302
Fund's Private Sector Facility 418

G

Geneva Convention 321
Geological Survey 344
Global 200 program 334
Global Forest Watch 179
Global stocktake 424
Global warming 24
Good Agricultural Practices 107
Great Plains of Northern India 366
Green cities 279
Green Climate 417
Green Economy Coalition 62
Green India Mission 364, 380
Green National Accounting system 381
Green Politics 38
Green Transport 265
Greener Destination Planning 274
Greener Vehicles 263
Greenhouse Gases 260
Greening Aquaculture 150
Greenpeace 28
Ground Water 354

H

Habitat Preservation 333
Health-care Waste 243
Hot Spots Approach 334

I

India's Inter-state Water Disputes 357
Indian Bureau of Mines 344
Indian Council of Agricultural Research 347
Indian Penal Code 392
Industrial Ecology 281
Industrial Revolution 410
Information and Communication Technology 256
Intact Forest Landscapes 177
Integrated Solid Waste Management 239
Intended Nationally Determined Contributions 375
International Ecotourism Society 271
International Energy Agency 175
International Energy Agency 2
International Energy Agency's 213
International Food Policy Research Institute 136
International Halibut Commission 331
International Maritime Organisation 310
International Transfer of Mitigation outcomes 425
International Tropical Timber Organisation 182
International Union for Conservation 8
International Water Management Institute 162
International Whaling Commission 332
Investing in Agroforestry 203
Issue Management Group 292
Issue of Land Acquisition 383

J

Johannesburg Declaration 61
Joint Crisis Initiatives 292
Just Transition 62

K

Kuznets curve 394
Kyoto Protocol 373

L

La Niña 414
LARR 385
Law of Non-Navigational Uses of International Watercourses 336
Law of the Sea 366
Leasehold Forestry Policy 302
life Cycle Approach 238
Long-Range Transboundary Air Pollution 321

M

Maloti Drakensberg Mountains 303
Malthusian catastrophe 394
Marine Mammals 332
Marine Pollution 329
Master Plan for the Forestry Sector 302
Millennium Development Goals 2
Millennium Ecosystem Assessment 8
Mines and Minerals (Development and Regulation) Bill 2011 368

N

NASA 395
National Agricultural Policy 411
National Air Monitoring Program 420
National Commission on Agriculture 361
National Commission on Environment and Ecology 409
National Conservation Strategy and Policy 411
National Environment Assessment and Monitoring Authority 371
National Environment Policy 2006 (NEP) 370
National Environment Policy 2006 (NEP) 410
National Forest Policy 411
National Law on Water 387
National Oceanic and Atmospheric Administration 413
National Population Policy 411
National Water Policy 411
National Waterways Bill 359
Natural Resource Investment Trust 404
Natural Resource Deficit 404
New Economics Foundation 62
New Mineral Policy, 2008 367
North American Acid Rain Control Program 321
Noxious Substances Convention 330

O

Occupational Safety and Health Administration 258
Ocean Dumping Convention 330
Ocean Governance 328
Oil and Natural Gas Commission 344
Oil pollution 330
Oil Pollution Liability and Compensation Scheme 330
One Planet Living Principles 61
Open Access 402
Organic Agriculture 296
Organic nutrient recycling 112
Organic/Biodynamic Agriculture 107
Our Common Future 21
Ozone Depleting Substances 244
Ozone Depletion 322
Ozone Deterioration 25

P

Payments for Ecosystem Services 187
Power Purchase Agreements 224
Principle of Common Concern 316

R

Regional Fisheries Management Organisations 155
Rio Declaration 61

S

Scientific Grounding 27
Shared Environmental Information System 49
Sloping Land Conversion Programme 200
Soil Land Use Survey 345
Solar Capacity 224
Species Extinction 26
State of Forest Report 345
State of World Fisheries and Aquaculture 148
Stockholm Declaration 61
Supreme Court 396

Surface water 354
Sustainable development 21
Sustainable Forest Management 182
Sustainable Forest Management 197
Sustainable Tourism 267

T

The Foresight Report 107
The PROSOL Programme 304
The Rodale Institute 136
Threshold 21 16
Tools for Delivering on Green Growth 85
Trail Smelter case 314
Treaty for Amazonian Cooperation 335

U

U.N. Economic, Scientific, and Cultural Organisation 311
Uganda Organic Agriculture Policy 297
Uganda Organic Standard 297
UN Climate Conference in Marrakech 416
UN Habitat 399
UN Laws on the Exclusive Economic Zone 366
UNEP-led Green Economy Initiative 291
United Nations Environment Programme 28
United Nations Framework on Climate Change Convention 216

V

Vehicle Emissions 396
Vienna Convention for the Protection of the Ozone Layer 324
Vision 2050 235

W

Waste Management Hierarchy 240
Water Accounting 159
Water Cycle 157
Water Pollution 163
Water Resources Group 163
Wetland Shield 272
World Business Council on Sustainable Development 235
World Commission on Environment and Development 6
World Commission on the Environment and Development 21
World Environment Organisation 309
World Green Economy Organisation 416
World Green Economy Summit 416
World Health Organisation 221
World Summit on Sustainable Development 152
World Wide Fund for Nature 28
World Wide Fund for Nature 8
World Wildlife Fund 334

Y

Yellow River 165

For Product Safety Concerns and Information please contact our EU representative GPSR@taylorandfrancis.com
Taylor & Francis Verlag GmbH, Kaufingerstraße 24, 80331 München, Germany

www.ingramcontent.com/pod-product-compliance
Ingram Content Group UK Ltd.
Pitfield, Milton Keynes, MK11 3LW, UK
UKHW020146200926
475702UK00005B/946
9781032073996